Second International Workshop 'SANS-YuMO User Meeting at the Start-up of Scientific Experiments on the IBR-2M Reactor 2011'

Devoted to the 75th Anniversary of Yu M Ostanevich's Birth
Journal of Physics: Conference Series Volume 351

Moscow, Russia
27-30 May 2011

Editors:

V. Gordeliy
A. Kuklin
M. Balasoiu

ISBN: 978-1-62276-768-7
ISSN: 1742-6588

Printed from e-media with permission by:

Curran Associates, Inc.
57 Morehouse Lane
Red Hook, NY 12571

Some format issues inherent in the e-media version may also appear in this print version.

Printed by Curran Associates, Inc. (2013)

For permission requests, please contact the Institute of Physics
at the address below.

Institute of Physics
Dirac House, Temple Back
Bristol BS1 6BE UK

Phone: 44 1 17 929 7481
Fax: 44 1 17 920 0979

techtracking@iop.org

Additional copies of this publication are available from:

Curran Associates, Inc.
57 Morehouse Lane
Red Hook, NY 12571 USA
Phone: 845-758-0400
Fax: 845-758-2634
Email: curran@proceedings.com
Web: www.proceedings.com

TABLE OF CONTENTS

SPECIAL AND INVITED LECTURES

BIOLOGICAL SYSTEMS

MAGNETIC NANODISPERSED STRUCTURES

POLYMERS AND SURFACTANTS

NANOSTRUCTURED ADVANCED MATERIALS

THEORY AND MODELING

INSTRUMENTAL AND SOFTWARE DEVELOPMENT

Preface

The Second International Workshop '**SANS-YuMO User Meeting at the Start-up of Scientific Experiments on the IBR-2M Reactor**', devoted to the 75th anniversary of the birth of Professor Yu M Ostanevich (1936–1992), an outstanding neutron physicist and the founder of small-angle neutron scattering (field, group, and instrument) at JINR FLNPh, was held on 27–30 May at the Frank Laboratory of Neutron Physics.

The first Workshop was held in October 2006. Research groups from different neutron centers, universities and research institutes across Europe presented more than 35 oral and poster presentations describing scientific and methodological results. Most of them were obtained with the help of the YuMO instrument before the IBR-2 shutdown in 2006.

For the last four years the IBR-2 reactor has been shut down for refurbishment. At the end of 2010 the physical launch of the IBR-2M reactor was finally realized.

Nowadays the small-angle neutron scattering (SANS) technique is applied to a wide range of scientific problems in condensed matter, soft condensed matter, biology and nanotechnology, and despite the fact that there are currently over 30 SANS instruments in operation worldwide at both reactor and spallation sources, the demand for beam-time is considerably higher than the time available. It must be remembered, however, that as the first SANS machine on a steady-state reactor was constructed at the Institute Laue Langevin, Grenoble, the first SANS instrument on a 'white' neutron pulsed beam was accomplished at the Joint Institute for Nuclear Research at the IBR-30 reactor, beamline N5.

During the meeting Yu M Ostanevich's determinative and crucial contribution to the construction of spectrometers at the IBR-2 high-pulsed reactor was presented, as well as his contribution to the development of time-of-flight (TOF) small-angle scattering technique, and a selection of other scientific areas. His leadership and outstanding scientific achievements in applications of the Mossbauer effect in physics and chemistry, in SANS studies of polyelectrolytes, small molecules, fractals, metallic glasses, macromolecules, polymers, etc., were recognized by a number of awards including the State Prize of the Russian Federation in 2000.

The scientific program of the workshop focused on fundamental and methodical research at the **YuMO** spectrometer and developments of the SANS instrument at the modernized IBR-2M reactor. We recall that the acronym YuMO of the small-angle neutron scattering spectrometer (MURN) was given in honor of Yu M Ostanevich.

One of the most important objectives of this user meeting was to discuss the further development possibilities of the YuMO spectrometer with experts, in the frame of a *SANS YuMO Round Table*, taking into account the specific performance of the modernized YuMO SANS instrument, and the scientific and technical requests of the instrument's users.

Highlights on modern achievements in nanoscience, polymers and biology were other significant goals of the meeting. The plenary invited talks were presented by leading scientists in small-angle neutron scattering and soft condensed matter, including members of the

Russian Academy of Sciences: Prof. Heinrich Stuhrmann, Prof. Alexei Khokhlov, Prof. Jose Teixeira, Prof. Alexander Ozerin, Prof. Albrecht Wiedenmann, etc.

There were 27 oral talks given and 32 posters presented by 92 participants from 12 countries: Czech Republic, Egypt, France, Germany, Hungary, Moldova, Mongolia, Poland, Romania, Russian Federation, Slovak Republic, and Ukraine.

The workshop was organized with the financial support of the Frank Laboratory of Neutron Physics (Joint Institute for Nuclear Research), Horia Hulubei National Institute of Physics and Nuclear Engineering – IFIN HH (Romania), Institute of Macromolecular Chemistry AS CR (Czech Republic) and Comenius University (Slovakia).

V. Gordeliy, A. Kuklin and M. Balasoiu

Participants of the Meeting

During sessions

Journal of Physics: Conference Series **351** (2012) 011001 doi:10.1088/1742-6596/351/1/011001

During sessions

A discussion between sessions

Organizers

Frank Laboratory of Neutron Physics, Joint Institute for Nuclear Research

Horia Hulubei National Institute of Physics and Nuclear Engineering - IFIN HH

Romanian Neutron Scattering Society

Sponsors

Frank Laboratory of Neutron Physics, Joint Institute for Nuclear Research

Horia Hulubei National Institute of Physics and Nuclear Engineering - IFIN HH

Comenius University in Bratislava, Slovakia

Institute of Macromolecular Chemistry AS CR, Czech Republic

Programme Committee

Valentin Gordely - *chairman*	Joint Institute for Nuclear Research, Russia
Heinrich Stuhrmann	Germany
Jose Teixeira	Laboratoire Leon Brillouin, France
Pavel Apel	Joint Institute for Nuclear Research, Russia
Pavol Balgavy	Comenius University in Bratislava, Slovakia
Alexander Belushkin	Joint Institute for Nuclear Research, Russia
Georg Bueldt	Institute of Structural Biology and Biophysics (ISB), Germany
Leonid Bulavin	Taras Shevchenko National University of Kyiv, Ukraine
Emil Burzo	Babes-Bolyai University, Romania
Vadim Cherezov	The Scripps Research Institute, Department of Molecular Biology, USA
Ion Ionita	Romanian Society of Neutron Scattering, Romania
Alexei Khokhlov	Moscow State University, Russia
Aziz Muzafarov	Institute of Synthetic Polymeric Materials, Russian Academy of Sciences, Russia
Alexander Ozerin	Institute of Synthetic Polymeric Materials, Russian Academy of Sciences, Russia
Gerard Pepy	Research Institute for Solid State Physics and Optics, Hungary
Josef Plestil	Institute of macromolecular chemistry CAS, Czech Republic
Aurel Radulescu	Juelich Centre for Neutron Science JCNS, Germany
Maria Balasoiu	Joint Institute for Nuclear Research, Russia
Alexander Kuklin	Joint Institute for Nuclear Research, Russia

Local Organizing Committee

Alexander Kuklin – *Chairman*
Maria Balasoiu – *Co-chairman*
Tatiana Murugova – *Secretary*
Natalia Malysheva
Natalia Dokalenko
Julia Gorshkova
Andrey Rogachev
Oleksandr Ivankov
Dmitry Soloviev
Lilia Anghel
Erhan Raul

Journal of Physics: Conference Series **351** (2012) 011002 doi:10.1088/1742-6596/351/1/011002

Conference Programme

SANS-YuMO User Meeting at the start-up of scientific experiments on the IBR-2M reactor, devoted to the75th anniversary of the birth of Yu M Ostanevich.

Thursday, 26.05.2011	
18:00 – 20:00	**REGISTRATION** (Hotel Dubna "Grill", building 3, str. Moskovskaya 2)

Friday, 27.05.2011	
10:00 – 11:00	**REGISTRATION**
11:00 – 11:20	**OPENING AND WELCOME**
11:20 – 11:50	Yury M. Ostanevich and science at Frank Laboratory of Neutron Physics ***Balagurov A. M.****, Joint Institute for Nuclear Research, Dubna, Russia*
11:55 – 12:10	Cser László:"Fragments of the history of the first small angle time-of-flight spectrometer established at IBR-2 reactor" ***Balagurov A. M.****, Joint Institute for Nuclear Research, Dubna, Russia*
12:10 – 12:40	COFFEE BREAK
12:40 – 13:10	The modernized reactor IBR-2M: history and prospects ***Belushkin A. V.****, Joint Institute for Nuclear Research, Dubna, Russia*
13:15 – 14:00	Nanostructures in polymer solutions: study by neutron scattering technique ***Khokhlov A.R.****, Moscow State University, Moscow, Russia.*
14:00 – 15:30	LUNCH
15:30 – 15:50	Status and perspectives of cold moderator at the IBR-2 reactor ***Kulikov S.*** *Joint Institute for Nuclear Research, Dubna, Russia*
15:55 – 16:20	SAS applications in soft condensed matter ***Funari S.****, HASYLAB, Hamburg, Germany*
16:25 – 16:55	Nuclear magnetic resonance as a complementary method to SANS: status and perspectives ***Chupin V.****, M.M. Shemyakin and Yu.A. Ovchinnikov Institute of bioorganic chemistry RAS, Moscow, Russia*
17:00 – 17:30	COFFEE BREAK
17:30 – 19:10	**EXCURSION TO YUMO INSTRUMENT**
20:00 – 22:00	**WELCOME PARTY**

Saturday, 28.05.2011	
10:00 – 10:45	Polymer-nanoparticles composites: perspectives and problems ***Ozerin A.** , ISPM RAS, Moscow, Russia*
10:50: – 11:35	Ordering and Relaxation Dynamics of Magnetic Nanomaterials studied by SANS **Wiedenmann A.**, *Institute Laue-Langevin, Grenoble, France*
11:40 – 12:05	Time-resolved polarized neutron scattering from dynamic polarized nuclear spin targets **Stuhrmann H.**, *GKSS Research Center, Germany and IBS, France, till July 2007*
12:05 – 12:30	COFFEE BREAK
12:30 – 12:55	Small–angle polarized neutron scattering equipment at WWR-M today **Runov V.**, *Petersburg Nuclear Physics Institute, Gatchina, Russia*
13:00 – 13:25	Spin-echo encoding of the momentum transfer in grazing incidence scattering ***Vorobiev A.**, ESRF, Grenoble, France*
13:30 – 13:55	2D, I=f(Qx,Qy), data treatment related to SANS: status and perspectives **Pepy G.**, *Research Institute for Solid State Physics and Optics, Budapest, Hungary*
14:00 – 15:30	LUNCH
15:30 – 15:55	Status and perspectives of YuMO spectrometer **Kuklin A.**, *Joint Institute for Nuclear Research, Dubna, Russia*
15:55 – 16:20	New configuration of YuMO spectrometer and moderator at modernized IBR-2 reactor **Rogov A.** , *Joint Institute for Nuclear Research, Dubna, Russia*
16:25 – 16:50	Monte Carlo simulations and proposal of an instrument configuration for YuMO SANS spectrometer using a cold moderator **Erhan R.**, *Joint Institute for Nuclear Research, Dubna, Russia*
16:50 – 17:20	COFFEE BREAK
17:20 – 18:00	Main features of the software control system for YuMO instrument **Kirilov A.**, *Joint Institute for Nuclear Research, Dubna, Russia*
18:05 – 19:00	**Round table** "Possibilities and perspectives of YuMO spectrometer"

Sunday, 29.05.2011	
10:00 – 10:25	Modern aspects of small-angle scattering in biomembrane research ***Gordeliy V.**, IBS, Grenoble, France, Bionanocenter, MIPT, Dolgoprudny, Russia, JINR, Dubna, Russia*
10:30 – 10:55	Application of small angle neutron scattering for investigation of structure of biological membrane systems ***Yaguzhinsky Lev**, A.N. Belozersky Institute of Physico-Chemical Biology, MSU, Moscow, Russia*
11:00 – 11:25	Lipid bilayer – DNA interaction mediated by divalent metal cations: SANS and SAXS study ***Uhrikova D.**, Comenius University in Bratislava, Bratislava, Slovakia*
11:25– 11:50	Prospectives of applying neutron scattering techniques in study of structure and dynamics of nucleoprotein complexes ***Lebedev D.**, Petersburg Nuclear Physics Institute, Gatchina, Russia*
11:50 – 12:10	COFFEE BREAK
12:10 – 12:35	Small Angle Neutron Scattering Programme at the Budapest Research Reactor ***Szekely N.**, Research Institute for Solid State Physics and Optics, Budapest, Hungary*
12:40 – 13:05	KWS-2, the high intensity/wide-Q small-angle neutron diffractometer for soft-matter and biological systems at FRMII: current status and perspectives ***Radulescu A.**, Forschungszentrum Juelich, Juelich, Germany*
13:10 – 13:55	Isotope effects on the hydration of cements ***Teixeira J.**, Laboratory Leon Brillouin, Saclay, France*
14:00 – 15:30	LUNCH
15:30 -16:00	Contrast variation application in small-angle neutron scattering experiments ***Stuhrmann H.**, GKSS Research Center, Germany and IBS, France, till July 2007*
16:05 – 16:35	Investigations of ferrofluids and magnetic elastomers at the YuMO-SANS instrument ***Balasoiu M.**, Joint Institute for Nuclear Research, Dubna, Russia, Horia Hulubei National Institute of Physics and Nuclear Engineering, Bucharest, Romania*

16:40 – 17:05	SANS and SAXS study of extended defects in diamonds ***Shiryaev A.*** *Institute of Physical Chemistry and Electrochemistry RAS, Moscow, Russia*
17:10 – 17:35	The alteration of micellar nanoreactor during the hydrolysis reaction proceedings. SANS study ***Bakeeva R.F.***, *Kazan State Technological University, Kazan, Russia*
17:40– 18:00	**COFFEE BREAK**
17:40 – 18:00	***Poster session***
20:00 – 23:00	***Boat excursion***

	Monday, 30.05.2011
10:00 – 10:30	Fluctuations and critical phenomena in soft matter: structure investigations ***Bulavin L.***, *Taras Shevchenko National University of Kyiv, Kyiv, Ukraine*
10:30 – 11:00	**Open discussion**
11:00	**CONFERENCE CLOSING AND EVALUATION**

All papers published in this volume of *Journal of Physics: Conference Series* have been peer reviewed through processes administered by the proceedings Editors. Reviews were conducted by expert referees to the professional and scientific standards expected of a proceedings journal published by IOP Publishing.

Past and present of time-of-flight small-angle neutron scattering at IBR-2

A Kuklin[1,2], A Islamov[1], M Balasoiu[1,3], T Murugova[1,2] and V Gordeliy[1,2,4,5]

[1] Frank Laboratory of Neutron Physics, JINR, Joliot-Curie-6, Dubna 141980 Russia

[2] Bionanocenter of Moscow Institute for Physics and Technology, Dolgoprudny, Russia

[3] Horia Hulubei National Institute of Physics and Nuclear Engineering, Bucharest, Romania

[4] Institute of Complex Systems (ICS), ICS-5: Molecular Biophysics, Research Centre Juelich, 52425 Juelich, Germany

[5] Institute of Structural Biology J.P.Ebel, Grenoble, France

E-mail: kuklin@nf.jinr.ru

Abstract

A historical overview of MURN-YuMO instrument (including some technical and scientific development prospects) is presented. Numerous references are cited which demonstrate a tireless activity on the first in the world SANS machine functioning at a neutron pulsed reactor. The relevant scientific activity trajectories of Yu.M.Ostanevich and his colleagues are retraced and put into perspective.

It is a great pleasure for us to give a brief introduction to Prof. Yu.M. Ostanevich scientific activity related to small-angle neutron scattering. Three of the authors of this introduction (AK, AI and VG) knew and worked with him, two of us (TM and MB) came to the Laboratory when he has left us, but we all appreciate very much the beautiful science, the YuMO instrument and scientific spirit, which Yu.M. Ostanevich created for us and our colleagues. Recall that the acronym YuMO of small angle neutron scattering spectrometer (MURN) was given in honour of Yu.M. Ostanevich.

Prof. Yu.M. Ostanevich had a broad knowledge and deep understanding of physics. This allowed him to enter quickly into a new scientific field and work there efficiently. He started his research in the sixties with the studies of Mössbauer effect in condensed matter. One of his activities was investigation of the effect of gravity on self-diffusion at critical point of water and neutron resonances. Since 1974 he was concentrated on creating a first time-of-flight small-angle spectrometers at IBR-30 and IBR-2 pulsed reactors. We and our colleagues, who are working at the YuMO SANS instrument at IBR-2, admire outstanding quality and longevity of Yu.M. Ostanevich's scientific and technical ideas lying in the basis of the instrument. Already first SANS studies done by Yu. Ostanevich and his colleagues were impressive and showed how wide were his scientific interests. The publications on the studies of biological macromolecules [1-3], polymers [4-6], lipid membranes [7,8], small molecules [9,10] and micelles [11-13] profoundly illustrate this point.

In our opinion, the series of articles, which have been written quite long time ago by N.I. Gorskiy and Yu.M. Ostanevich, are still very interesting today [11-13]. N. Gorskiy and Yu. Ostanevich applied remarkable ability of the surfactants to form invert micelles in hydrophobic solvents for study of size effects at nanoscale.

Another example of fruitful activity of Yu.M. Ostanevich is the review written together with I.N. Serdyuk [14]. It has been written before SANS spectrometer at the IBR-2 was commissioned, was very timely and appreciated by the biological scientific community interested in applications of SANS to their biological objects. In that review the authors examined theoretical and practical problems of application of small-angle scattering in biology.

Yu.M. Ostanevich actively applied SANS to studying small molecules [15,16]. Such experiments are not trivial because the scattering from small objects is weak and such experiments require excellent experimental skills, which certainly were inherent to Yu.M. Ostanevich.

Investigations of polymers at the MURN spectrometer have been started by J. Plestil and Yu.M. Ostanevich with polyelectrolytes. Hydration and local structure of polyions, binding of counterions to polymers were the first problems addressed in [17,18]. The interpolyion correlations and size of polyions were obtained in [19,20]. The Nylon-6 in swollen state [21] and thermosensitive gels [22,23] were studied by SANS at the MURN instrument.

More detailed account about the investigations of polymers as well as block copolymer micelles (corona micelles, three-layer micelles and nanoparticles) can be found in the papers and reviews by J. Plestil, who was one of the closest co-workes of Yu.M. Ostanevich [24,25] and short review [26].

At present, the investigation of polymers at the YuMO spectrometer at the IBR-2 is one of the most

Figure 1. Detector system of the CsOK spectrometer.
On the photo: Yu. Ostanevich (right) and L. Cser (left).

active fields. It is also worth to mention here important studies done by the known polymer schools of Prof. A.R. Khokhlov and Prof. A.N. Ozerin. Among these works are investigations of self-assembly polyelectrolyte rods in polymer gel and in solution [27] and a new type of polymers – dendrimers [28-34], the charge-induced microphase separation in polyelectrolyte phenomena [35-37], poly(ethylene oxide)/poly(propylene oxide) copolymers in aqueous solution [38].

Small angle neutron scattering is the powerful method to study different types of objects with typical sizes of nanometers in biology, polymers, colloids and materials science field.

It is known that the first SANS machine on steady-state reactor was constructed at ILL. First machine on "white" neutron beam was constructed at JINR at reactor IBR-30 on beam N5 [39-43]. Neutron wavelength on such reactors is determined using time-of-flight (TOF) method based on start (pulse) - stop (registration of events on the detector). The idea of spectrometer belongs to Yuri M. Ostanevich and Laszlo Cser. They suggested that TOF spectrometer must have a lot of advantages in comparison with fixed-wavelength machines. In particular, using white neutron beams with high wavelength resolution allows reducing uncertainty of neutron momentum transfer at scattering processes to geometrical parameters of the spectrometer. It is interesting to point out that the first spectrometer got the name "CsOK", after the names of its principal creators – Cser Laszlo, Ostanevich Yuriy and Kozlov G.A. Word "CsOK", by the way, is conformable with Hungarian "csok", which means "kiss". Imagine, the spectrometer had been build from "zero" level in short period of a couple of months! CsOK consisted of detector system, built from 8 boron counters (NMI-52 type), situated on the left and right sides from the direct beam (see Picture 1). Such geometry of SANS installation is called "rectangular". The central part of the detection plane with central beam passing through it, remains free from counters and any construction materials. Using this type of construction, an effective reduction of the direct beam background in the detector space can be achieved. It was the idea that could be used for ring-wire detectors for spectrometer MURN (YuMO) at IBR-2. In addition to these detectors, the moved counter for measurements of the direct beam and symmetry center of the detectors (prototype of direct beam detector at MURN spectrometer) was used. The first experiments on the CsOK were done with biological objects, namely, hemoglobin solution, 50S ribosome and collagen (taken from a rat's tale moistened with light water and heavy water). For hemoglobin and 50S ribosome, an average value for the radius of gyration was obtained. For collagen with 400 fibers was obtained the value of 65.4 nm for wet samples and 64 nm for dry ones. The first experiments with proteins in solutions confirming the method as well as the basic characteristics of the spectrometer were briefly described in the published work. It is interesting to note that a new type of detector of thermal neutrons for the SANS instrument, circular multi-wires He^3 detector with a central hole, was presented before this publication. Namely, such a detector has been the main part of data acquisition system of SANS instrument at IBR-2 during more than 20 years. Later, this type of detectors allowed us to propose and experimentally realize in 2000 a new approach to the collection of time-of-flight SANS data by using two detector systems at the YuMO [44]. The SANS group tried to preserve this good tradition performing modernization of the instrument, installation of a two-detector system as well as a new type two dimensional position sensitive detector at the YuMO [45]. First experiments with magnetic fluids realized at MURN using contrast variation method [46-48] give the perspective to improve the sample enviroment, in particulary, using magnetic field [49].

Yu.M. Ostanevich devoted continuous attention to the development of time of-flight methodology and the instruments at IBR-30 and IBR-2 pulsed reactors. The first ideas and experimental results of Yu.M. Ostanevich and his colleagues on time-of-flight SANS have been published in 1976 [50].

Yu.M. Ostanevich was an outstanding scientist and he understood well the users of the instrument, being deeply involved into nearly all experiments done on the machine. He always was trying to make improvements of the instrument without shutting down the machine when the reactor was in operation. The SANS group was saving this good tradition during modernization of the instrument, installation of two detector system as well as a new type two dimensional position sensitive detector at the YuMO [45] and during the shutdown of IBR-2, in particular through preparing the conditions for its modernization in the nearest future.

We know now how much of additional efforts Yu.M. Ostanevich had to apply to keep the instrument in operation, continuously modernizing it.

Yu.M. Ostanevich has created one of the most efficient scientific schools at the Laboratory working for a very wide scientific society in Russia, JINR member countries and others. We hope that the participants of the Workshop will devote an additional time in order to suggest some interesting new ideas about principal modernization of YuMO.

We also hope that one of the most productive methods and instruments at the IBR-2, YuMO, launched by Yu.M. Ostanevich and further developed by his team, will get a sufficient support and will be ready for the future successful work when the IBR-2 will be in operation again.

Acknowledgements

The sponsors of the Workshop: Plenipotentiary member of Romania, Czech Republic, Slovak Republic, Frank Laboratory of Neutron Physics, Joint Institute for Nuclear Research.

The Workshop attracted many leading scientists from all over the world, who are exploiting and developing SANS. We are very grateful to A. Balagurov, A. Belushkin, P. Balgavy, V. Chupin, V. Isaev-Ivanov, A. Khokhlov, D. Lebedev, A. Ozerin, G. Pepy, J. Plestil, A. Radulescu, H. Stuhrmann, J. Teixera, A. Wiedenmann, L. Yaguzhinsky for kindly agreeing to participate in our user meeting.

We are especially thankful to L. Czer for his help in creation of chronological events related to CsOK.

We are very indebted to the members of the Organizing Committee: N.A. Malisheva, N. Dokalenko, A.V. Rogachev, O.I. Ivankov, Yu.E. Gorshkova, D.V. Soloviev, Yu.S.Kovalev as well as A.V. Belushkin, D.P. Kozlenko, S.V.Kozenkov,, S.F. Yarovikov, V.G. Simkin for permanent help in preparing the meeting.

References

[1] Cser L, Franek F ,Gladkikh I A, Nezlin R S, Novotny J, Ostanevich Yu M,1978 Febs lett **93** 312

[2] Cser L, Franek F, Gladkikh I A, Nezlin R S, Novotny J, Ostanevich Yu M 1980 Immunol Lett, **1**, 185

[3] Cser L, Franek F , Gladkikh I A, Nezlin R S, Novotny J, Ostanevich Yu M 1981 Eur J Biochem, **116**, 109

[4] Plestil J , Ostanevich Yu M , Bezzabotnov V Yu , Hlavata D 1986 Polymer, **27**,1241

[5] Plestil J, Mikes J , Dusek K, Ostanevich Yu M, Kuncenko A B 1981 Polym Bull, **4**, 225

[6] Plestil J , Hlavata D , Polymer 1988 **29**, 2216

[7] Klose G, Bruckner S, Bezzabotnov V Yu, Borbely S, Ostanevich Yu M 1986 Chemistry and Physics of Lipids, 41293-307

[8] Gordeliy V I, Ivkov V G, Ostanevich Yu M, Yaguzhinskij L S 1991 Biochimica et Biophysica Acta, **1061**, 39-48

[9] Cser L, Farago B, Grosz T, Jancso G, Ostanevich Yu M 1992 Physica B **180&181** 848-850

[10] Bezzabotnov V Yu, Cser L, Crosz T, Jancso G Ostanevich Yu M 1992 The Journal of Physical Chemistry, **96**

[11] Gorski N 1989 Z Phys Chem Leipzig, **270**, 817 (1989)

[12] Gorski N , Ostanevich Y M 1990 Ber Bunsenges Phys Chem, **94**, 737

[13] Gorski N , Ostanevich Y M 1993 J de Physique, **3**, 149

[14] Ostanevich Y M, Serdyuk I N 1982 Neutronographic investigations of biomacromolecular structures Uspekhi fizicheskikh nauk, **137**, 1, 85-116

[15] Bezzabotnov V Yu, Cser L, Grosz T, Jancso G, Ostanevich Yu M 1992 J of Physical Chemistry, **96**, 976-982

[16] Plestil J, Hlavata D 1988 Polymer, **29**, p.2216
[17] Plestil J , Ostanevich Yu M , Bezzabotnov V Yu , Hlavata D 1986 Polymer, **27**, p 1241
[18] Plestil J , Mikes J , Dusek K , Ostanevich Yu M , Kuncenko A B, 1981 Polym Bull, **4**, 225
[19] Plestil J , Ostanevich Yu M , Bezzabotnov V Yu , Hlavata D , Labsky J 1986, Polymer, **27**, 839
[20] Plestil J , Hlavata D , Labsky J Ostanevich Yu M , Bezzabotnov V Yu 1986 Polymer, **28**, 213
[21] Plestil J, Baldrian J, Ostanevich Yu M, Bezzabotnov V Yu 1991 Polym Sci Part B:Polym Phys, **29**, 509
[22] Plestil J , Ostanevich Yu M , Bordely S, Stejskal J, Ilavsky M 1987 Polym Bull, 17, 465
[23] Plestil J , Ilavsky M , Pospisil H 1993 Polymer **34**, 4846
[24] Pospisil H, Plestil J, Tuzar Z 1993 Collect Czech Chem Commun **58**, 2428
[25] Kriz J , Masar B , Pospisil H 1996 Macromolecules, **29**, 7853
[26] Plestil J Small-Angle Neutron Scattering Investigation of Polymer Systems Communication of the JINR, Dubna 2002 E14-2002-60
[27] Zaroslov Yu D, Gordeliy V I, Kuklin A I, Islamov A H, Philippova O E, Khokhlov A R, Wegner G 2002 Macromolecules, ***35***, 4466-4471
[28] Kuklin, A I, Ignat'eva G M, Ozerina L A, Islamov A Kh, Mukhamedzyanov R I, Shumilkina, N A, Myakushev V D, Sharipov E Yu, Gordeliy V I, Muzafarov A M& Ozerin, A N 2002 *Polym Sci* **A44** 12, 1-10
[29] Ozerin Alexander N, Muzafarov Aziz M, Gordeliy Valentin I, Kuklin Alexander I, Ignat'eva Galina M, Krykin Mikhail A, Ozerina Lyudmila A, Shumilkina Natalia A, Islamov Akhmed Kh, Sharipov Eugene Yu, Mukhamedzyanov Ruslan I 2003 Macromol Symp ***195***, *171-178*
[30] Ozerin A N, Muzafarov A M, Kuklin A I, Islamov A Kh , Gordelyi V I, Ignat'eva G M, Myakushev V D, Ozerina L A, Tatarinova E A 2004 Doklady Chemistry, **395**, Part 2, 59–62
[31] Kuklin Alexander I, Ozerin Alexander N, Islamov Akhmed Kh, Muzafarov Aziz M, Gordeliy Valentin I, Rebrov Eugeniy A, Ignat'eva Galina M, Tatarinova Elena A, Mukhamedzyanov Ruslan I, Ozerina Lyudmila A and Sharipov Eugeniy Yu J 2003 Appl Cryst **36**, 679-683
[32] Ozerin A N, Svergun D I, Volkov V V, Kuklin A I, Gordeliy V I, Islamov A Kh, Ozerina L A and Zavorotnyuk D S 2005 J Appl Cryst **38**, 996–1003
[33] Rogachev A V, Kuklin A I., Cherny A Yu, Ozerin A N, Muzafarov A M, Tatarinova E A, Gordeliy V I 2010 Physics of the Solid State 2010 **52** 5 1045–1049.
[34] Rogachev A.V., Cherny A.Yu., Ozerin A.N., Gordeliy V.I., Kuklin A.I 2007 Crystallography Reports **52** 3 500–504
[35] Philippova Olga E, Andreeva Assol S, Khokhlov Alexei R, Islamov Akhmed Kh, Kuklin Alexander I and Gordeliy Valentin I,2003, Langmuir, **19**, 7240-7248
[36] Andreeva A S, Fomenkov A I, Islamov A Kh, Kuklin AI, Filippova O E and Khokhlov A R 2005 Polymer Science, Ser A, **47**, 2 194–201
[37] Andreeva Assol S; Philippova, Olga E; Khokhlov, Alexei R, Islamov Akhmed Kh; Kuklin Alexander I 2005 Langmuir: The ACS Journal Of Surfaces And Colloids **21** 4, 1216-1222
[38] Plestil J, Pospisil H, Sikora A, Krakovsky I and A I Kuklin 2003 J Appl Cryst **36**, 970-975
[39] Ostanevich Yu M 1977 Journal of Polymer Symposium **61**, 359-368
[40] Gladkih I A, Kozlov Zh A, Ostanevich Yu M and L Cser 1974 The Neutron Small Angle Scattering by the Time-of-Flight Method (in Russian), JINR Communications, (Dubna) 3-7655
[41] Gladkih I, Ostanevich Yu M, Cser L A, Spectrometer for Investigation of the Neutron Small Angle Scattering by the Time-of-Flight Method (Theory), 1975 (in Russian) JINR Communications (Dubna) P3-8389

[42] Cser L Investigation of biological macromolecular systems with a pulsed neutron source in Neutron Scattering for the Analysis of Biological Structures, 1975 Brookhaven Symposia in Biology No 27 VII 1 -VII 29

[43] Kozlov Zh, Ostanevich Yu M, Cser L A remark about Small-Angle Scattering of Neutrons by the Time-of-Flight Method, in:Proc Neutron Diffr Conf , Petten (1975), RCN Report, RCN 234, p 101

[44] Kuklin A I, Islamov A Kh, Gordeliy V I Neutron News, **16**, 3, 16-18

[45] Kuklin AI, Kutuzov S A, Gabriel A, Eckold G, Utrobin P K, Smirnov A A, Ivankov A, Islamov A Kh, Kovalev Yu S, Feoktistov A, Rogachev A V, Kirilov A S, Gordeliy V I Measurements with New Type of Position Sensitive Detector in Backscattering and Small-angle Setup 2007 Annual report FLNP 129-132, LLB Scientific Report 2003-2004 165

[46] Grabcev B, Balasoiu M, Bica D and Kuklin A I 1994 *Magnitnaia Gidrodinamica* **10**(2) 156

[47] Balasoiu M, Grabcev B and Bica D 1995 *Rom. Rep. Phys.* **47**(3-5) 319

[48] Grabcev B, Balasoiu M, Tarziu A, Kuklin A I and Bica D 1999 J.Magn.Magn. Mater. **201** 140

[49] Kuklin A.I. et al 2011. J.Phys:Conference Series 291(1) 012013

[50] Voronov B I, Gladkih I, Luzanov N I, Kunchenko A B, Ostanevich Yu M and Cser L A 1976 Spectrometer for Investigation of the Neutron Small Angle Scattering by the Time-of-Flight Method (in Russian), JINR Communication, (Dubna) 14-9451

Contrast Variation Application in Small-Angle Neutron Scattering Experiments.

H. B. Stuhrmann

Helmholtz-Forschungszentrum, Geesthacht, Germany, and Institut de Biologie Structurale Jean-Pierre Ebel, CEA/CNRS/UJF, Grenoble, France.

heinrich.stuhrmann@orange.fr

Abstract. The mathematical formalism of contrast variation is presented in terms of an expansion of spherical harmonics. Early attempts of contrast variation in X-ray small-angle scattering are compared with the corresponding more versatile techniques of neutron small-angle scattering. Some applications in life sciences illustrate the power of nuclear spin contrast variation.

1. Introduction

In the early times contrast variation aimed at the elimination of intra-particle background scattering in X-ray small-angle scattering [1]. X-ray small-angle scattering from solutions of myoglobin in solvents of different electron density allowed the determination of three basic scattering functions, one of them being due to the scattering from the shape of the protein molecule [2]. As the shape of a molecule is described by much less parameters than its total structure the probability to find a unique molecular shape from its scattering function is greatly enhanced [3].

2. The spherical harmonics as natural co-ordinates in small-angle scattering.

Spherical harmonics can be considered as the natural co-ordinates for the description of the structure of spherical viruses [4]. Another reason for the use of spherical harmonics is the question of uniqueness of structure determination from small-angle scattering (SAS). In order to discuss this point a reminder of some basic equations is mandatory.

We start from an expansion of $\rho(\mathbf{r})=\rho(r,\omega)$ and of its Fourier transform, the scattering amplitude $A(\mathbf{Q}) = A(Q,\Omega)$, as a series of spherical harmonics, $Y_{l,m}$.

$$\rho(\mathbf{r}) = \sum_{l=0}^{\infty} \rho_{l,m}(r)\, Y_{l,m}(\omega) \quad \Leftrightarrow \quad A(\mathbf{Q}) = \sum_{l=0}^{\infty} A_{l,m}(Q)\; Y_{l,m}(\Omega) \qquad (1)$$

Q is a vector in the Fourier space, and Q is its modulus. ω and Ω are unit vectors in the real space and in the Fourier space, respectively. The intensity of small-angle scattering, $I(Q)$, then is obtained as the sum of absolute squares of the coefficients $A_{l,m}(Q)$

$$I(Q) = 2\pi^2 \sum_{l=0}^{\infty} \sum_{m=-l}^{l} \left|A_{l,m}(Q)\right|^2 \equiv 2\pi^2 \sum_{l=0}^{\infty} I_l(Q) \qquad (2)$$

The coefficients $A_{l,m}(Q)$ are the Hankel transforms of $\rho_{l,m}(r)$. They may be calculated from the polar co-ordinates (r_n, θ_n, φ_n) of the N atoms of a molecule.

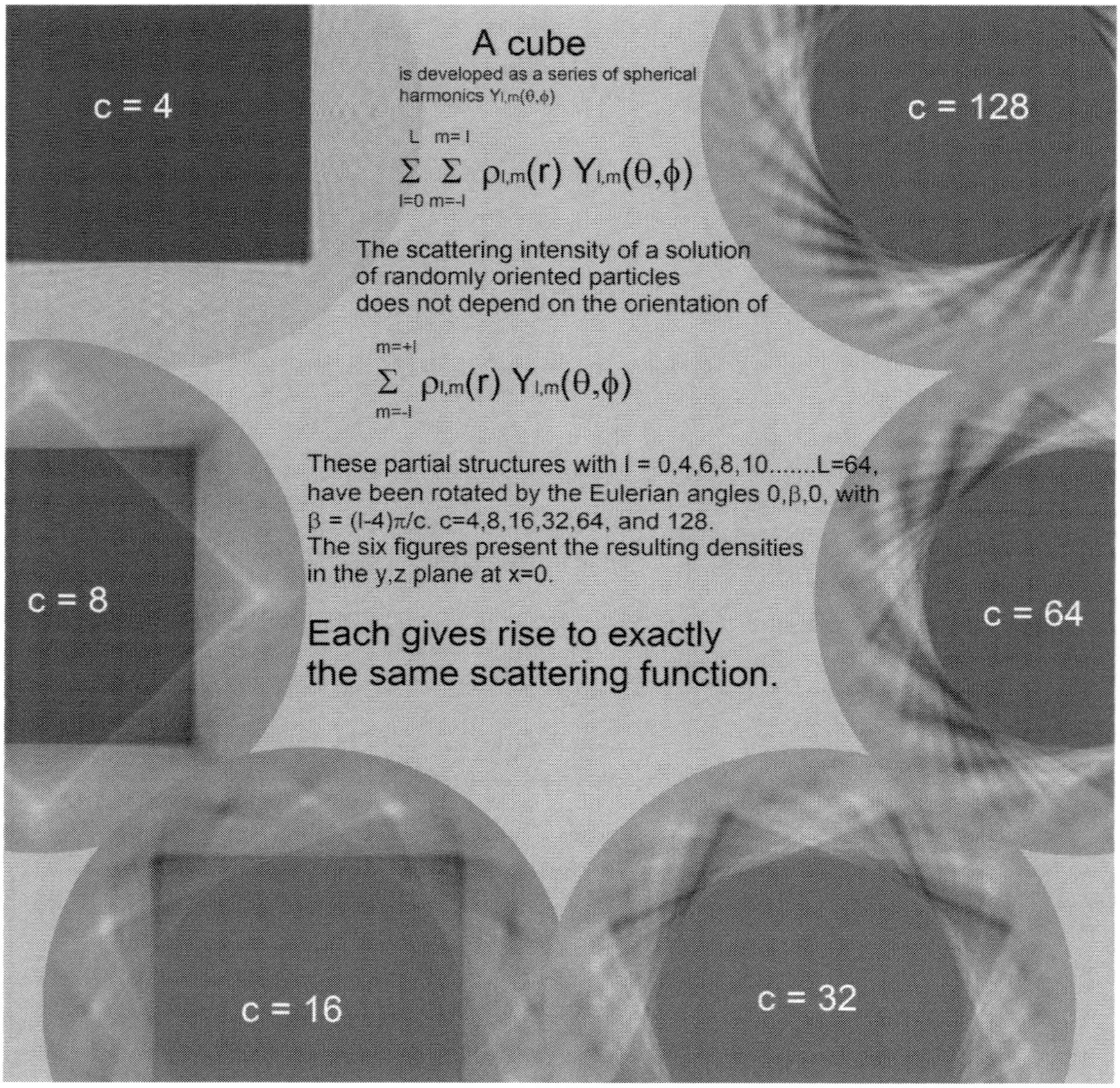

Fig. 1. Various density maps giving rise to the scattering function of a cube.

$$A_{l,m}(Q) = \sqrt{\frac{2}{\pi}}\, i^l \int_{r=0}^{\infty} \rho_{l,m}(r)\, j_l(Qr)\, r^2 dr = \sqrt{\frac{2}{\pi}}\, i^l \sum_{n=1}^{N} j_l(Qr_n) Y^*_{l,m}(\theta_n, \varphi_n) \qquad (3)$$

j_l are the spherical Bessel functions. $Y^*_{l,m} = (-1)^m Y_{l,-m}$ is the conjugate complex of $Y_{l,m}$. Each partial structure $\rho_{l,m}(r) Y_{l,m}(\theta,\varphi)$ gives rise to a partial scattering function $|A_{l,m}(Q)|^2$.

The rotation of a single set of partial structures, $\sum_{m=-l}^{l} \rho_{l,m}(r) Y_{l,m}(\theta,\varphi)$, by an arbitrary angle with respect to the rest of the structure has no influence on $I_l(Q)$ defined in (2), while $\rho(\boldsymbol{r})$ may change considerably [5]. Some density maps of $\rho(\boldsymbol{r})$ giving rise to the scattering function of a cube are shown in Fig. 1. The rotation of the partial structures of the cube by l-dependent angles blurs the sharp contours of the original structure. Hence, the guess may be allowed that there is only one shape which may be associated with the scattering function of a cube.

Let us turn to the molecular shape determination as it has been proposed 40 years ago [3]. It was assumed that the surface of the shape could be described by a unique function. $F(\omega) = \sum_{l=0}^{L} \sum_{m=-l}^{l} f_{l,m} Y_{l,m}(\omega)$. The shape scattering function is developed as a power series of Q^2 [3].

$$I_C(Q) = 2\pi^2 \sum_{k=0}^{\infty} b_k Q^{2k} \text{ where}$$

$$b_k = \sum_{l=0}^{k} \sum_{p=0}^{k-l} \sum_{m=-l}^{l} \frac{d_{l,p}\, d_{l,k-p}\, f^{(l+p+3)}_{l,m}\, f^{*(2k-l-p+3)}_{l,m}}{(l+p+3)(2k-l-p+3)}, \qquad f^{(q)}_{l,m} = \int_{\omega} (F(\omega))^q\, Y^*_{l,m}(\omega)\, d\omega \qquad (4)$$

(7)

Computer simulations with model bodies indicate that the low resolution shape determination from error-free data is unique, even when very limited ranges are used in the simulated curves [6].

3. Solvent contrast variation Introduction

In a first paper, the idea of contrast variation has been illustrated by a simulated model, consisting of a cube with a non-uniform density [1]. The structure of the model is described by two terms, the shape $\rho_C(\boldsymbol{r})$ multiplied by the contrast ρ, and the internal structure $\rho_S(\boldsymbol{r})$

$$\rho(\boldsymbol{r}) = \rho\, \rho_C(\boldsymbol{r}) + \rho_S(\boldsymbol{r}) \qquad (5)$$

The contrast ρ is the difference between the average scattering density of the dissolved particle and that of the solvent. The shape $\rho_C(\boldsymbol{r})$ has the value 1 inside the volume excluded to the solvent. The intensity diffracted by $\rho(\boldsymbol{r})$ is the absolute square of the corresponding amplitude, $|A(\boldsymbol{Q})|^2 = |\rho A_C(\boldsymbol{Q}) + A_S(\boldsymbol{Q})|^2$. For randomly oriented particles we obtain from (2):

$$I(Q)=\sum_{l=0}^{\infty}\sum_{m=-l}^{l}\left|\rho A_{l,m}^{(C)}(Q)+A_{l,m}^{(S)}(Q)\right|^{2}\equiv\rho^{2}I_{C}(Q)+\rho I_{CS}(Q)+I_{S}(Q) \qquad (6)$$

Among the three basic scattering functions of solvent contrast variation, $I_S(Q)$ is the only one which in favorable cases can be measured directly at zero contrast, $\rho = 0$. The shape scattering function $I_C(Q)$ is dominant at high contrast. The cross term $I_{CS}(Q)$ reflects the convolution between the shape with the internal structure. As $A_{0,0}^{(S)}(0)=\int_V \rho_S(\boldsymbol{r})dV=0$, both $I_{CS}(0)$ and $I_S(0)$ are zero.

Sperm whale myoglobin was the first protein studied by X-ray small-angle scattering using solvent contrast variation. The electron density of the solvent was changed by adding glycerol to water. The seven small-angle scattering curves shown in Fig.2 were analyzed in term of the basic scattering functions of (6).

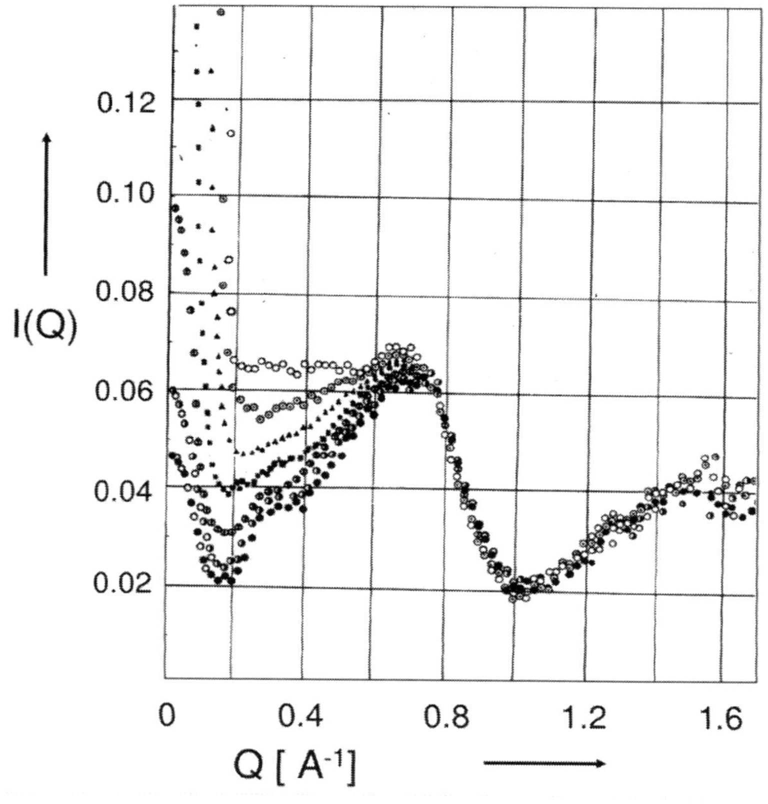

Fig. 2. X-ray small-angle scattering of myoglobin in glycerol/water mixtures. The weight concentration of glycerol varies between 0 (open circles) and 94% (full circles). At large Q small-angle scattering is mainly due to the internal structure $\rho_S(\boldsymbol{r})$.

The shape scattering function $I_C(Q)$ of myoglobin is developed as a power series of Q^2. Up to 9 coefficients of the power series were used for the determination of the shape of the myoglobin molecule. The result from small-angle scattering [5] shown in Fig. 3 is in good agreement with the model from crystallographic studies [8]. More recent examples of protein shape determination have been reviewed by Vachette and Svergun [9].

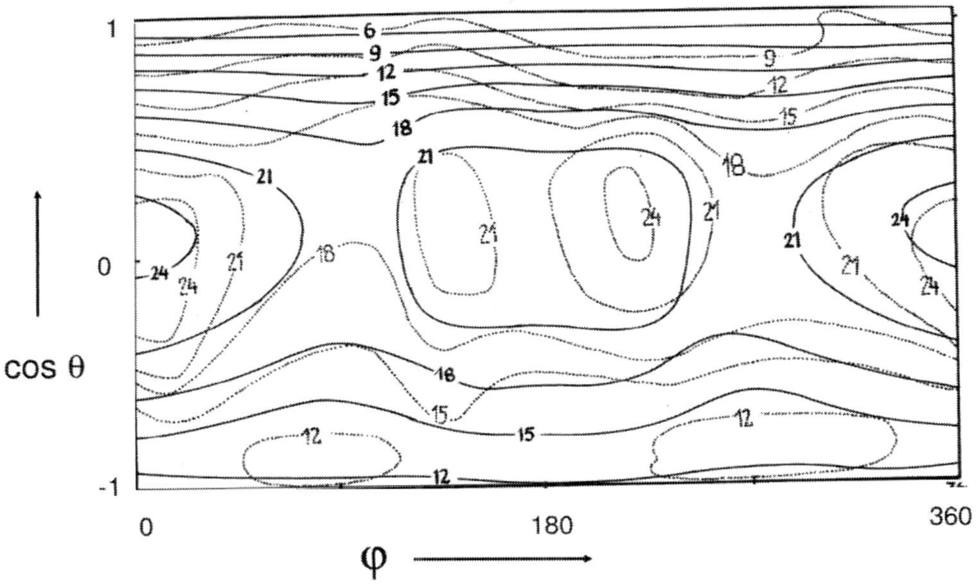

Fig. 3. The shape function $F(\omega) = F(\cos\theta, \varphi)$ of myoglobin determined from the shape scattering function $I_C(Q)$. The coefficients $f_{l,m}$ up to l=3 of $F(\omega)$ have been determined. The distances of the isohypse lines are given in Å.

The internal structure is more complicated than the shape. Its elucidation is often restricted to the analysis of the variation of the apparent radius of gyration with the contrast ρ [10].

$$R^2 = R_C^2 + \frac{\alpha}{\rho} - \frac{\beta}{\rho^2} \tag{7}$$

The coefficients α and β provide an unmistakable feature of the low-resolution of complex particles in solution. Particles with a high density core (e.g. ribosomes, ferritin) will give rise to a negative α whereas particle with a low density core (e.g. nucleosome core particle, low density lipoprotein) give rise to positive α, in a spherical approximation. A non-vanishing β will be due to a dipolar structure, i.e. the centers of mass of different components do not coincide.

Having said this, we find ourselves in the field of neutron scattering. In fact, there are only very few examples of contrast variation using X-ray small-angle scattering [2,11]. But there are considerably more applications of contrast variation in neutron scattering.

4. Neutron scattering from hydrogen

Nearly all applications of contrast variation in neutron scattering use the extraordinary properties of the interaction of neutrons with the hydrogen nuclei ^{1}H (=H) and ^{2}H (=Deuterium). More recent applications use dependence of neutron scattering length $b_{coh.}$ of these hydrogen isotopes on their nuclear polarization P(H) and P(D), respectively.

$$b_{coh}(H) = \left[-0.374 + 1.456\, pP(H)\right] 10^{-12}\ cm$$
$$b_{coh}(D) = \left[+0.667 + 0.27\ \ pP(D)\right] 10^{-12}\ cm \tag{8}$$

This equation holds if a completely polarized neutron beam is used. The polarization p of the neutron beam then will be +1 or -1. The nuclear polarization may assume values between -1 and +1. The methods for achieving neutron and nuclear polarization will be presented in the other paper.

Similarly, there is a strong variation of the cross section of incoherent scattering with polarization.

$$\sigma_{inc.}(H) = \left[105\left(\frac{3}{4} - \frac{1}{2}pP - \frac{1}{4}P^2\right)\right] 10^{-24}\ cm^2$$
$$\sigma_{inc.}(D) = \left[1.0\ \left(2\ - pP\ -\ P^2\right)\right] 10^{-24}\ cm^2 \tag{9}$$

There is no incoherent scattering for P=1.

5. Solvent contrast variation by isotopic substitution

Most of the applications of neutron small-angle scattering rely on the difference between the lengths of coherent scattering of H and D. This difference is large compared with scattering lengths of other elements. Another fact is also important: hydrogen is abundant in organic matter. The variation of the scattering density of organic solvents due to isotopic substitution in neutron scattering is much larger than equivalent methods in X-ray scattering. Solvent contrast variation by isotopic substitution is almost exclusively used in neutron scattering.

Neutron small-angle scattering in H_20/D_20 distinguishes clearly between lipids material protein and RNA/DNA (Fig. 4). Most of the applications aim at the internal structure of complex macromolecules. As an example we cite the studies on the large subunit of E.coli ribosomes, which consists of rRNA by two third of its mass and more than thirty different ribosomal proteins. The variation of the apparent radius of gyration with the D_2O content of H_20/D_20 mixtures provides a low resolution model, with a high RNA content close to the centre and proteins preferring the surface region [12,13]. Deuteration of the rRNA confirmed this model [14] (Fig. 5). The shape of this ribosomal subunit being asymmetric [15], the elucidation of a more detailed internal structure had to be left to labeling techniques using specific deuteration (see below).

The analysis of the internal structure is greatly simplified with spherical particles, because of the absence of spherical harmonics with $l > 0$ in (2).

$$I(Q) \approx \left| \int_{r=0}^{\infty} \rho_{0,0}(r)\, j_0(Qr)\, r^2 dr \right|^2 = \left| \int_{r=0}^{\infty} \rho_{0,0}(r) \frac{\sin Qr}{Qr} r^2\, dr \right|^2 \tag{10}$$

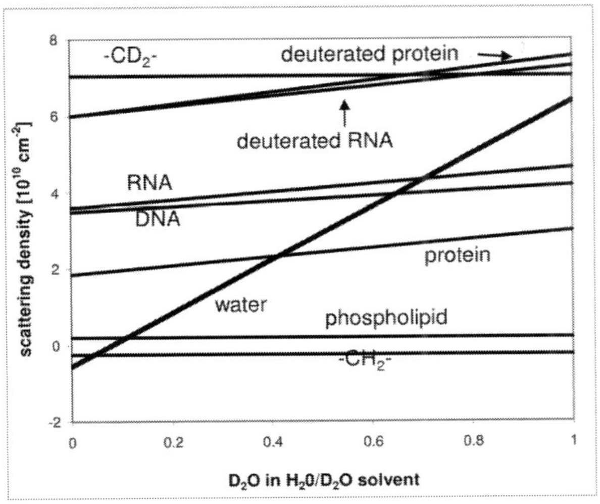

Fig.4. The scattering density of water and of some components of biological macromolecules. The scattering density of H_20/D_20 mixtures varies between -0.56 10^{10} (H_20) to 6.4 10^{10} cm^{-2} (D_20) covering the scattering densities of the main components of living cells.

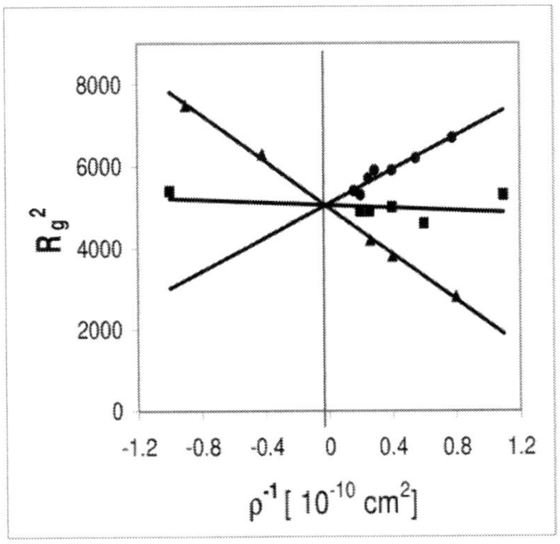

Fig. 5. The variation of the radius of gyration, Rg, of the large subunit of E. coli ribosomes with the contrast ρ. The deuteration of the ribosomal proteins has been varied from 0 (Δ) through 40% (□) to about 80% (O). After [14].

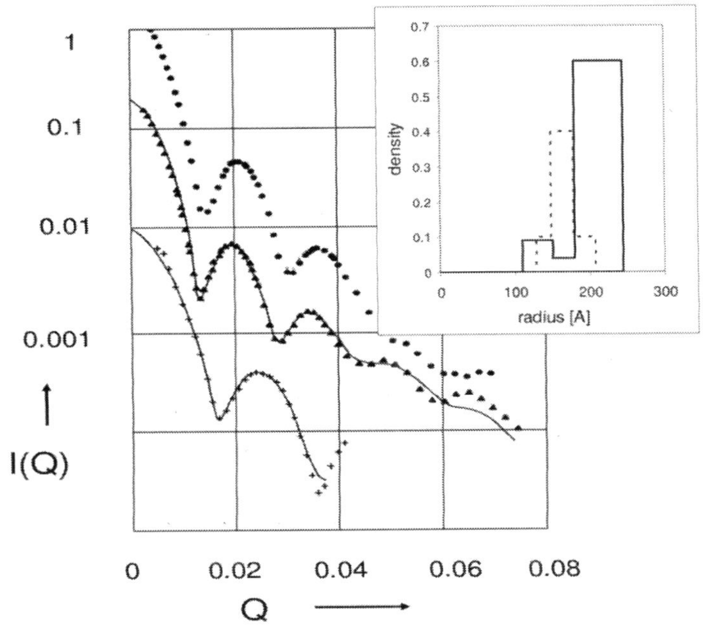

Fig. 6 Angular distribution of neutrons scattered by cauliflower mosaic virus. Solution in water (●), solution with 65% D_2O in the solvent (▲), solution with 42% D_2O in the solvent (+). The insert shows the radial distribution of DNA (broken line) and of protein (line) [16]..

As an example we shall cite the studies on spherical virus, the cauliflower mosaic virus [16]. The virus has a diameter of 50 nm with a molecular weight of 22.8 10^6 dalton. Its double stranded DNA has a molecular weight of about 4 10^6 dalton. Fig. 6 shows the intensity of neutron small-angle scattering from the virus in H_20, 42% D_20 (protein masked), and 65% D_20 (DNA masked). The very high intensity of the first side maximum in all buffers is due to the scattering from a hollow sphere indicating that both the protein and the DNA are essentially distributed on shells on the outer part of the virus [16]. A more quantitative analysis is based on a concentric spherical shell model shown in the insert of Fig.6. The solid curves in Fig. 6 are calculated from this model [16].

6. Specific isotopic labeling

Specific deuteration has led to unique applications of neutron scattering. With this technique, single polymer chains can be studied amongst others, which are identical except for their hydrogen isotope content. The determination of the radius of gyration of polymethylmethacrylate in a glassy matrix elucidated the conformation of a polymer in solid state [17]. Similar studies were done with dilute and semidilute solutions of polystyrene [18].

Biological macromolecules often are complexes of several unique components (subunits) of identical or different chemical composition. Sometimes, their dissociation and reconstitution offers en elegant way of selective deuteration. Let $\rho_O(\boldsymbol{r})$ be the structure of non-spherical particle. A region $\rho_L(\boldsymbol{r})$ inside this particle will have an isotopic composition that is different from that of the rest of the particle. The contrast of the label due to the different isotopic composition is ρ. The intensity of neutron scattering is written as

$$I(Q)=\sum_{l=0}^{\infty}\sum_{m=-l}^{l}\left|A_{l,m}^{(O)}(Q)+\rho\, A_{l,m}^{(L)}(Q)\right|^{2}\equiv I_O(Q)+\rho I_{OL}(Q)+\rho^2 I_L(Q) \tag{11}$$

The scattering intensity of the unlabelled particle being known from a separate measurement, we are left with the sum of a cross term $I_{0L}(Q)$ and the scattering function $I_L(Q)$ of the label. The determination of the latter function requires the preparation of a third sample with an intermediate deuteration of $\rho_L(\boldsymbol{r})$. The set of three scattering functions is a basis for internal or label contrast variation.

$$\begin{aligned} I_1(Q) &= I_O(Q) \qquad \textit{no labeling} \\ I_2(Q) &= I_O(Q) + \rho I_{OL}(Q) + \rho^2 I_L(Q) \\ I_3(Q) &= I_O(Q) + \rho/2\, I_{OL}(Q) + \rho^2/4\, I_L(Q) \end{aligned} \tag{12}$$

The sum of $I_1(Q)+I_2(Q)$ diminished by $2I_3(Q)$ yields the scattering function $\frac{\rho^2}{2}I_L(Q)$ of the label. In practice, the intensity of neutron scattering, $I_{1,2}(Q)$, from a mixture of the solutes giving rise to $I_1(Q)$ and $I_2(Q)$ is measured. The difference $I_{1,2}(Q)-2I_3(Q)$ provides the scattering function of the label [19]. The latter procedure largely eliminates interparticle scattering and scattering from contamination and aggregates provided these effects are equal for each sample [20].

The determination of the in situ structure of components of a complex is one of the numerous applications of the method of triple isotopic substitution (TIS) [20].

Biological macromolecules are often composed several well-defined constituents. An important approach to determine their architecture starts with the measurement of the spatial correlation function of two selected components. The number of different distinct pair correlation functions is increased to a point where the total structure emerges. The pair of selected regions $\rho_{L1}(\boldsymbol{r})$ and $\rho_{L2}(\boldsymbol{r})$ is embedded in the total structure $\rho_O(\boldsymbol{r})$. The corresponding amplitudes are $A_1(\boldsymbol{Q}), A_2(\boldsymbol{Q})$ and $A_O(\boldsymbol{Q})$, respectively. Each of the two regions may be labeled, i.e. it may have its isotope ^{1}H substituted by ^{2}H. Each of the labeled region may have an excess spatial scattering length distribution described by $[b(D)-b(H)]\sum_n \delta(\mathbf{r}-\mathbf{r}_n)$. In fact, four different solutions have to be prepared in order to obtain the inter label scattering function from neutron scattering. 1: the unlabelled particle, 2: $\rho_{L1}(\boldsymbol{r})$ is labeled, 3: $\rho_{L2}(\boldsymbol{r})$ is labeled, 4: $\rho_{L1}(\boldsymbol{r})$ and $\rho_{L2}(\boldsymbol{r})$ are labeled. Each of these samples gives rise to the following scattering functions.

$$\begin{aligned}
I_1 &= A_O^2 \\
I_2 &= (A_0+A_1)^2 = A_O^2 + 2A_OA_1 + A_1^2 \\
I_3 &= (A_0+A_2)^2 = A_O^2 + 2A_OA_2 + A_2^2 \\
I_4 &= (A_0+A_1+A_2)^2 = A_O^2 + A_1^2 + A_2^2 + 2A_0A_1 + 2A_0A_2 + 2A_1A_2
\end{aligned} \tag{13}$$

The samples giving rise I_1 and I_4 are mixed and so are the samples giving rise to I_2 and I_3. From the difference between I_1+I_4 and I_2+I_3 one obtains the cross term $2A_1A_2$. This method of mixing the samples has the great advantage that it eliminates interparticle scattering. Using (3) and the development of $\frac{\sin Q|r-r'|}{|r-r'|}$ as a series of spherical harmonics we obtain

$$\begin{aligned}
A_1A_2 &= \sum_{l=0}^{\infty}\sum_{m=-l}^{l} A_{l,m}^{(1)}(Q)\,A_{l,m}^{*(2)}(Q) \\
&\approx \sum_{l=0}^{\infty}\sum_{m=-l}^{l}\sum_{n=1}^{N}\sum_{n'=1}^{N'} j_l(Qr_n^{(1)})\,j_l(Qr_{n'}^{(2)})\,Y_{l,m}(\omega_n^{(1)})\,Y_{l,m}^{*}(\omega_{n'}^{(2)}) \approx \sum_{n=1}^{N}\sum_{n'=1}^{N'} \frac{\sin Q\left|\mathbf{r}_n^{(1)}-\mathbf{r}_{n'}^{(2)}\right|}{Q\left|\mathbf{r}_n^{(1)}-\mathbf{r}_{n'}^{(2)}\right|}
\end{aligned} \tag{14}$$

The super scripts (1) and (2) refer to the hydrogen atoms of the labels $\rho_{L1}(\boldsymbol{r})$ and $\rho_{L2}(\boldsymbol{r})$, respectively. Representing the labels by their center of gravity, one obtains.

$$I_X(Q) = [(I_1(Q)+I_4(Q)] - [I_2(Q)+I_3(Q)] \approx \frac{\sin Qd}{Qd} \tag{15}$$

$I_X(Q)$ images the spatial relationship of the pair of components in question [21]. The Fourier transform of $I_X(Q)$ is the distribution of lengths of all possible vectors connecting the non-exchangeable hydrogen sites in the two labeled regions. The second moment of the length distribution related to a pair of labels is equal to the sum of the squares of their radii of gyration *in situ* plus the square of the separation between their centers [22]. A table of such distances yields the spatial arrangement of the components. In total, $n(n-1)/2$ distances exist between n components. In order to reconstruct their spatial arrangement by triangulation at least $4n-10$ distance values must be known for $n \geq 4$ [23].

The three-dimensional model of the 21 proteins of the small ribosomal subunit is shown in Fig. 7. It has been constructed by triangulation from 105 distances, appreciably more than the minimum of 78 required for a 21 protein structure [23].

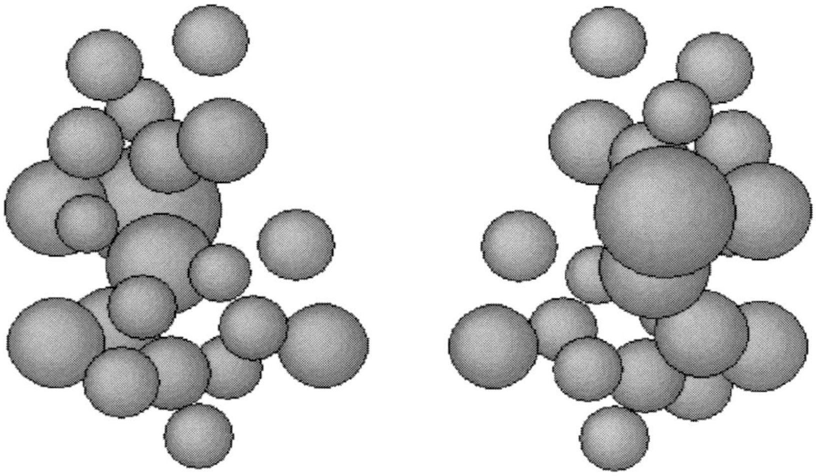

Fig. 7. Two views of the neutron map of the small ribosomal subunit from Escherichia coli. Each protein is represented by a sphere whose volume is the same as that of the protein. The maximum linear dimension of the array is about 190 Å. The two views are related to each other by a 180° rotation about an axis oriented vertically in the plane of the figure. After [23].

7. Nuclear spin contrast variation

In contrast to the methods of isotopic substitution, which require the preparation of several samples, a single sample will sufficient for contrast variation by nuclear spin polarization. Using a completely polarized neutron beam, the variation of the length of coherent scattering and of the intensity of coherent scattering with the polarization of the hydrogen isotopes ^1H and ^2H is given by (11) and (12), respectively.

A high polarization of protons or deuterons is obtained by the method of dynamic nuclear spin polarization (DNP). Without entering into details of this method, we mention that we use the method proposed by Abragam, which works in insulators [24][25]. In this case the nuclear spins are polarized in the presence of a small amount of paramagnetic centers. The temperature is kept between 0.1 and 1 K, and a high magnetic field is used. Microwave irradiation at a frequency slightly below or above the paramagnetic resonance will polarize the nuclear spins in positive or negative direction with respect to the external magnetic field direction, respectively. With solutions of biomolecules in a mixture of deuterated glycerol and heavy water the final values of nuclear polarization will be around 70% (or -70%) for protons and 20% (or -20%) for deuterons after several hours of microwave irradiation. At a temperature of 0.1 K, the polarization of the protons and deuterons remains nearly constant for days.

A proton spin polarized sample is obtained by destroying the deuteron polarization. A deuteron spin sample is obtained by destroying the proton polarization. At a temperature of 0.1 K, differently polarized nuclear spin systems will coexist for many hours without significant change of polarization.

A typical experiment of polarized neutron scattering takes 5 days [26]. Loading the frozen sample and cooling the refrigerator to 1 K takes 6 hours. After calibration of the proton NMR signal and neutron scattering from the sample at P=0, microwave irradiation will polarize the nuclear spins of the sample at a temperature well below 1 K. The microwaves are switched off and the temperature of the sample will drop to 0.1 K. The deuterons are depolarized. Neutron scattering from the proton spin polarized sample is measured for two days. The direction of the neutron spin polarization is changed each ten minutes. Then the same procedure is repeated for the deuteron spin polarized sample for another two days. The sample is unloaded on a Friday afternoon. The data set consists of five spectra of neutron small-angle scattering. The spectra of the sample at P=0, and the spectra of the proton polarized target and deuteron polarized target at two neutron beam polarizations, in direction and opposite to the external field [26].

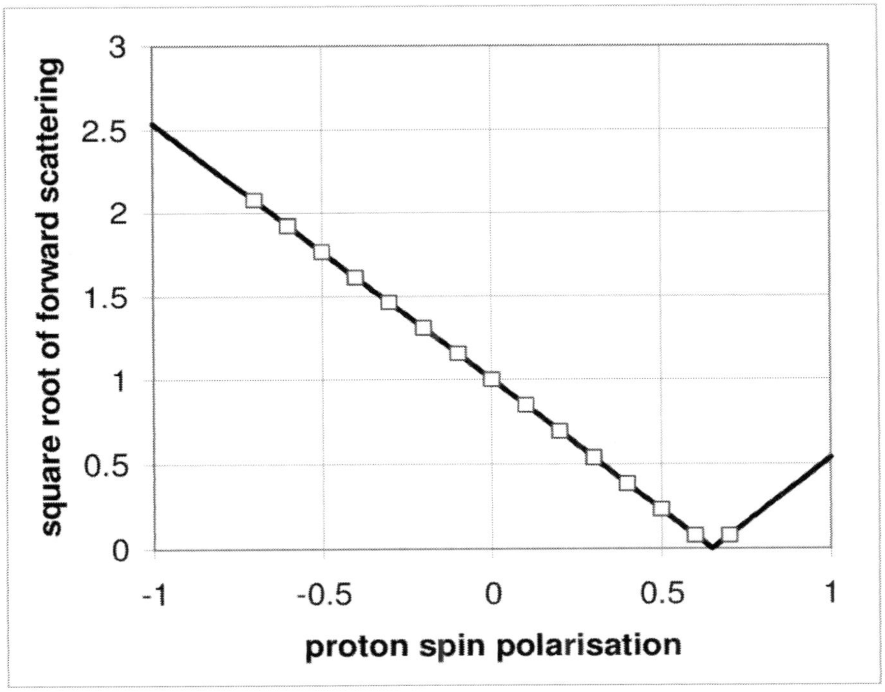

Fig. 8. The variation of zero-angle of a protein in a mixture of deuterated solvent and heavy water (1/1) with proton polarization.

Nuclear spin contrast variation is mainly used as a method of internal contrast variation.

$$I_{(\pm)}(Q) = \sum_{l=0}^{\infty} \sum_{m=-l}^{m=l} \left| U_{l,m}(Q) + p\,P V_{l,m}(Q) \right|^2 \equiv I_U(Q) + p\,P\,I_{UV}(Q) + P^2\,I_V(Q) \qquad (16)$$

$I_{(+)}(Q)$ and $I_{(-)}(Q)$ are obtained with positive and negative polarization p of the neutron beam, respectively. The nuclear polarization P stands for the proton polarization P(H) or deuteron polarization P(D). $U(\boldsymbol{Q})$ is the amplitude of the sample at P=0. With deuterated solvents $U(0)$ will be negative. With increasing proton spin polarization P(H) the amplitude P(H)V(0) will decrease the intensity of forward scattering (Fig. 8). At P(H) = 0.65 the nuclear spin contrast matches the native

Journal of Physics: Conference Series **351** (2012) 012002 doi:10.1088/1742-6596/351/1/012002

contrast of the protein. The matching point changes hardly with RNA or DNA as solute. This means that nuclear spin contrast is an excellent amplifier of an existing contrast, but unlike the method of isotopic substitution it is not suitable for the distinction between chemically different components.

Assuming that the extreme polarizations P are equal, the basic scattering functions are easily obtained from $I_{(+)}(Q)$ and $I_{(-)}(Q)$.

$$I_{UV}(Q) = \frac{1}{2P}\left[I_{(+)}(Q) - I_{(-)}(Q)\right]$$
$$I_V(Q) = \frac{\left[I_{(+)}(Q) + I_{(-)}(Q)\right] - 2I_U}{2P^2} \qquad (17)$$

The gain in contrast due to nuclear polarization considerably extends the application of contrast variation. The neutron scattering studies on ribosomes at the end of the last century may serve as an illustration. While the small 30S subunit of the E.coli ribosome has been studied in great detail (Fig. 7), a similar approach for the large 50S subunit of the ribosome seemed to be prohibitively difficult. Attempts were made to reduce the incoherent scattering by massive deuteration of the solvent and of the solute. The idea of the contrast free 'glassy ribosome' should facilitate the in situ structure determination of the labeled (protiated) component [27]. However, the interference of any residual internal contrast between deuterated rRNA and the deuterated proteins with the strong contrast of the label could not be entirely excluded. It was at that time when first experiments of nuclear spin contrast variation were started. It happened that the optimal composition of samples for this kind of study were those developed for the 'glassy ribosome'. In fact, the scattering density of a completely deuterated ribosome is almost equal to that of a mixture of deuterated glycerol and D_20. The first experiments of proton spin contrast variation on the deuterated 50S subunit with one of its proteins left protiated showed that the forward scattering from the high contrast of the label and the weak contrast of the deuterated ribosome were comparable in size. Thus the analysis of the data had to take into account the shape and internal structure in order to determine the site (and if possible) the shape the labeled ribosomal component [26].

The four additional basic scattering functions from nuclear spin contrast variation of (20) increased considerably the structural information. Clearly, the basic scattering functions from proton spin contrast variation, and in particular the cross term $I_{UV}(Q)$, were most important. Valuable additional structural information came from electron microscopy [28]. A number of ribosomal proteins of the large subunit have been studied by this method. We point out to the fact that the results concerning the co-ordinates of the label with respect to the large ribosomal subunit depend on the knowledge of the low resolution structure of the ribosome known at that time [29].

The study of labeled components in the functional E. coli ribosome with its two protiated tRNAs and a protiated mRNA fragment in a completely deuterated environment was a still larger challenge for two reasons:

1: The contrast of RNA in a deuterated solvent is lower by a factor 1.5 with respect to proteins due to the lower hydrogen content. The same holds for the proton spin added contrast.

2: the occupation density of the ribosomal binding sites by tRNAs does not exceed 40%.

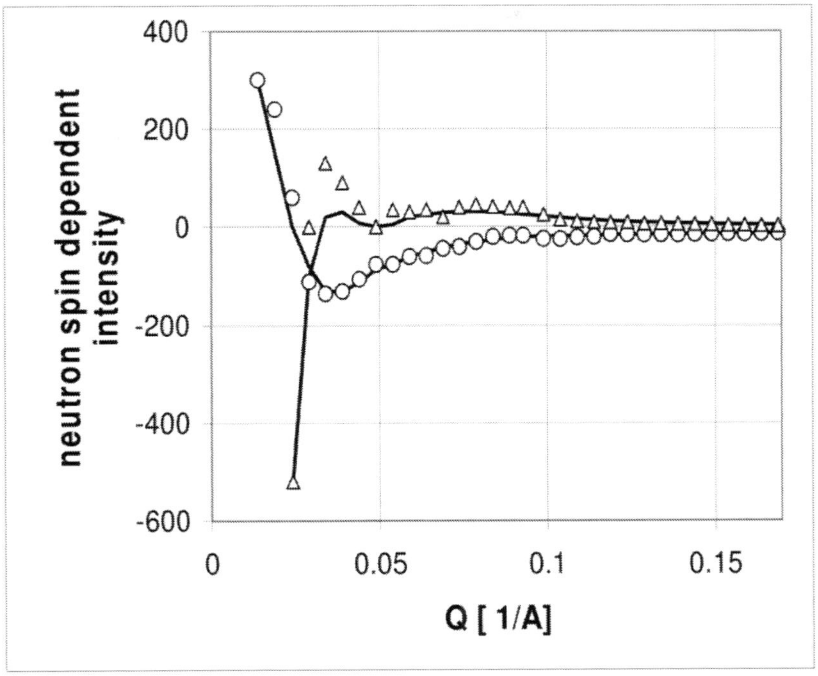

Fig. 9. The spin dependent scattering function $I_{UV}(Q)$ of protiated tRNA bound to the functional complex of the deuterated E.coli ribosome. The polarizations of the proton spins and deuteron spins were 0.7 and 0.15, respectively. For the intensity scale: a solution of the protiated ribosome in the same deuterated solvent would have an I(0) = 8 10^6, concentrations being equal. After [30].

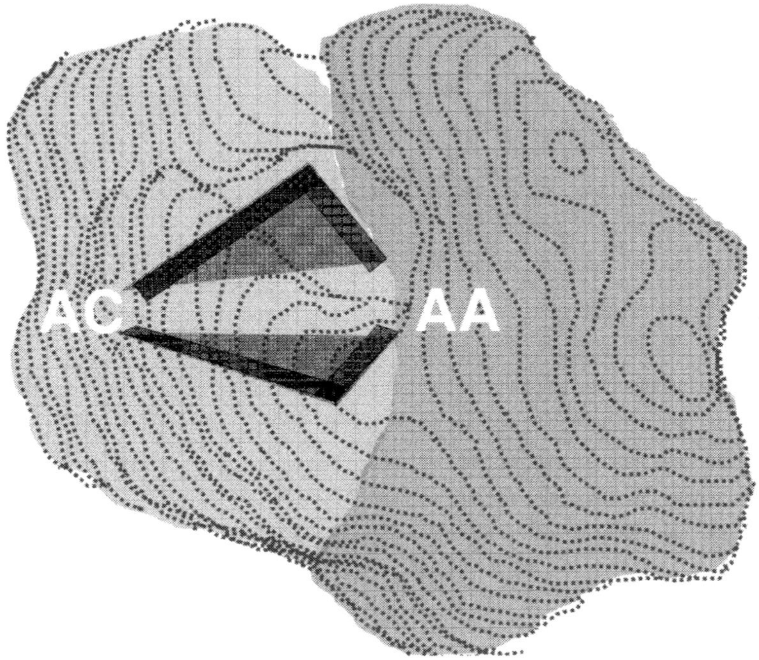

Fig. 10. The site and the orientation of two tRNAs in the E. coli ribosome. The tRNAs are given in the low resolution description of triangles. The anticodon (AC) of the tRNA is close to the neck of the small subunit whereas the aminoacyl group (AA) of the tRNA approaches the central protuberance of the large subunit After [30].

Nevertheless, polarized neutron scattering from the nuclear spin polarized sample clearly revealed the basic scattering functions of proton and deuteron spin contrast variation [30]. The change of the neutron scattering intensity with the polarization direction of the incident neutron beam is shown in Fig. 9 for both the proton spin polarized and the deuteron spin polarized sample.

The cross terms shown in Fig. 9 differ in the sign at very small scattering angles, and so do the contrasts induced by the polarization of protons and deuterons, respectively. As the density of deuterons in the solvent was higher than in the solute, the deuteron spin contrast was negative. Hence the proton spin contrast was positive and so was U(0). At $Q > 0.05$, the intensity is mainly due to the amplitude of the protiated tRNAs.

The determination of the in situ structure of the two tRNAs was done in two steps. In a first step the centre of gravity of the $(tRNA)_2$-mRNA complex was determined. In a next step, the orientation of this complex with respect to the ribosome had to be found. Two facts were helpful: (i) the structure of the tRNA is known to atomic resolution, (ii) the extremes of the structure of tRNA must be rather close to each other. There is also an estimation of the angle between the planes of the tRNAs, which may be as large as 90° [31]. The orientation of the tRNA complex was varied over the whole range of Eulerian angles while the center was kept. The best fit of the experimental data was obtained with orientation of the tRNAs shown in Fig. 10.

8. Conclusion

Contrast variation techniques have found many applications in neutron small-angle scattering both in life sciences and in condensed matter research in general. As the structural aspect prevails in biological applications, we have developed the concept and the mathematical formalism of contrast variation in this frame. Isotopic substitution is still the mostly used technique. It discriminates between chemically different components of complex structures. Nuclear spin contrast variation remains an excellent amplifier of an existing contrast

References

[1] H.B. Stuhrmann, R.G. Kirste, Z. f. physik. Chemie, Neue Folge **46**, 247-250 (1965).
[2] R.G. Kirste, H.B. Stuhrmann, Z. f. physik. Chemie, Neue Folge, **56**, 338-341 (1967).
[3] H.B. Stuhrmann, Z.f. physik. Chemie, Neue Folge, **72**, 177-184 (1970).
[4] Y. Zheng, P.C. Doerschuk, J.E. Johnson, Biophysical Journal 69, 619-639 (1995).
[5] H.B. Stuhrmann, Acta Cryst, 72, 177-184 (1970).
[6] D.I. Svergun, V.V. Volkov, M.B. Kozin, H.B. Stuhrmann, Acta Cryst **A52**, 419-426 (1996).
[7] H.B. Stuhrmann, Z.f. physik. Chemie, Neue Folge, **72**, 185-198 (1970).
[8] G. Bodo, H.M. Dintzis, J.C. Kendrew, W.H. Wyckoff, Proc. Royal Soc. A 253 70 (1959).
[9] P. Vachette, D.I. Svergun, Structure and Dynamics of Macromolecules, HERCULES, Vol. IV, edited by E. Fanchon, E. Geissler, J.-L. Hodeau, J.-R. Regnard, P.A. Timmins, pp 199-237, Oxford University Press (2000)
[10] K. Ibel, H.B. Stuhrmann, J. Mol. Biol. **93**, 255-265 (1975)
[11] L. Mateu, A. Tardieu, V. Luzzati, L. Aggerbeck, A.M. Scanu, J. Mol. Biol. **70**, 105-116 (1972)
[12] I.N. Serdyuk, A.K. Grenader, FEBS Letters, 59, 133 (1975).
[13] H.B. Stuhrmann, J. Haas, K. Ibel, M.H.J. Koch, B. de Wolf, R. Parfait, R.R. Crichton, Proc. Nat. Acad. Sci. USA, 73, 2379-2383 (1976).
[14] R.R. Crichton, D.M. Engelman, J. Haas, M.H. Koch, P.B. Moore, R. Parfait, H.B. Stuhrmann, Proc. Nat. Acad. Sci. USA, 12, 5547-5550 (1977).
[15] H.B. Stuhrmann, M.H.J. Koch, R. Parfait, J. Haas, K. Ibel, R.R. Crichton, Proc. Nat. Acad. Sci. USA, 74, 2316-2320 (1977).
[16] C. Chauvin, Thesis, Etude de la structure interne de virus sphériques par diffusion neutronique, University of Grenoble (1979).

[17] R.G. Kirste, W.A. Kruse, J. Schelten, Die Makromolekulare Chemie, **162**, 299-303 (1973).
[18] J.P. Cotton, B. Farnoux, G. Jannink, J. Chem. Phys. **57**, 290-294 (1972).
[19] I.N. Serdyuk, M. Yu. Pavlov, Makromol. Chem. Macromol. Symp. **15**, 167-184 (1988).
[20] M. Yu. Pavlov, I.N. Rublevskaya, I.N. Serdyuk, G. Zaccai, R. Leberman, Yu. M. Ostanievich, J. Appl. Cryst. **24**, 243-254 (1991).
[21] D.M. Engelman, P.B. Moore, Proc. Nat. Acad. Sci. USA, 69, 1997 (1972)
[22] V. Ramakrishnan, M. Capel, M. Kjeldgaard, D.M. Engelman, P.B. Moore, J. Mol. Biol. 174, 265 (1984).
[23] Capel M.S. et al. Science, 238, 1403 (1988).
[24] A. Abragam, Phys. Rev. **98**, 1729.
[25] A. Abragam, M. Goldman, Nuclear Magnetism, Order and Disorder, Oxford: Clarendon ((1982).
[26] H.B. Stuhrmann, Rep. Prog. Phys. **67**, 1073-1115 (2004).
[27] R.P. May, H.B. Stuhrmann, K.H. Nierhaus, Neutrons in Biology, edited by B.P. Schoenborn, Basic Life Sciences Vol. 27, A. Hollaender, General Editor, pp; 25-45, Plenum Press (1984).
[28] J. Frank, P. Penczek, R. Grasucci, S. Srivastave, J. Cell. Biol. 115, 597-605 (1991).
[29] R. Willumeit et. al., Biochim. Biophys. Acta, **1520**, 7-20 (2001).
[30] H.B. Stuhrmann, K.H. Nierhaus, Neutrons in Biology, edited by Schoenborn and Knott, pp. 397-413, Plenum Press (1996).
[31] V. Lim, C. Venclovas, A. Spirin, R. Brimacombe, P. Mitchell, F. Müller, Nucleic Acids Research, 20, 2627-2637 (1992).

Time-resolved Polarized Neutron Scattering from Dynamic Polarized Nuclear Spin Targets.

H.B. Stuhrmann

Helmholtz-Forschungszentrum, Geesthacht, Germany and Institut de Biologie Structurale Jean-Pierre Ebel, CEA/CNRS/UJF, Grenoble, France

heinrich.stuhrmann@orange.fr

Abstract. The experiments of time-resolved polarized neutron scattering from dynamically polarized nuclei of the EHBA-Cr(V) complex are reviewed. The analysis of the data is extended to protiated samples, and the existence of polarized proton spin domains in the absence of deuteration is shown. It is concluded that pulsed neutron sources may gain considerable importance in time-resolved neutron scattering.

1. Introduction

It is the huge pseudomagnetic moment of the hydrogen isotope ^1H which justifies its use in polarized neutron scattering. The concomitant changes of the coherent scattering length, b, and of the cross section of incoherent scattering $\sigma_{inc.}$ of the hydrogen isotopes ^1H (=H) and ^2H (=Deuterium) with nuclear polarization P ($-1 \le P \le +1$) and neutron polarization p (± 1) are given by the following relations [1,2]:

$$b(H) = (-0.374 + 1.456\, pP(H)\ )\, 10^{-12}\, cm$$
$$\sigma_{inc.}(H) = 105\ (\ 0.75 - 0.5p\, P(H) - 0.25\, P^2(H))\ 10^{-24}\ cm^2$$

$$b(D) = (0.667 + 0.27\, pP(D)\ )\ 10^{-12}\, cm$$
$$\sigma_{inc.}(D) = \ (2 - p\, P(D) - P^2(D))\ 10^{-12}\ cm^2 \qquad (1)$$

P(H) and P(D) are the polarization of protons and deuterons, respectively. While almost completely polarized neutron beams are obtained routinely, a high nuclear polarization is achieved less readily.

In terms of its neutron scattering length, deuterium behaves very much like other nuclei, whereas normal hydrogen is 'exotic', with a scattering length half the size of deuterium but negative. Moreover, the neutron scattering length varies strongly with the relative orientation of the neutron spin interacting with the proton spin, which led to the following explicit formulation:

"Lorsque la cible diffractante contient des noyaux de spin non nul, les meilleures conditions d'expérimentation sont en principe réalisées lorsque simultanément les neutrons incidents et les noyaux de la cible sont polarisés" [3].

Moreover, it has been argued that for very complex crystals it is beneficial to minimize in some way the amount of data to be refined in order to get the atomic positions of interest [4]. One way is to selectively depolarize nuclei whose Larmor frequencies differ either because of their magnetic moments or because of their proximity to paramagnetic ions in the crystal. In this way, it is concluded that one can control and vary the contribution to the structure factor of individual atoms within the unit cell [4].
The first step in this direction has been taken in the late nineties, when the possibility to use tyrosyl radicals as labels in catalase was studied. The idea was and still is that the formation of polarized proton spin domains near the tyrosyl radicals would enhance their scattering power to such a degree that a radical density map of catalase might be established. In a joint effort of the Institut de Biologie Structurale, Grenoble and of the Paul-Scherrer Institut (PSI), Villigen, it has been shown, that tyrosyl radicals do support DNP. This finding led to an enlarged collaboration which included the Service de Physique de l'Etat Condensé, CEA Saclay, and the Universities of Mainz and Munich. It was agreed that the study of the build up of proton polarization by DNP in organic radicals of different molecular weight would help to analyze the mechanism of DNP with radicals in more complex biological structures. The experiments of time-resolved neutron scattering were done at the Institute Laue-Langevin (ILL), Grenoble, at the Orphée reactor of the Laboratoire Léon Brillouin (LLB), Saclay, and at the PSI, Villigen. In the following we will recall some methods of neutron and nuclear polarization.

2. Neutron polarization

In small-angle scattering, cold neutrons with a wavelength $\lambda > 4$ Å and band width $\Delta\lambda/\lambda \approx 0.1$ are polarized by a magnetic super mirror, which reflects only one orientation of the spin [5]. A magnetic guide field between the neutron polarizer and the sample preserves their polarization.

Another important ingredient is the neutron spin flipper. Various kinds exist: an adiabatic spin flipper relies on the method of adiabatic fast passage (AFP). It requires a radio frequency (rf) coil to be placed into a magnetic field gradient such the neutron will be resonant while it crosses the flipper [6]. Another way of reversing the direction of neutron polarization is provided by a flat coil spin flipper, which is due to F. Mezei [7]. The polarization of the beam measured at the sample is better than 0.96.

3. Nuclear polarization

As the magnetic moment of a nuclear spin is very small, the natural polarization even under the conditions of a strong magnetic field and low temperatures remains negligibly small. In a field of 2.5 T, a proton polarization of 0.25% is reached at a temperature of 1K. In order to obtain higher nuclear polarization, the method of dynamic nuclear polarization is used. As we are interested in the nuclear polarization of organic hydrogenous material, the method of Abragam is used [8]. In this case, the nuclear spins are polarized in the presence of paramagnetic centres that have been added in small amount. The temperature is kept between 0.1 and 1 K, and a magnetic field of 2.5 or 3.5 T is used.

The macroscopic aspects of DNP are well understood in the frame of spin temperature theory [9]. The thermodynamic model assigns heat reservoirs to various degrees of freedom of the electronic and the nuclear spin system that are coupled via mutual and external interactions. The mechanism of DNP then is described as a two step process: the cooling of the non-Zeeman reservoir by a non-saturating microwave field and the subsequent transfer of entropy to the nuclear spin system via thermal mixing. Irradiation of the system at a frequency below the electron paramagnetic resonance,

$\omega < \omega_e$, with $\omega_e - \omega \approx \omega_I$ (nuclear Zeeman transition) leads to positive nuclear polarization. Irradiation at a frequency $\omega > \omega_e$ leads negative nuclear polarization. The magnitude of the mixing is strongly influenced by the microscopic structure of the material and in particular the nuclear spins close to the unpaired electrons [10].

In a simple microscopic picture of DNP, the nuclear polarization develops near the paramagnetic centers through the electron nuclear dipolar interaction decreasing with the third power of the distance between electron and nuclear moments. More distant bulk protons are polarized by dipolar interactions between nuclei (spin diffusion). The same mechanism in the reverse order is responsible for nuclear relaxation in most insulating solids.

In spite of considerable theoretical and experimental work, the mechanism of DNP is not fully understood. The problem is complicated by the fact that the local field created by the paramagnetic moment shifts the Larmor frequencies of close nuclei with respect to that of bulk nuclei, thus suppressing spin diffusion. Furthermore, interaction between the paramagnetic centers (the electron non-Zeeman reservoir in the thermodynamic model) plays an important role in the establishment of a common spin temperature between close and bulk nuclei and also between isotopes in a sample.

4. Experimental

A facility for DNP meeting the requirements of neutron diffraction has the following features:

1. The sample is cooled by liquid helium (^{4}He).

2. For temperatures below 1 K, the ^{4}He bath of the sample cell is coupled to the mixing chamber of the dilution refrigerator, the ^{3}He/^{4}He mixture of which reaches temperatures slightly below 0.1 K.

3. The liquid helium consumption is moderate: between 1 and 2 liter/h.

4. The magnet is designed so as to allow a large solid angle for the scattered neutrons.

5. The sample exchange time is short: less than 0.5 h for 1 K cells, less than half a day for 0.1 K cells.

6. The sample volume is slightly less than 1 ml. A typical size of the platelets is 15 x 15 x 3 mm^3.

The NMR signal is detected by a continuous wave (cw) spectrometer. It consists of a radio frequency (rf) source, a NMR coil, a detection system called Q-meter, and a digitization unit. The system is sensitive to a change of the impedance of the sample due to polarization. For proton NMR with magnetic fields of 2.5 T and 3.5 T the system is operating at 106 MHz and 150 MHz respectively. Through a coaxial switch inserted in the (3/2)λ cable connecting the NMR coil with the Q-meter a separate dedicated rf system can be connected to the coil. It is used to manipulate selectively the nuclear spin systems of the samples, e.g. to perform AFP polarization reversals [11] or to destroy the nuclear polarization.

The microwave system consists of a source, a wave guide, and a multimode cavity. The high frequency of 70 GHz source is a carcinotron operated at B = 2.5 T. The frequency is easily modulated: a frequency jump of 300 MHz takes only a few microseconds [12]. The experiments with time resolved polarized neutron scattering were done with a different microwave system: Two IMPATT diodes of 100 mW output power, tuned to frequencies corresponding to positive and negative DNP, respectively (97.0 GHz and 97.5 GHz, with B= 3.5 T), can be connected alternatively to the sample cavity (Fig. 1) by an electro-mechanical wave guide switch. This is the microwave system from PSI (Fig. 1).

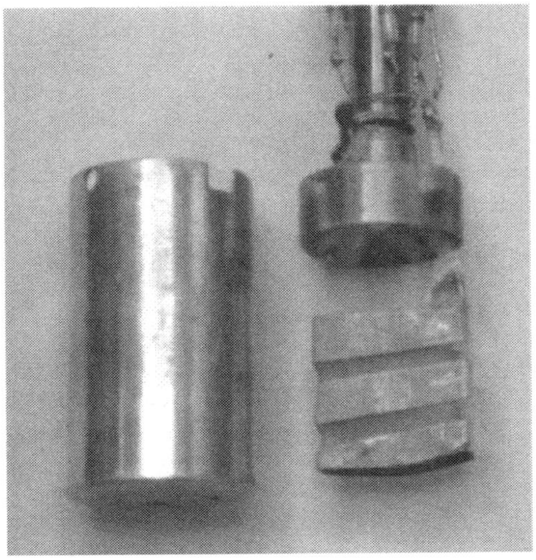

Fig. 1. The open sample cavity. The sample fits into the flat NMR coil fixed to the microwave guide.

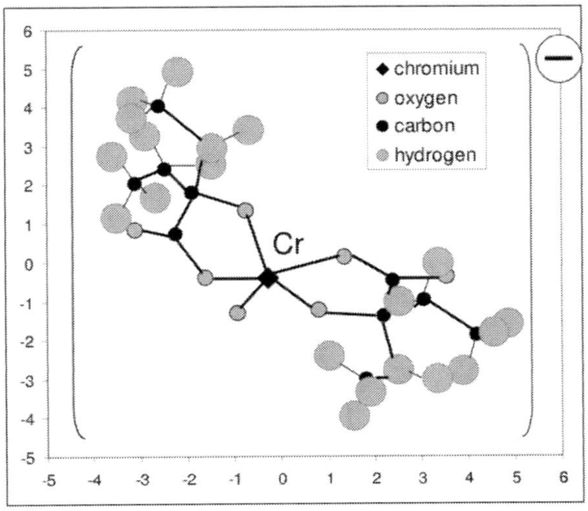

Fig. 2. The molecular structure of EHBA-Cr(V) anion. The 20 hydrogen atoms of the four ethyl groups are shown as large grey spheres. Units in Å.

In many cases DNP has been performed in the presence of a chromium-(V) complex, $C_{12}H_{22}CrO_7Na.H_2O$ (EHBA-Cr(V)) (Fig 2), dissolved in a mixture of glycerol and water. The synthesis of EHBA-Cr(V) is done in one step [13]. EHBA-Cr(V) is very soluble in water. It is stable in acid pH. Glassy slabs of 3 mm thickness are obtained by injecting the solution into a copper mould cooled to by liquid nitrogen. This is done in a dry nitrogen atmosphere of a glove box.

Dynamic nuclear spin contrast

By dynamic nuclear spin contrast we mean a time-dependent nuclear spin contrast, which is observed in time-resolved polarized neutron scattering. The nuclear polarization obtained by DNP is far from the polarization at thermal equilibrium. It will decay more or less readily depending on the temperature of the sample. At a temperature of 1 K and in magnetic field of 3.5 T, the relaxation time is about 10 to 20 minutes. The build up of a local proton polarization near paramagnetic centres or its decay is expected to occur in a much shorter time. An abrupt and selective change of the polarization leads to a strong polarization gradient near paramagnetic centres. The observation of the subsequent relaxation process by neutron scattering tells us how the perturbed polarization distribution finds its way to equilibrium.

Among the various ways of creating a strong spatial non-equilibrium polarization, the method of AFP seems to be the obvious choice. An rf scan restricted to the central peak of the proton NMR signal, will lead to a reversal of proton spins not too close to paramagnetic centers, i.e. of the bulk protons. The polarization of the protons close to the paramagnetic centre, i.e. of the 'close' protons, will hardly be affected. An rf scan across the wings of the proton NMR signal will reverse the

polarization of the ‘close’ protons. This method requires a reasonable high initial nuclear polarization that is achieved after prolonged microwave irradiation.
The other way is to address to the paramagnetic centers directly. Again the method of AFP can be considered. The more promising way is a periodic change in the direction of DNP. This is most easily done by changing the microwave frequency. Typically, the direction of dynamic polarization was changed each 10 seconds. During a half-cycle 100 pictures were taken. The cycle was repeated several thousand times [14]. Most of the data shown below have been obtained in that way.

4.1. EHBA-Cr(V) in deuterated solvents.

We recall some of the results on this compound which have been published some years ago [14][15]. Then we will extend the analysis to that part of experimental data which remained unexploited for reasons which will be outlined below.

Time-resolved neutron scattering from a solution of EHBA-Cr(V) in a mixture of deuterated glycerol and D_2O shows most clearly the build up of proton polarization (Fig. 3). Within 10 seconds the polarization of the protons of the EHBA-Cr(V) molecules changes from -0.35 to +0.22, i.e. by 57%, whereas the polarization of the few protons of the solvent (< 0.02) as monitored by NMR varies by 12 %, only. This difference in the evolution of the polarization between the close and the bulk protons reflects the mechanism of DNP: a strong initial gradient develops due to the fast polarization of the protons close to the paramagnetic centre that then spreads out to the bulk with a slower rate [14].

At this point, the immediate build up of a proton polarization domain near paramagnetic centers may be attributed to two circumstances: (i) The most evident one is the gradient of isotopic composition. The scarcity of protons in the strongly deuterated solvent impedes nuclear spin diffusion from the high proton density region of the radical molecule to the deuterated solvent almost completely deprived of its protons. (ii) There may be an influence of the local magnetic field of the paramagnetic centre which changes the Larmor frequency of the close protons rendering them less apt for nuclear spin state exchange. This spin diffusion barrier might also contribute to a confinement of nuclear polarization near paramagnetic centers. A crude estimate yields 8 Å for the radius of an isotropic diffusion barrier. This is the distance at which the difference in the z-component of the field created by the paramagnetic centre at neighboring nuclei is comparable with the bulk NMR line width. The field gradient is of the order of $3\mu_B/r^4$, which gives 250 $GÅ^{-1}$ at 5 Å, roughly outside the EHBA-Cr(V) molecule [14].

In order to show the efficiency of a spin diffusion barrier in creating a polarization gradient near paramagnetic centers the time resolved neutron scattering experiments on proton spin polarized samples should done with protiated samples, i.e. in the absence of deuteration. In fact the French-Swiss-German collaborative team (see Acknowledgements) started experiments on EHBA-Cr(V) in glycerol/H_2O at the very beginning, i.e. in 2000. The hope to see a neat contrast from a local proton polarization was drowned by the apparent impossibility to extract any structural information from time-resolved neutron scattering intensity data dominated by incoherent scattering. We will come back to this point further below.

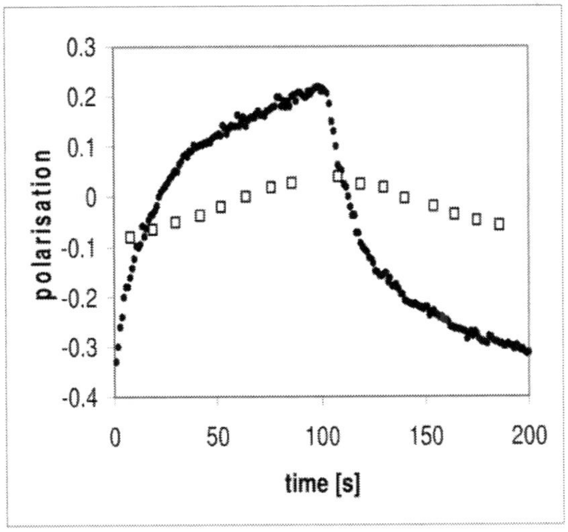

Fig.3. Close proton polarization deduced from the time-resolved neutron scattering data (o) and bulk proton polarization measured be NMR (□). The deuteration of the solvent is 0.98 [14].

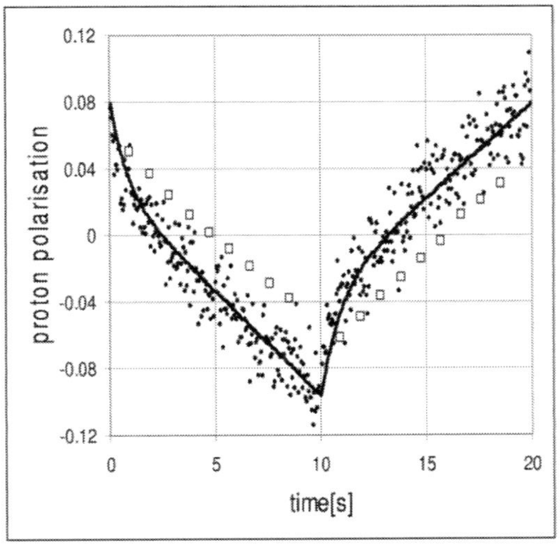

Fig. 4. Close proton polarization deduced from the coherent neutron scattering intensity of EHBA-Cr(V) and bulk proton polarization from NMR (□). The deuteration of the solvent is 0.8.

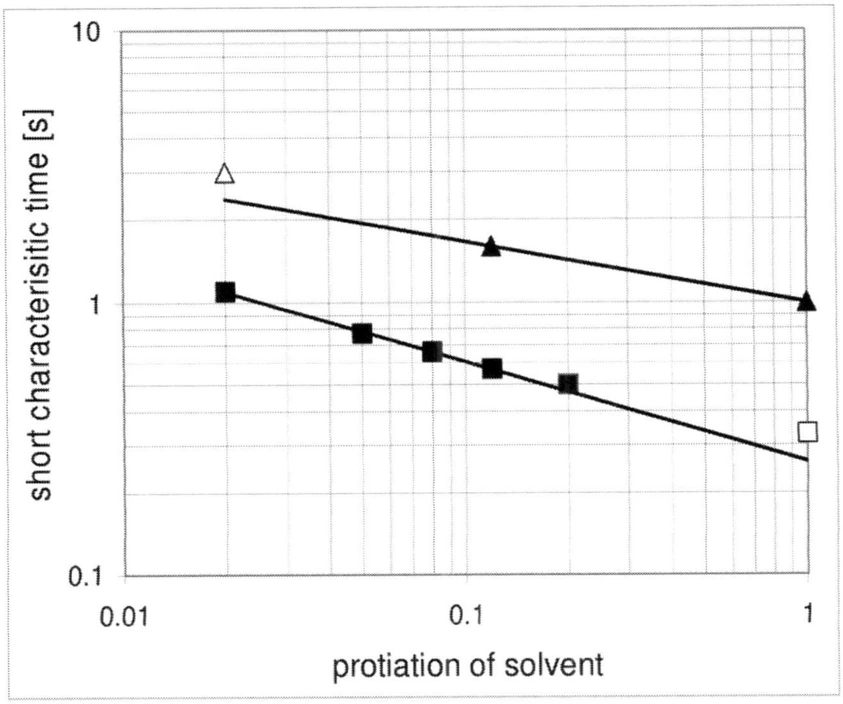

Fig. 5. The variation of the short time constant τ_1 of the build up of proton polarization with the deuteration of the solvent. The squares result from the periodic variation of the direction of DNP. The triangles present NMR data from the relaxation of the proton polarization in the absence of microwaves. Open symbols indicate less accurate values. The lines are fixed to black symbols.

The more prudent approach was to decrease the deuteration gradually, and to try to extrapolate to the state of a completely protiated sample [15]. The results of the experiments using alternating direction of DNP will be recalled briefly. The analysis of the data of time-resolved neutron scattering starts with the determination of the variation of the proton polarization with time and its analysis in terms of exponentials. From the data of EHBA-Cr(V) in a deuterated solvent shown in Fig.3, a short and a long relaxation time, τ_1 = 1.1 s and τ_2 = 5.5 s, are obtained. With decreasing deuteration of the solvent, the intensity of incoherent scattering from the protons increases, while the intensity of coherent scattering due to the contrast of the solute decreases. The analysis of the data from sample of lower deuteration shown in Fig. 4 is more difficult, also because τ_1 becomes shorter and τ_2 becomes much longer [15]. The variation of the characteristic time constant τ_1 with the deuteration of the solvent is shown in Fig. 5. From the double logarithmic plot we expect a characteristic time τ_1 of about 0.3 s in the absence of deuteration.

4.2. EHBA-Cr(V) solution in the absence of deuteration.

The study of the proton polarization build up in solutions of EHBA-Cr(V) in the absence of deuteration is most conveniently observed by proton NMR in the absence of microwave irradiation. In practice, a high proton polarisation is achieved after prolonged microwave irradiation. Immediately thereafter, the polarization of the bulk protons is reversed by AFP. Fig. 6 shows the variation of the proton polarization after positive DNP and reversal of the bulk protons to negative polarization by AFP. The short relaxation time of 1 s may be attributed to the exchange of polarisation between positively polarized close protons near the unpaired electron of an EHBA-Cr(V) molecule and the more distant negatively polarized bulk protons. This short relaxation time of τ_1 = 1 s measured in the absence of microwave irradiation is about three times longer than the expected relaxation time from cyclic change of the direction of DNP, i.e. in the presence of microwave irradiation (Fig. 5).

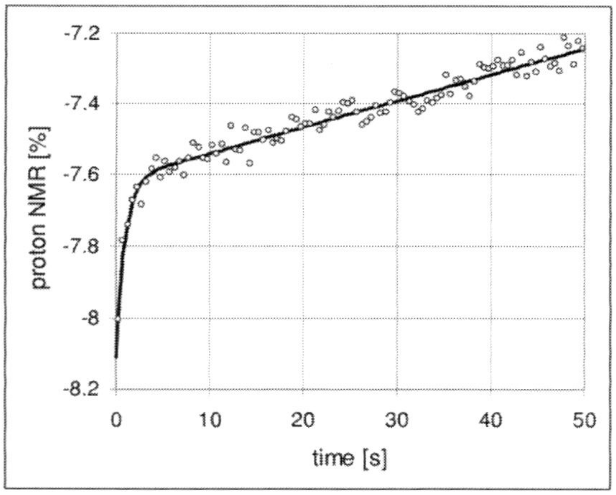

Fig.6. Time-resolved proton NMR from EHBA-Cr(V) in glycerol-water (1/1) after positive DNP and subsequent polarization reversal of the bulk protons by AFP. The time of fast relaxation is 1 s. Microwaves OFF.

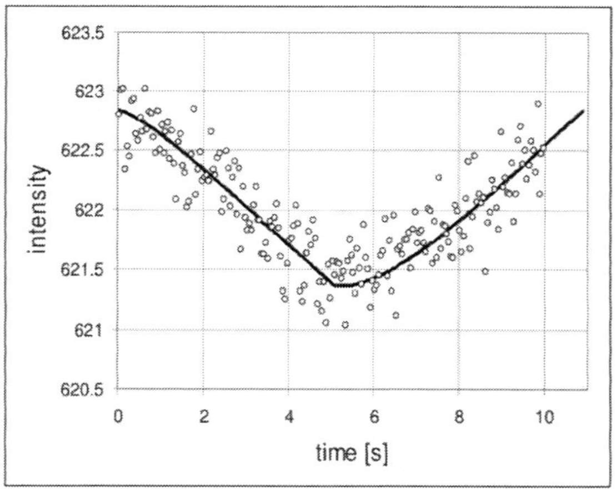

Fig. 7. Time-resolved neutron scattering from EHBA-Cr(V) during one cycle of positive and negative DNP. The solid line represents the approximation by (2), with the restriction that the periodic function is continuous. Microwaves ON.

Journal of Physics: Conference Series **351** (2012) 012003 doi:10.1088/1742-6596/351/1/012003

Let us turn to time-resolved neutron scattering using alternating direction of DNP. From Fig. 5 a fast characteristic time τ_1 = 0.3 s is expected. The variation of the neutron scattering intensity shown in Fig. 7 seems to confirm the estimated value.

The variation of the intensity of time-resolved neutron scattering is analyzed for intervals of the modulus of momentum transfer Q of different total $\Delta Q \geq 0.04$ Å^{-1}. As we expect a linear variation with time of the intensity of neutron scattering with a short non-linear variation after each change of the direction of polarization, the following approximation was used

$$I(Q,t) = a(Q) + b(Q)t + c(Q)e^{t/\tau_1} \qquad (2)$$

The decrease of the intensity at the onset of positive DNP shown in Fig. 7 seems to be slightly slower at the start of the cycle, which leads to a negative coefficient c(Q). At the onset of negative polarization a similar non-linear change of the scattering intensity with time is observed.

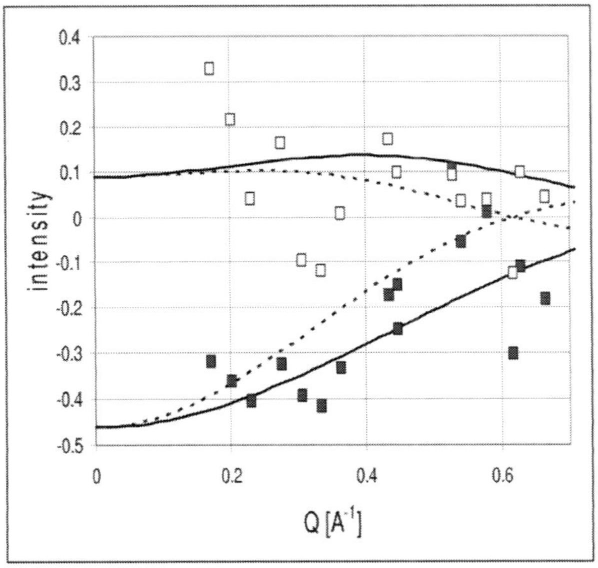

Fig. 8. The Q-dependence of the change c of the neutron scattering intensity characterized by τ_1. The solid and open squares present c(Q) at the start of positive (lower part) and negative (upper part) DNP, respectively. Results obtained with $\Delta Q \geq 0.09$ Å^{-1} are shown. Lines are calculated from the EHBA-Cr(V) model surrounded by a spherical distribution of selectively polarized protons of two different widths close to that of the protons of the EHBA-Cr(V) molecule (see text).

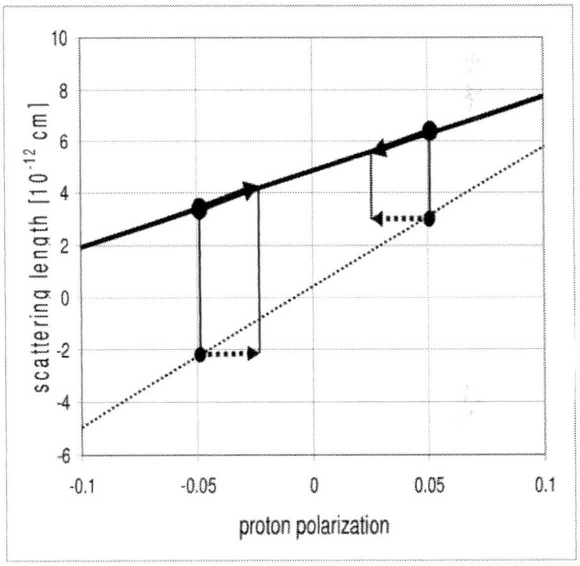

Fig.9. The variation of the scattering length of the EHBA-Cr(V) molecule (full line) and of the solvent molecules displaced the EHBA-Cr(V) molecule (dashed line) as function of proton polarization. The horizontal arrows present the nearly constant scattering density of the solvent at the onset of positive DNP (right) and negative DNP (left). The vertical bars present the contrast.

The coefficient c(Q) of the exponential term in (2) varies with Q as shown in Fig. 8. c(Q) is negative at the start of positive DNP, and its modulus decreases with Q. Changing to negative DNP at the start of the second half cycle, c(Q) tends to be slightly positive.

We assign the exponential term in (2) to the polarization of the protons close to the unpaired electron of the EHBA-molecule. As the solvent is not deuterated and neglecting the angular dependence of the electron spin proton spin interaction [16], the domain of preferentially polarized close protons is assumed to be spherical. In other words, the domain will comprise close hydrogen atoms mostly of the EHBA-Cr(V) molecule and those hydrogen atoms of the solvent molecules which are close to the unpaired electron of the EHBA-Cr(V) molecule.

The change of the neutron small-angle scattering intensity from spherical domains of selectively polarized protons is most conveniently described by a development of the scattering amplitude, $A(\mathbf{Q})$, as a series of spherical harmonics, $Y_{l,m}(\Omega)$.

$$A(\mathbf{Q}) = \sum_{l=0}^{\infty} \sum_{m=-l}^{l} A_{l,m}(Q) Y_{l,m}(\Omega) \qquad (3)$$

where $\mathbf{Q} = (Q,\Omega)$. Ω is a unit vector in the momentum space. With the amplitude V(**Q**) of the selectively polarized close nuclei and U(**Q**) as the amplitude of the residual structure including the slowly polarizing bulk nuclei, the intensity of neutron scattering will vary with the nuclear polarization P and the neutron polarization p as follows.

$$I(Q,P) = \sum_{l=0}^{\infty} \sum_{m=-l}^{l} \left| U_{l,m}(Q) \right|^2 + 2pPU_{o,o}(Q)\, V_{0,0}(Q) + P^2 V_{0,0}^2(Q) \qquad (4)$$

The spherical structure of the polarization domain implies that the polarization dependent scattering intensity functions will be dominated by the monopole term (l=0). In a first approximation, a change of the intensity of neutron scattering due to the build up of the polarization domain is given by $U_{0,0}(Q)V_{0,0}(Q)$, i.e. this cross term is proportional to c(Q) in (2).

At Q = 0, c(Q) is proportional to the product of the contrast of the polarized proton spin domain and the contrast of the EHBA-Cr(V) molecule. The effective scattering length (= contrast * volume) of the solute as a function of the proton polarization is obtained from Fig. 9 as the difference between the scattering length of the EHBA-Cr(V) molecule and the scattering length of the solvent molecules displaced by the latter. The proton spin contrast vanishes at P=0.2.

For a further analysis we make the simplifying assumption that the variation of the polarization of the close protons during one cycle of DNP hardly exceeds that of the bulk protons ($\Delta P = 0.10$). The polarization of both the close protons and those of the bulk then varies between -0.05 and +0.05. What is going to happen during one cycle? At the start of positive DNP (t=0 in Fig. 7, left side of Fig. 9) the close protons may have a polarization of P = -0.05, the same as that of the solvent (left side of Fig. 9). DNP will lead to a fast polarization of close protons in a fraction of a second. During this time the polarization of the bulk protons will have hardly changed (horizontal arrow in Fig. 9). The contrast (vertical bars in Fig. 9) increases rapidly for a very short time. The concomitant increase of coherent scattering will compensate for a short time the decrease if the intensity due to incoherent scattering (Fig. 7). At the end of the first half cycle, i.e. after 5 seconds of DNP, the polarization of the bulk protons will have come close to that of the close protons. The next half cycle starts from a positive polarization of both close protons and bulk protons. For a short time negative DNP will decrease the polarization and the scattering density of the close protons more rapidly than those of the bulk. During this short period the contrast of the solute will decrease compensating to some extent the increase of the intensity due to incoherent scattering (right side of Fig. 7 and of Fig. 9).

The cycle of dynamic proton spin contrast variation allows an estimate of the change of the scattering curves of EHBA-Cr(V) at very short times after the start of each half cycle. Assuming an isotropic distribution of polarized close protons with a range similar to that of the protons of EHBA the change of the scattering intensity due to the preferential polarization of the close protons has been calculated for the contrast at P= -0.05 and P= +0.05. The widths of the radial distribution of the close protons were those of the EHBA-Cr(V) molecule multiplied by 0.8 (full line in Fig. 8) and 1.2 (dashed line in Fig. 8). The corresponding scattering intensity profiles $U_{0,0}(Q)V_{0,0}(Q)$ agree with the experimental data (Fig. 8). From the present stage of the data analysis it may be assumed that the size of the spherical magnetic diffusion barrier is close to the size of the EHBA-Cr(V) molecule. A more rigorous treatment of the data using a thermodynamic model for the dynamics of nuclear polarization [15] of the data is in progress.

Conclusions

The analysis of time resolved neutron scattering from EHBA-Cr(V) in a protiated solvent shows the existence of a proton spin polarization gradient near the paramagnetic centres. This observation can be taken as a clear manifestation of a magnetic spin diffusion barrier near paramagnetic centres.

The low contrast is a real technical challenge for time-resolved neutron scattering experiments from dynamic polarized nuclei in hydrogenous material. Similar experiments with polyethylene illustrate this [17]. Apart from the rather complex set up for dynamic polarization a very powerful source of thermal neutrons is needed. The reactor of the Institute Laue-Langevin is most appropriate for this purpose. At this point the use of a pulsed neutron source might offer some additional advantages. One them is the fact that fast neutron of a pulse are the first to arrive at the sample. These neutrons with a short wavelength are well suited for probing the short range order, e.g. the onset of dynamic nuclear polarization near paramagnetic centers. The slower neutrons arriving slightly later are more apt to see the expansion of the polarized proton spin domain. Other aspects concern the way how a stroboscopic experiment is conducted. The cyclic repetition of DNP in different directions, eventually coupled with AFP, offer new possibilities of a pulsed neutron source. The pulse width and the pulse repetition rate of IBR2 [18] matches well the characteristic times of nuclear polarization build up in dynamic polarized targets.

5. Acknowledgements

This work was performed by the French-Swiss-German collaboration on the polarization build up by DNP, the team members of which are B. van den Brandt, H. Glättli, I. Grillo, H. Jouve, J. Kohlbrecher, J.A. Konter, E. Leymarie, S. Mango, R.P. May, A. Michels, the Author, and O. Zimmer. We thank the Paul-Scherrer Institut (PSI) for the facilility of dynamic nuclear polarization and the Institut Laue Langevin for allocating beam time at the small-angle diffractomer D22.

References

[1] A. Abragam, M. Goldman, Nuclear Magnetism, Order and Disorder, (Oxford: Clarendon) (1982)
[2] H. Glättli, M. Goldman, Methods of Exp. Physiscs, C **23**, 241-286 (1987).
[3] J. des Cloizeaux, G. Jannink, *Les Polymères en Solution: leur Modélisation et leur Structure*, p.184, les éditions de physique, Les Ulis, France (1987).

[4] J.B. Hayter, G.T. Jenkin, J.W. White, Phys. Rev. Lett. **33**, 696-698 (1974).
[5] O. Schärpf, Physica B, **156&157**, 631-638, see also from the same author http://82.135.31.182/neutronpol.pdf, *The spin as a measuring probe* (2010).
[6] V.I. Lushikov, Yu; V. Taran, Nucl. Instrum. Methods Phys. Res. A **228**, 159-160 (1984)
[7] O. Schärpf, AIP Conf. **89**, 182-189 (1982).
[8] A. Abragam, Phys. Rev. **98**, 1729-1735 (1955).
[9] A. Abragam, M. Goldman, Rep. Prog. Phys. **41**, 395 (1978).
[10] S.F.J. Cox, S.F.J. Read, W. Th; Wenckebach, J. Phys. C: Solid State Phys. **10**, 2917-2936 (1977).
[11] P. Hautle, W. Grübler, B. van den Brandt, J.A. Konter, S. Mango, M. Wessler, Phys. Rev. B, **46**, 6596-6599 (1992).
[12] E. Leymarie, Thesis, Université Paris XI, Orsay, No. 7052.
[13] M. Krumpolc, J. Rozek, J. Am. Chem. Soc. **101**, 3206-3209 (1979).
[14] van den Brandt, H. Glättli, I. Grillo, P. Hautle, H. Jouve, J. Kohlbrecher, J.A. Konter, E. Leymarie, S. Mango, R.P. May, H.B. Stuhrmann, O. Zimmer, Europhys Lett. **59**, 62 (2002).
[15] B. van den Brandt, H. Glättli, I. Grillo, P. Hautle, H. Jouve, J. Kohlbrecher, J.A. Konter, E. Leymarie, S. Mango, R.P. May, A. Michels, H.B. Stuhrmann, O. Zimmer, European Phys. J. B **49,** 157-165 (2006).
[16] H. B. Stuhrmann, Acta Cryst. A **63**, 455-464 (2007).
[17] T. Kumada, Y. Noda, S. Koizumi, T. Hashimoto, The Journal of Chemical Physics **133**, 054504 (2010)
[18] Yu. M. Ostanievich, Die Makromolekulare Chemie, Macromolecular Symposia, edited by B. Sedláček, Editor in chief Hartwig Höcker, **15**, 91-103 (1988).

SAXS Study of Sterically Stabilized Lipid Nanocarriers Functionalized by DNA

Borislav Angelov[1,*], Angelina Angelova[2], Sergey Filippov[1], Göran Karlsson[3], Nick Terrill[4], Sylviane Lesieur[2] and Petr Štěpánek[1]

[1]Institute of Macromolecular Chemistry, ASCR, Prague, Czech Republic,
[2]CNRS UMR8612, Physico-Chimie-Pharmacotechnie-Biopharmacie, Univ Paris Sud, Châtenay-Malabry, France,
[3]Department of Physical and Analytical Chemistry, Uppsala University, Uppsala, Sweden,
[4]Diamond Light Source Ltd, Chilton, Didcot, Oxfordshire OX11 0DE, UK

E-mail: angelov@imc.cas.cz

Abstract. The structure of novel spontaneously self-assembled plasmid DNA/lipid complexes is investigated by means of synchrotron radiation small-angle X-ray scattering (SAXS) and Cryo-TEM imaging. Liquid crystalline (LC) hydrated lipid systems are prepared using the non-ionic lipids monoolein and DOPE-PEG$_{2000}$ and the cationic amphiphile CTAB. The employed plasmid DNA (pDNA) is encoding for the human protein brain-derived neurotrophic factor (BDNF). A coexistence of nanoparticulate objects with different LC inner organizations is established. A transition from bicontinuous membrane sponges, cubosome intermediates and unilamelar liposomes to multilamellar vesicles, functionalized by pDNA, is favoured upon binding and compaction of pBDNF onto the cationic PEGylated lipid nanocarriers. The obtained sterically stabilized multicompartment nanoobjects, with confined supercoiled plasmid DNA (pBDNF), are important in the context of multicompartment lipid nanocarriers of interest for gene therapy of neurodegenerative diseases.

1. Introduction

Small-angle X-ray scattering (SAXS) and cryogenic transmission electron microscopy (cryo-TEM) are powerful structural methods [1-20] for assessment of the changes in the inner and overall supramolecular organizations of self-assembled cationic lipid nanoparticles (NPs) upon complexation with plasmid DNA (pDNA) in non-viral gene delivery vehicles [21-30]. The ionic interaction between positively charged lipid membranes and nucleic acids, to generate lipoplexes, may cause dramatic structural transformations in the initially formed lipid NPs. Depending on the degree of DNA compaction and uploading, the lipid phase may self-organize into periodic multilamellar, cubic or inverted hexagonal (H_{II}) structures as well as into non-periodic bilayer tubules or intermediate type aggregates. Studies have reported that nonbilayer type liquid crystalline vehicles, such as H_{II}-phase lipoplexes, display higher transfection activity [23]. Thus, investigations have been initiated to employ nonlamellar "helper" lipids for non-bilayer lipoplex preparation [2,3]. PEGylation of the lipid nanovehicles has been shown to provide longer circulation times in the biological fluid before gene delivery [26-29]. The understanding of the interaction of non-viral self-assembled nanocarriers with the cellular barriers and their cellular internalization pathways, as well as of the nucleic acid release mechanisms from the pDNA/lipid nanoparticles, requires more detailed knowledge about the inner structural organization of the nanocarriers as determined by the involved compaction mechanism.

In this paper, we report SAXS and Cryo-TEM structural studies that elucidate the structure of novel pDNA/lipid nanoparticles designed for application in gene therapy of neurodegeneration diseases. Monoolein-based positively charged PEGylated lipid nanocarriers of liquid crystalline nature are prepared by self-assembly, and subjected to complexation with plasmid DNA encoding for human brain-derived neurotrophic factor (BDNF). The protein BDNF is a potential therapeutic agent for treatment of neurodegenerative diseases such as multiple sclerosis (MS), Alzheimer's, Parkinson's and Huntington's diseases, and amyotropic lateral sclerosis (ALS) [31]. The deficiency of the neuroprotector BDNF is expected to be compensated by gene transfection strategies [32].

We aim at preparation of advanced LC nanovehicles for delivery of the BDNF gene (pBDNF). The cationic amphiphile, cetyltrimethylammonium bromide (CTAB), is used to supply positive charges in the generated liquid crystalline lipid NPs serving for compaction of the plasmid DNA. In excess aqueous medium, the nonionic lipid monoolein (MO) self-assembles into a bicontinuous cubic liquid crystalline phase [33-39], which can be dispersed into nanoparticles upon inclusion of PEGylated amphiphiles or polymer surfactants of the Pluronic type [19,20,33,35]. By changing the molar ratio between the neutral lipid (MO) and the cationic (CTAB) component in the self-assembled lipid mixture we vary the quantity of the positive charges in the lipid membrane available for complexation with pBDNF. The obtained results indicate successful formation of sterically stabilized gene delivery vehicles and a rich variety of self-assembled nanoparticulate structures as interesting potential candidates for neuroregenerative gene therapy studies.

2. Experimental

2.1 Materials

Liquid crystalline nanoparticles were prepared by the hydration of a lyophilized mixed lipid film followed by vortex shaking, agitation by sonication, and filtering through a 0.2 μm filter (Minisart High Flow, *Sartorius*). The mixed film consisted of a nonlamellar monoglyceride lipid 1-oleoyl-*rac*-glycerol (MO) (M_W 356.55, purity 99.5%, *Sigma*), a cationic amphiphile hexadecyltrimethyl-ammonium bromide (CTAB) (M_W 364.46, BioUltra, ≥99.0%, *Fluka*), and a PEGyated lipid 1,2-dioleoyl-*sn*-glycero-3-phosphoethanolamine-*N*-(methoxy(polyethylene glycol)-2000) ammonium salt (DOPE-PEG_{2000}) (M_W 2801.51, *Avanti Polar Lipids*). The PEGylated lipid was incorporated at a molar ratio MO/DOPE-PEG_{2000} 98/2 (mol/mol). The molar ratios between the nonionic lipid (MO) and the cationic amphiphile (CTAB), MO/CTAB, were varied in the sequence 90/10, 85/15, 80/20 and 75/25 mol/mol. The hydration and dispersion of the self-assembled lipid phase in excess phosphate buffer solution (NaH_2PO_4/Na_2HPO_4, 1.10^{-2}M, pH 7, p.a. grade, Merck) was performed at ambient temperature. An ultrasonication ice bath (Branson 2510 ultrasonic bath, "set sonics" mode) was applied during 2-4 minutes in order to get homogeneous dispersions of cationic nanoparticles. Endotoxin-free plasmid DNA (pBDNF), encoding for human BDNF, was prepared by custom gene synthesis, subcloning and purification (*GenScript Co.*). Sterically stabilized lipid/pBDNF nanocomplexes were assembled after dilution of the pBDNF solution in a phosphate buffer and subsequent incubation with a dispersion of cationic lipid nanocarriers MO/CTAB/DOPE-PEG_{2000}. The concentration of the positive charges in the NPs systems was varied through the MO/CTAB molar ratio, while keeping the pBDNF concentration constant. The negative charges of pDNA were maintained in deficiency with respect to the cationic charges in order to avoid the overall precipitation of the generated pBDNF/lipid complexes.

2.2 SAXS

Small angle X-ray scattering measurements with diluted lipid and pDNA/lipid samples were performed at the I22 beamline of the Diamond Light Source (Didcot, U.K). The liquid solutions of nanoparticles were filled in borosilicate glass capillaries (Hilgenberg Gmbh, Germany, diameter 1.5 mm, 10 μm wall thickness). The samples were mounted in a holder in air, without a special vacuum chamber. SAXS measurements were performed at room temperature (21 °C). The detector at the I22 station was a gas wire RAPID 2D detector [40], the area of which was divided into 512x512 pixels. The wavelength of the incident X-ray beam was 0.155 nm during the experiments and the beam size at the samples was ~350x250 μm^2. The accessible q-range (magnitude of the scattering vector) was from 0.09 nm^{-1} to 4.17 nm^{-1}. A metal attenuator was installed in order to prevent the liquid crystalline nanoparticle samples from X-ray damages. The exposure time was about 1 sec. The incident and transmitted beam intensities were simultaneously recorded. Silver behenate was used for the q-range calibration and glassy carbon for the intensity normalization. After the data acquisition, the obtained 2D images were integrated into 1D scattering curves by means of the *Fit2D* software (www.esrf.eu/computing/scientific/FIT2D/). The backgrounds, coming from the glass capillary and the solvent, were measured and subtracted using conventional procedures. One and the same capillary was used during the measurements with and without solvent. This allowed for very precise subtraction of the background.

The fitting of the experimental data was based on the multilamellar vesicle model [41]. The scattering intensity was modeled as a sum of the scattering from individual spherical shells:

$$I(q)=\left(\sum_{i=0}^{n-1}\left[K(q,R_c+i\,t_{sh}+i\,t_{sol},\Delta\eta)-K(q,R_c+(i+1)\,t_{sh}+i\,t_{sol},\Delta\eta)\right]\right)^2,$$

where

$$K(q,R,\Delta\eta)=\frac{4}{3}\pi R^3\Delta\eta\,3\frac{\sin(qR)-qR\cos(qR)}{(qR)^3}$$

is the scattering from a spherical shell, R is the radius of the shell, $\Delta\eta$ is the difference between the scattering length density of the surfactant and the solvent. The number of surfactant layers is denoted by n, and t_{sk} and t_{sol} are the thicknesses of the surfactant and the solvent layers, respectively.

2.3 CryoTEM

For cryo-TEM experiments, a drop of the NPs solution was placed on a pretreated copper grid, which was coated with a perforated polymer film. The excess solution was removed by blotting with a filter paper. The sample film preparation was done under controlled environment conditions at a constant temperature of 25°C. A chamber with a relative humidity of 98-99% was used in order to avoid evaporation of the liquid sample. Rapid vitrification of the thin film was achieved by plunging the grid into liquid ethane maintained just above its freezing point. The sample was then transferred to the electron microscope Zeiss 902A (*Carl Zeiss NTS*, Oberkochen, Germany). The instrument was operated at an accelerating voltage of 80 kV and in zero-loss bright-field mode. The temperature was kept below -165 °C. The sample was protected against atmospheric conditions during the entire procedure in order to prevent the formation of ice crystals. The resolution of the method was 3-5 nm. Digital images were acquired with a BioVision Pro-SM Slow Scan CCD camera (*Proscan Electronische Systeme*, GmbH, Germany). The image processing was done by means of the *iTEM* software (*Olympus Soft Imaging Solutions*, GmbH, Germany).

3. Results

Figure 1 shows the scattering curves from the investigated NPs samples prepared with pBDNF (Fig.1a) and without plasmid DNA (Fig. 1b). The position of the characteristic 1st peak is located around q=0.04 Å^{-1}. The corresponding repeat lamellar spacings, d, nearly coincide in values, which are in the range 15.4 – 15.6 nm for three of the NPs samples (see the SAXS patterns 2, 3 and 4 in Fig.1b). For the first sample (SAXS pattern 1 in Fig. 1b), the first peak is very broad and is positioned at q=0.037 Å^{-1}, which corresponds to a spacing of 16.8 nm. After the formation of pDNA/lipid complexes, the position of the characteristic Bragg peak shifts to q=0.106 Å^{-1} and the corresponding lamellar spacing becomes d = 5.9 nm. This result demonstrates the dramatic condensation of the plasmid DNA in the compacted MO/CTAB/DOPE-PEG_{2000}/pBDNF lipoplexes.

The SAXS patters (Fig. 1) and the cryo-TEM images (Fig. 2) suggest that the NPs system, obtained at a neutral-to-charged lipid ratio MO/CTAB 90/10 (mol/mol), contains liquid crystalline objects of mainly spongosome and intermediate cubosome types. Multilamellar vesicles are not a representative feature at this molar ratio. The small portion of positive charges (only 10 mol% CTAB) appears to provide lower complexation for pBDNF (see the weak peak of the pDNA/lipid complex in Fig.1a). The broad SAXS peaks indicate sponge-like membrane organization and a very small fraction of periodic lamellar formations.

The obtained structural results indicate a coexistence of multicompartment liquid crystalline NPs and vesicles in the MO/CTAB/DOPE-PEG_{2000} samples with MO/CTAB molar ratios 85/15 and 80/20 (mol/mol). They reveal that the increase of the number of positive charges, in the sequence from a molar ratio 85/15 to 80/20 (MO/CTAB, mol/mol), results in a corresponding augmentation of the percentage of the pBDNF complexation. Thus, the number of ordered multilamellar formations increases and the intensity of the characteristic SAXS peaks rises (Fig.1a).

The SAXS data and the cryo-TEM images evidence a structural transition (Fig. 3) from multicompartment, sponge-type NPs to multilamellar vesicular nanocarriers upon increasing pDNA complexation at rising positive charges. It is notable that the pBDNF/cationic lipid interaction leads to flattening of the lipid bilayer curvature in the sponge membranes and formation of multilamellar-type carriers. The SAXS peak of the multilamellar vesicles (MLV) becomes stronger and more narrow with the degree of the pBDNF complexation, indicating an enhanced multilamellar formation inside the generated lipoplex NPs at a MO/CTAB molar ratio 80/20.

The most positively charged sample, with a MO/CTAB ratio 75/25 (mol/mol), involves predominantly vesicular NPs (cryo-TEM images not shown here). The lower intensity of the SAXS patterns (Fig. 1a) suggests that the complexation of pBDNF with the closed bilayer vesicles seems to be weaker as compared to its complexation with the cationic liquid crystalline NPs involving flexible bilayers and an inner network of open channels. Another possible explanation for this observation is that the internal organisation of the NPs in sample 4 (MO/CTAB 75/25, mol/mol), is not so dense (the NPs interior is filled by aqueous phase) with regard to the sponge membrane NPs interior, which is built-up from a random lipid bilayer network forming channels (Fig. 3, left).

4. Discussion

The obtained SAXS results provide evidence that the interplay between the membrane charges and curvature plays a key role in the determination of the structural organization of the pBDNF/lipid nanocarriers. The functionalization of the NPs by supercoiled pDNA appears to lead to a flattening of the curved lipid bilayer membranes constituting the monoolein bicontinuous cubic structure. The same tendency is produced also by the CTAB and DOPE-PEG_{2000} co-lipids, which help to disperse the bicontinuous cubic phase of MO into nanoparticles. The cationic amphiphile CTAB has been known as an emulsifier providing high plasmid loading efficiency into particles. To ensure steric repulsion between the obtained nanocarriers, the investigated here pBDNF/lipid complexes were prepared in a deficiency of negatively charged pDNA with regard to the total number of cationic charges in the NPs system.

The SAXS results showed that, at increasing molar fractions of CTAB in the MO/CTAB/DOPE-PEG_{2000} self-assembled mixtures, the initially formed liquid crystalline NPs, lacking inner multilamellar organization, are progressively transformed into bilayer-based nanoobjects with lower interfacial curvature of the lipid membrane. According to the molecular shape concept, the average critical packing parameter of the mixture of a cationic lipid (CTAB), a helper lipid (MO) and a co-lipid (DOPE-PEG_{2000}) will be modified upon rising the CTAB molar fraction as well as upon interaction with pDNA. The reorganization of the lipid bilayer from a 3D sponge-type membrane with open channels (sample 1) to a lamellar-type nanocarrier, upon gene loading, is schematically presented in Figure 3. The SAXS results established the coexistence of cationic lipid NPs without associated DNA and lipid nanocarriers loaded with pBDNF. The latter involve periodicaly organized lipid/pBDNF complexes with more compact structures as evidenced by the strong Bragg diffraction peaks in the SAXS patterns (Fig. 1a).

The particles morphology, determined by cryo-TEM imaging, suggests that plasmid DNA can be associated on the nanocarrier surfaces but also entrapped between the lipid bilayer surfaces forming multicompartment nanocontainers (Figure 2). The performed EM imaging established a coexistence of nanovehicles of a liquid crystalline type, displaying dense inner membrane organisation, and nanovehicles of a vesicular membrane type, involving lipid lamellas that surround an aqueous core. Figure 2 reveals that the complexation of the cationic lipid NPs with pBDNF facilitates the bridging between the bilayer surfaces, i.e. favours the tendency of increasing the number of lamellas in the generated lipoplex NPs.

The reported here structural findings prove to be useful for the design of novel lipid nanocarriers of the BDNF gene. The choice of the MO/CTAB molar ratio will allow to directly determining the amphiphilic MO/CTAB/DOPE-PEG$_{2000}$ compositions, which could compromise the fusogenic and release properties of the NPs. A recent work [8] has stressed that highly compacted multilamellar lipopexes appear to be rather stable topologically-closed particles, which do not easily release the entrapped pDNA from the closed multilamellar lipid containers. However, a small number of lamellas in the NPs would not hamper the pDNA release. The delivery of pDNA from self-assembled lipid carriers should be facilitated also by open-channel structures such as membrane sponges and tubular assemblies. Some of the advanced nanovector systems, prepared in this investigation, correspond to these criteria.

5. Acknowledgement

The SAXS measurements at beamline I22 were carried out with the support of the Diamond Light Source Ltd. (Chilton, Didcot, Oxfordshire, UK). The Users support obtained at the synchrotron radiation facility DLS is gratefully acknowledged (Diamond Experiment SM3313). Also, we would like to thank Prof. Katarina Edwards, Uppsala University, Department of Physical and Analytical Chemistry for help with Cryo-TEM experiments.

References

1. Schmutz M Durand D Debin A Palvadeau Y Etienne A and Thierry A R 1999 *Proc. Nat. Acad. Sci.* **96** 12293.
2. Koltover I Salditt T Rädler J O and Safinya C R. 1998 *Science* **281** 78.
3. Zidovska A Evans H M Ewert K K Quispe J Carragher B Potter C S and Safinya C R 2009 *J. Phys. Chem. B* **113** 3694.
4. Rädler J O Koltover I Salditt T and Safinya C R 1997 *Science* **275** 810.
5. Leal C Bouxsein N F Ewert K K and Safinya C R 2010 *J. Am. Chem. Soc.* **132** 16841.
6. Zidovska A Evans H M Ahmad A Ewert K K and Safinya C R 2009 *J. Phys. Chem.* **113** 5208.
7. Marchini C Montani M Amici A Amenitsch H Marianecci C Pozzi D and Caracciolo G 2009 *Langmuir* **25** 3013.
8. Caracciolo G Pozzi D Capriotti A L Marianecci C Carafa M Marchini C Montani M Amici A Amenitsch H Digman M A Gratton E Sanchez S S and Lagana A 2011 *J. Med. Chem.* **54** 4160.
9. Bilalov A Olsson U and Lindman B 2011 *Soft Matter* **7** 730.
10. Dias R S Lindman B and Miguel M G 2002 *J. Phys. Chem. B* **106** 12600.
11. Rosa M Moran M C Miguel M G and Lindman B 2007 *Colloids and Surfaces A: Physicochem. Eng. Aspects* **301** 361.
12. Andreoli de Oliveira E Teixeira da Silva E R Fevrier A Grelet E Nallet F and Navailles L 2010 *EPL* **91** 28001.
13. McLoughlin D Imperor-Clerc M and Langevin D 2004 *ChemPhysChem* **5** 1619.
14. Uhrikova D Pullmannova P Kucerka N Funari S S Teixeira J and Balgavy P 2009 *Eur. Phys. J. Special Topics* **167** 191.
15. Filippov S K Konak C Kopeckova P Starovoytova L Spirkova M and Stepanek P 2010 *Langmuir* **26** 4999.
16. Sternberg B Sorgib F L and Huang L 1994 *FEBS Lett.* **356** 361.
17. Torchilin V P Levchenko T S Rammohan R Volodina N Papahadjopoulos-Sternberg B and D'Souza G G M 2003 *Proc. Nat. Acad. Sci.* **100** 1972.
18. Efrat R Kesselman E Aserin A Garti N and Danino D 2009 *Langmuir* **25** 1316.
19. Gustafsson J Ordd G Nyden M Hansson P and Almgren M 1998 *Langmuir* **14** 4987.
20. Gustafsson J Arvidson G Karlsson G and Almgren M 1995 *Biochim. Biophys. Acta* **1235** 305.
21. Guo X and Szoka F C 2003 *Acc. Chem. Res.* **36** 335.
22. Li W and Szoka Jr F C 2007 *Pharm. Res.* **24** 438.
23. Zuhorn I S Engberts J B F N and Hoekstra D 2007 *Eur. Biophys J.* **36** 349.
24. Mintzer M A and Simanek E E 2009 *Chem. Rev.* **109** 259.
25. Zuber G Dauty E Nothisen M Belguise P and Behr J-P 2001 *Adv. Drug Deliv. Rev.* **52** 245.
26. Chen T Palmer L R Fenske D B Lam A M L Wong K F and Cullis PR 2004 *J. Liposome Res.* **14** 155.
27. Ambegia E Ansell S Cullis P Heyes J Palmer L and MacLachlan I 2005 *Biochim. Biophys. Acta* **1669** 155.
28. Palmer L R Chen T Lam A M I Fenske D B Wong K F MacLachlan I and Cullis P R 2003 *Biochim. Biophys. Acta* **1611** 204.
29. Tam P Monck M Lee D Ludkovski O Leng E C Clow K Stark H Scherrer P Graham R W and P.R. Cullis P R 2000 *Gene Therapy* **7** 1867.

30. Maurer N Wong K F Stark H Louie L McIntosh D Wong T Scherrer P Semple S C and Cullis P R 2001 *Biophysical J.* **80** 2310.
31. Nagahara A H Merrill DA Coppola G Tsukada S Schroeder BE Shaked GM Wang L Blesch A Kim A Conner J M Rockenstein E Chao M V Koo E H Geschwind D Masliah E Chiba A A and Tuszynski M H 2009 *Nature Medicine* **15** 331.
32. Di Polo A Aigner L J Dunn R J Bray G M and Aguayo A J 1998 *Proc. Nat. Acad. Sci.* **95** 3978.
33. Angelova A Angelov B Mutafchieva R Lesieur S and Couvreur P 2011 *Accounts of Chemical Research* **44** 147.
34. Angelov B Angelova A Mutafchieva R Lesieur S Vainio U Garamus V M Jensen G V and Pedersen J S 2011 *Physical Chemistry Chemical Physics* **13** 3073.
35. Angelov B Angelova A Vainio U Garamus V M Lesieur S Willumeit R and Couvreur P 2009 *Langmuir* **25** 3734.
36. Angelov B Angelova A Garamus V M Le Bas G Lesieur S Ollivon M Funari S S Willumeit R and Couvreur P 2007 *J. Am. Chem. Soc.* **129** 13474.
37. Angelov B Angelova A Papahadjopoulos-Sternberg B Lesieur S Sadoc J.-F Ollivon M and Couvreur P 2006 *J. Am. Chem. Soc.*, 2006 **128** 5813.
38. Angelova A Angelov B Lesieur S Mutafchieva R Ollivon M Bourgaux C Willumeit R Couvreur P 2008 *Journal Drug Delivery Science and Technology* **18** 41.
39. Angelova A Angelov B Papahadjopoulos-Sternberg B Ollivon M Bourgaux C 2005 *Journal Drug Delivery Science and Technology* **15** 108.
40. Lewis R A Berry A Hall C J Helsby W I and Parker B T 2000 *Nucl Instrum Meth A* **454** 165.
41. Lesieur P Kiselev M A Barsukov L I and Lombardo D 2000 *J. Appl. Cryst.* **33** 623.

Figures

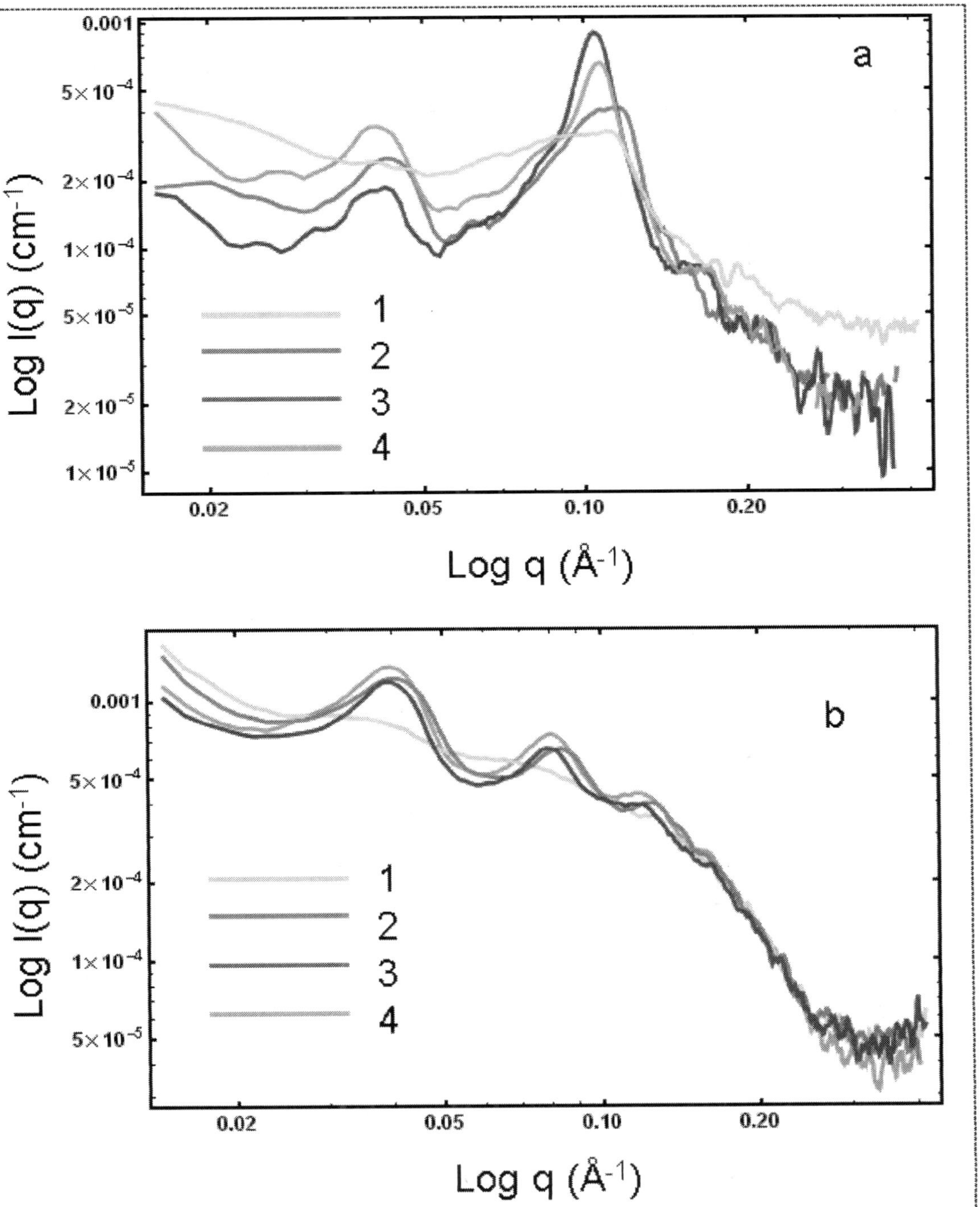

Figure 1. Synchrotron radiation SAXS patterns obtained with diluted samples of PEGylated cationic nanoparticles (MO/CTAB/DOPE-PEG_{2000}) complexed with pBDNF (a) or lacking plasmid DNA (b). The neutral-to-cationic lipid molar ratio, MO/CTAB, in the self-assembled mixtures is: 90/10 (line 1, green), 85/15 (line 2, red), 80/20 (line 3, blue), and 75/25 (line 4, orange) mol/mol.

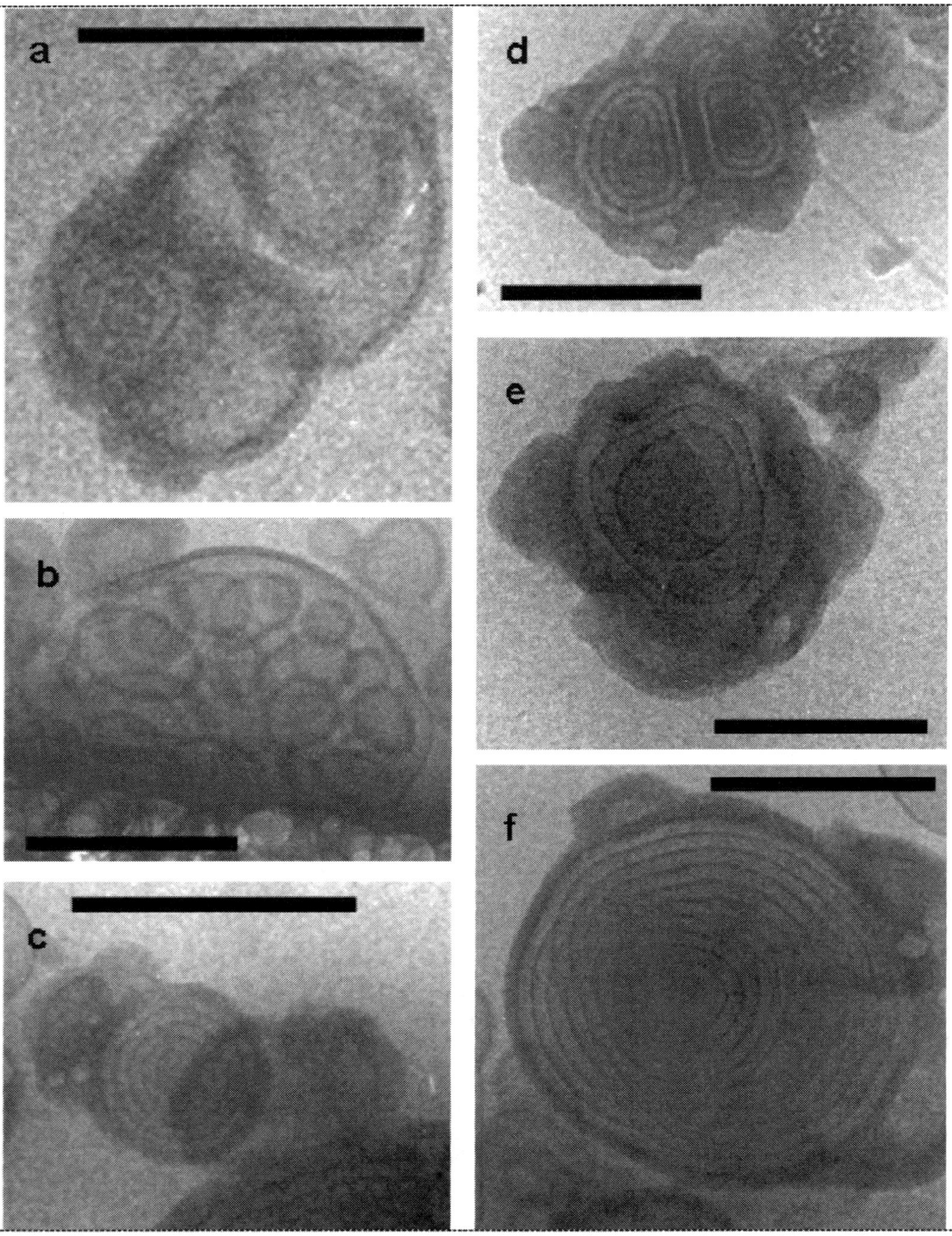

Figure 2. Cryo-TEM micrographs of coexisting self-assembled PEGylated lipid nanoparticles (MO/CTAB/DOPE-PEG_{2000}) interacting with negatively charged plasmid DNA encoding for human BDNF (pBDNF). The nonionic-to-cationic (MO/CTAB) amphiphile molar ratio in the nanocarriers is 80/20 mol/mol. The left column represents nanoobjects with a low degree of compaction, while the right column shows multicompartment nanoparticles with an extended multilamellar organization favoured by the pBDNF compexation. The scale bar corresponds to 200 nm in every image.

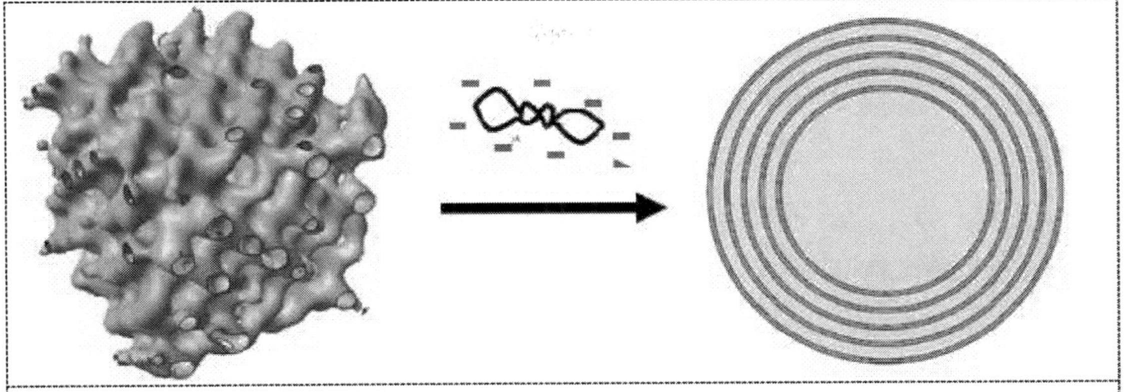

Figure 3. Structural transformation of the inner organisation of self-assembled liquid crystalline cationic lipid nanoparticles induced by the complexation with negatively charged plasmid DNA and revealing the morphological change from a sponge membrane type to a multilamellar vesicular type nanocarrier.

Journal of Physics: Conference Series **351** (2012) 012005 doi:10.1088/1742-6596/351/1/012005

Characterization of bio-synthesized nanoparticles produced by *Klebsiella oxytoca*

L Anghel[1,2,a], M Balasoiu[1,3,b], L A Ishchenko[4], S V Stolyar[4,5], T S Kurkin[6], A V Rogachev[1], A I Kuklin[1], Yu S Kovalev[1], Yu L Raikher[7], R S Iskhakov[4,5], G Duca[8]

[1]Joint Institute of Nuclear Research, Dubna, 141980, Russia
[2]Institute of Chemistry of ASM, Chisinau, Republic of Moldova
[3]Horia Hulubei National Institute of Physics and Nuclear Engineering, Bucharest, Romania
[4]Siberian Federal University, 660041, Krasnoyarsk, Russia
[5]Kirensky Institute of Physics, Siberian Branch of RAS, Krasnoyarsk, Russia
[6]Institute of Synthetic Polymer Materials RAS, 117393, Moscow, Russia
[7]Institute of Continuous Media Mechanics, Ural Branch of RAS, Perm, Russia
[8]Academy of Science of Moldova, Chisinau, Republic of Moldova

Email: [a]angel@nf.jinr.ru, [b]balasoiumaria@yahoo.com

Abstract. Structural and morphological properties of biogenic ferrihydrite nanoparticles produced by bacteria *Klebsiella oxytoca* are investigated. The stability of water dispersions of biomineral particles produced by *Klebsiella oxytoca* was monitored by UV-Vis spectroscopy. Their chemical composition was determined by FT-IR spectroscopy. The vibrational spectra of biogenic ferrihydrite nanoparticles revealed typical absorption peaks of exopolysaccharides. Morphological analysis based on Raman spectroscopy indicated the presence of exopolysaccharides on the surface as well as inside the pores of the ferrihydrite nanoparticles. Structural investigations of ultrasonic assisted samples of different concentration of water dispersed particles were performed using small angle X-ray scattering analysis. Model calculations and fitting procedures revealed scattering objects of an elongated shape with 6.73±0.16 nm radius of gyration.

1. Introduction

New applications of nanomaterials are emerging rapidly. The synthesis of nanoparticles is a cornerstone of nanotechnology. New methods to study synthesis of nanoparticles are an important area of research. The methods currently being used encompass chemical routes. The by-products associated with metal production by these methods rise problems with respect to environmental pollution. Additionally, some of these processes are expensive, the synthetic procedures involving conditions such as high temperature, pressure and environmental inertness. Many chemical routes are known to use toxic chemicals for the synthesis of nanoparticles. The need-of-the-hour; however is to evolve the procedures for nanoparticles synthesis through environmental benign routes. Researchers in this field, therefore, have been eagerly looking at biological systems as alternative eco-friendly systems.

Microbial cells are highly-organized units, regarding morphology and metabolic pathways, capable of synthesizing well size-calibrated and well-structured particles. Furthermore, biogenic nanoparticles often are water-soluble and biocompatible, which is essential for many applications. Some well-known examples of microorganisms synthesizing inorganic materials include magnetotactic bacteria (which synthesize magnetic nanoparticles) [1], diatoms (which synthesize siliceous materials) [2], and S-layer bacteria (which produces gypsum and calcium carbonate layers) [3].

Bacteria *Klebsiella oxytoca* produce two types of ferrihydrite nanoparticles depending on the microorganisms' growth conditions [4,5]. Earlier investigations have quite well identified these two fractions by means of Mossbauer spectroscopy [6,7], static magnetic measurement analysis [8,9], scanning electron microscopy and small angle X-ray scattering methods [10] on dry powder samples. SAXS studies on concentrated and non-ultrasonicated samples of water dispersions of ferrihydrite have revealed scattering objects with the form factor of a cylinder of radius R=4.87±0.02 nm and length L=2.12±0.04 nm [11]. It was shown also by HRTEM analysis that nanoparticles or their clusters bear on their surfaces some adsorbed protein molecules [10]. However, in order to take full advantage of the obtained nanoparticles in establishing new technologies one needs to investigate more precisely structural characteristics and morphological properties of the biogenic ferrihydrite particles. In this paper, UV-Vis spectroscopy, FT-IR spectroscopy, Raman spectroscopy and small-angle X-ray scattering analysis were used to investigate the structure of bio-mineral nanoparticles produced by bacteria *Klebsiella oxytoca*.

2. Materials and methods

2.1. Sample preparation

Aqueous samples of biogenic particles of ferrihydrite were provided by Siberian Federal University, Krasnoyarsk, Russia. Initial concentration of biomineral nanoparticles in aqueous solutions was 12.5 g/l (5g of biomineral powder dissolved in 400 ml double distilled water).

Ferrihydrite concentration in the analyzed sample was determined using a spectrophotometric method based on the interaction of 1,10-phenantroline, a heterocyclic organic compound, and metal ions. Spectrophotometric measurements revealed the concentration of ferrihydrite in the analyzed sample of 2.68 g/l.

2.2. UV-Vis spectroscopy

UV-Vis spectra of the prepared samples were registered in the wavelength range of 400—650 nm at room temperature (Figure 1).

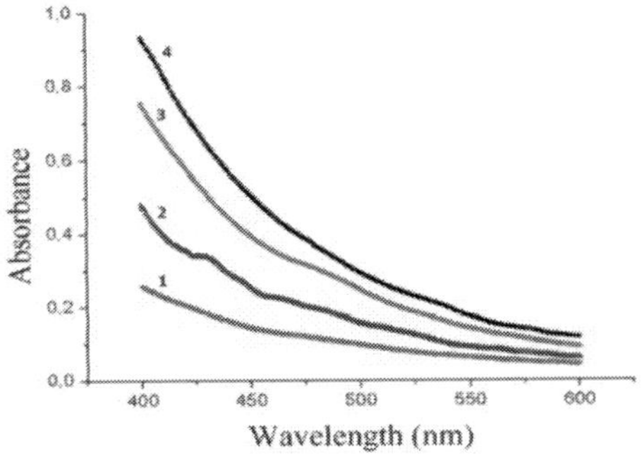

Figure 1. The absorption spectrum versus concentration of ferrihydrite: 26.8×10^{-2}g/l (1), 20.1×10^{-2}g/l (2), 13.4×10^{-2}g/l (3), 6.7×10^{-2}g/l (4).

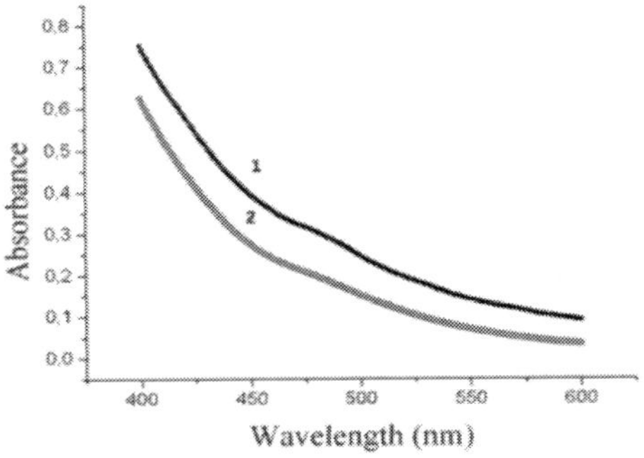

Figure 2. The absorption spectrum of ferrihydrite sample with concentration 13.4×10^{-2}g/l: before ultrasonic treatment (1) and after ultrasonic treatment (2).

The spectra are rather nondescript with gradual absorbance decline from 400—650 nm and do not allow to determine the fraction of ferrihydrite in the analyzed samples at a specific wavelength.

Journal of Physics: Conference Series **351** (2012) 012005 doi:10.1088/1742-6596/351/1/012005

Therefore, the changes produced by ultrasonic treatment of the samples were determined by comparison of the absorption spectra of the sample before and after ultrasonic treatment (Figure 2).

The decrease of absorbance values might be related to the partial destruction of the particle clusters and by the decrease of the amount of organic material on the surface of the particles. Time stability of ultrasonic-assisted samples of different concentrations was monitored using UV-Vis spectroscopy by comparison of the spectra, which were registered at 24h, 36h, 72h and one week after the ultrasonic treatment of the samples. The obtained data shows that the samples remained stable during one week. This study was useful for optimization the samples condition for small-angle X-ray scattering analysis.

2.3. FT-IR spectroscopy

In order to elucidate the structure and to identify the organic compounds which cover the ferrihydrite nanoparticles, a Fourier transform infrared (FT-IR) spectroscopic analysis was performed.

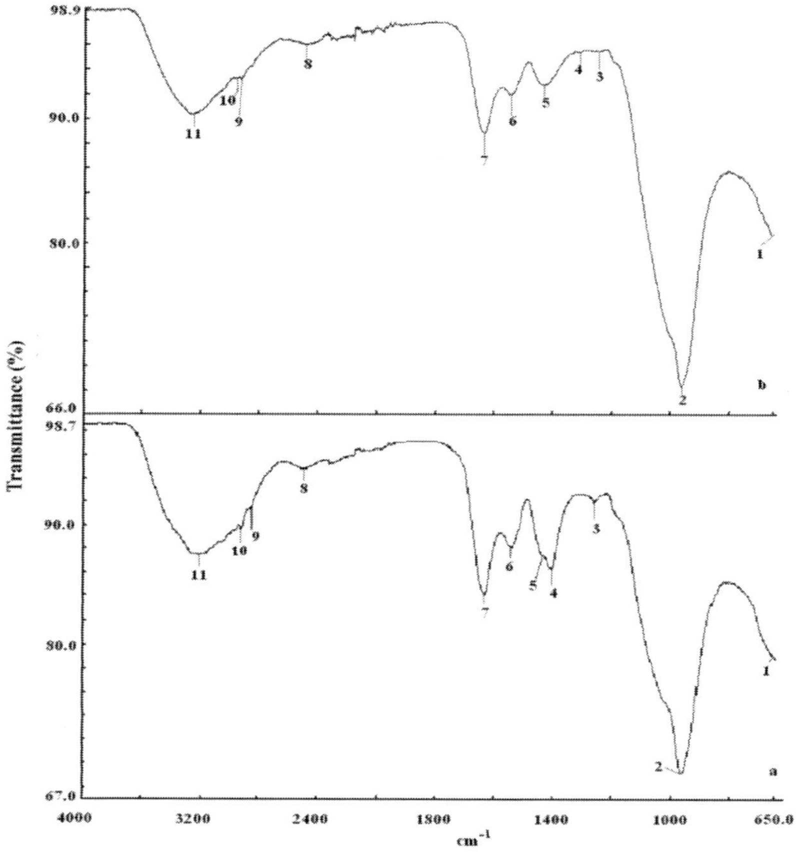

Figure 3. FT-IR analysis of biogenic ferrihydrite nanoparticles produced by *Klebsiella oxytoca* cultivated in dark growth conditions (a) one week and (b) 3 weeks, showing similar spectra. Peaks: 1 and 2 correspond to alkenes =CH bond, 3 carboxylic and/or hydroxyl C-O; 4 indicates the presence of OCH, COH, CCH groups; 5 and 6 are characteristic to amide I and II of proteins; 9 corresponds to CH vibrations of Cl; 10 is characteristic of OH stretching vibrations.

Dry samples of ferrihydrite nanoparticles produced by bacteria *Klebsiella oxytoca* were subjected to FT-IR analysis using Perkin Elmer IR spectrometer at the Institute of Chemistry of ASM, Chisinau. The infrared spectra of biogenic ferrihydrite nanoparticles obtained from bacteria *Klebsiella oxytoca* cultivated in dark growth conditions one week and three weeks are shown in Figure 3. (a) and (b), respectively.

FT-IR spectra of the ferrihydrite bio-mineral particles show a peak at 3255.0–3216.2 cm^{-1} characteristic of OH stretching vibrations. The peak at 2929.5–2926.8 cm^{-1} corresponds to CH vibrations of Cl, 1406.2 cm^{-1} indicates the presence of OCH, COH, CCH groups. These peaks clearly indicate the presence of glucose [12]. Further, the band at 1311.1 cm^{-1} indicates the CO bond of a polysaccharide. Strains of bacteria *Klebsiella oxytoca* are known to produce exopolysaccharides [13] and among these, several polysaccharides are specifically involved in the process of iron-binding [14]. Therefore, we assume that ferrihydrite nanoparticles isolated from bacteria *Klebsiella oxytoca* are wrapped by iron-binding exopolysaccharides. Furthermore, the bands at 1636.3 cm^{-1} and 1546.6 cm^{-1} confirmed the presence of amine I and II of proteins [15]. This suggests that iron-binding exopolysaccharides found in the analyzed samples are associated with the proteins residue.

2.4. Raman spectroscopy

The Raman spectroscopic characterization of biogenic ferrihydrite nanoparticles, from practical viewpoint, is useful for studying the trace deposits on substrates and is able to provide information regarding the particle chemical composition.

The Raman spectra of water dispersed biogenic ferrihydrite nanoparticles were obtained with a Laser Confocal Scanning Microscop SOLAR TII. The samples were studied at the wavelength of 441.6 nm, with the laser power of 100 MW.

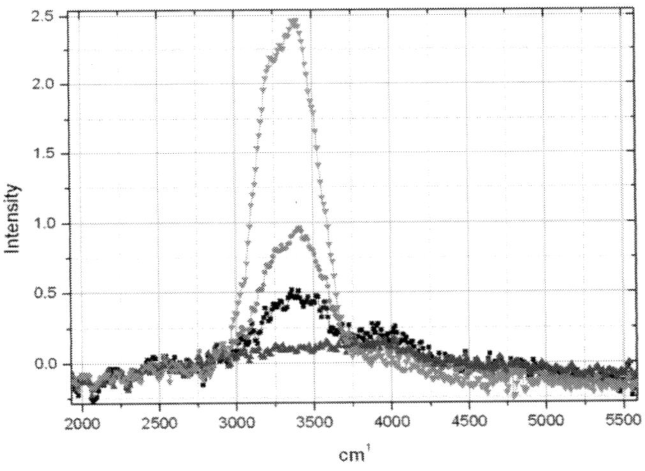

Figure 4. Raman spectra of distilled water (■) and of the multiple scan of a water dispersed sample of biogenic ferrihydrite with the concentration of 2.68 g/l: (▲) the scan of the nanoparticles in the sample; (▼) near to the surface of nanoparticles and (●) the scan deep in the sample.

The Raman spectrum of biogenic ferrihydrite nanoparticles in a single scan with the power set to 100 MW is shown in Figure 4. The spectrum consists of a broad weak band between 3650 cm^{-1} and 3000 cm^{-1}. In Raman spectroscopy this is characteristic of OH bonds stretching vibrations in the exopolysaccharides covering the ferrihydrite nanoparticles and of the water molecules. The comparison of the spectrum of ferrihydrite nanoparticles obtained near the surface of the particles and the spectrum obtained deep in the sample revealed a small difference in the absorption band of OH. The spectra obtained for measurement deep in the sample presents a relatively greater intensity of the 3650 cm^{-1} - 3000 cm^{-1} broad band. Taking into account the fact that ferrihydrite has a high-specific surface area (approximately 200-840 m^2/g) [16], and that because of its pores and arrangement of the nanoparticles, it has a large and accessible inner surface, the enhanced inner absorption can be related to the augmented quantity of exopolysaccharides absorbed in the ferrihydrite pores.

2.5. Small-angle X-ray scattering

The samples were also studied by small-angle X-ray scattering (SAXS) method on a Brucker Nanostar SAXS spectrometer available at Institute of Synthetic Polymer Materials RAS, Moscow. The experimental setup covered the q range 0.007 – 0.23 Å^{-1}. The experimental scattering curves were analyzed using model calculations included in ATSAS Package (PRIMUS, GNOM and GASBOR) [17,18]. In order to eliminate concentration or aggregation influence on the modeled experimental data, a linear extrapolation to zero concentration was done for the scattering data obtained on the samples with different concentrations of ferrihydrite particles.

The radius of gyration, R_G, which characterizes the distance of the scattering object parts from its centre of gravity, was calculated in the q range of 0.024-0.034Å^{-1} using the Guinier approximation:

$$I(q) = I(0)\exp\left(q^2 R_G^2 / 3\right). \quad (1)$$

This gave 6.41±0.13nm as an estimate for the radius of gyration independently of the shape of the investigated particles was obtained.

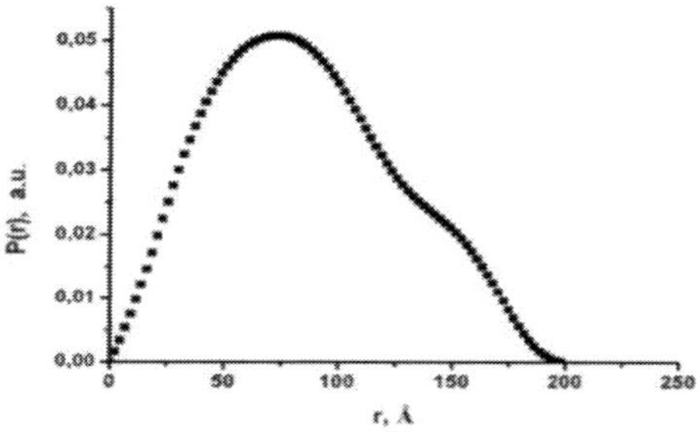

Figure 5. Function $p(r)$ calculated from the scattering curve obtained from extrapolation to zero concentration

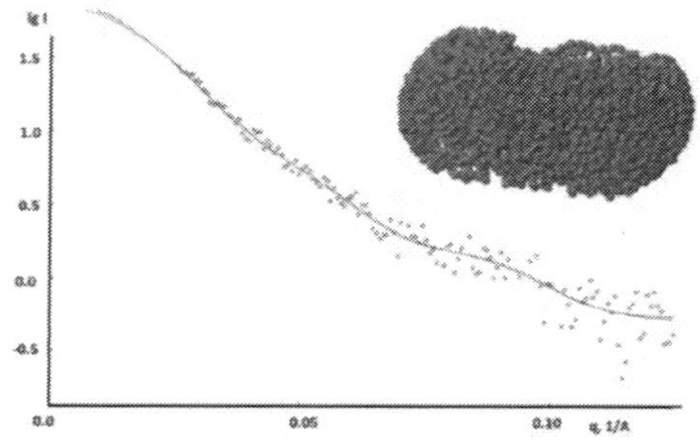

Figure 6. *ab initio* reconstruction of water dispersed ferrihydrite nanoparticles.

In order to calculate more precisely the values of R_G, the pair distribution function of biomineral nanoparticles was computed using the fitting procedures included in GNOM software from ATSAS package. The $p(r)$ function or *pair-distance distribution function* describes the pairewise set of all distances between points within the scattering. In SAXS, $p(r)$ is used to describe the spatial distribution of the electrons within the macromolecular structure, and is a useful tool for visibly detecting conformational changes within an ensemble of macromolecules. Typically, the particle distance distribution function, $p(r)=\gamma(r)r^2$, where $\gamma(r)$ is the characteristic function of the particles, is calculated by an indirect Fourier transformation to avoid problems due to the discrete sampling of the $I(q)$ curve over a finite range [19]. The indirect Fourier transform essentially constructs trial $p(r)$ functions that are Fourier transformed and evaluated in comparison with the experimental scattering. In the GNOM program [18], a regularizing multiplier is used to balance the smoothness of the trial $p(r)$ function with the goodness of the fit to the data.

The radius of gyration is obtained from the p(r) function using formula:

$$R_G^2 = \int_0^{D_{max}} r^2 p(r)\, dr \,/ \int_0^{D_{max}} p(r)\, dr. \quad (2)$$

In Eq. (2), D_{max} denotes the maximum distance inside the scattering particle. This method of determining R_G takes into accounts all the collected experimental data, not only those limited to small q domains, as is used in the Guinier approximation [19]. Therefore, the obtained real space parameters are likely better estimated for the samples, where small amounts of aggregation can influence the accuracy of the information regarding the scattering profile. Calculated $p(r)$ distribution function in the q range 0.02 – 0.26 $Å^{-1}$ is presented in Figure 5.

There was not found any resemblances between the obtained $p(r))$ function and those corresponding to the objects of a specific shape (sphere, rode, core-shell, etc.) described in the literature [19]. The elongated tail of the $p(r)$ function within the r-range 120 – 200 Å indicates the presence of macromolecules in the scattering solutions. This fact confirms the earlier results that the biogenic nanoparticles removed from bacterium *Klebsiella oxytoca* are still wrapped in an organic sheath [11].

We remark that the value of radius of gyration of 6.73±0.16 nm calculated from the pair-distribution function is close to the value estimated using the Guinier approximation.

The overall shape of the particles was determined with the aid of the program GASBOR. This software performs an *ab initio* reconstruction of molecular structure by a chain-like ensemble of *dummy residues* [18]. Figure 6 displays the resulting plot for the scattering curve obtained from a sample of biogenic ferrihydrite nanoparticles. The identified elongated 3D object resembles quite closely the rod-like model reported earlier [11]. As any other method that generates 3D structures from the 1D scattering data, the resulting pattern is not unique and might be optimized with supplementary experimental data.

3. Conclusions

Dry samples of ferrihydrite nanoparticles produced by bacteria *Klebsiella oxytoca* were subjected to FT-IR analysis. The FT-IR spectra revealed typical absorption peaks of exopolysaccharides. Additionally, the Raman spectra showed the presence of exopolysaccharides on the surface and inside the pores of the ferrihydrite nanoparticles. Characterization of ultrasonic assisted biogenic ferrihydrite particles dispersed in water was performed by SAXS. Analysis of the scattering curves using ATSAS software package suggests ferrihydrite particles of an elongated shape with the radius of gyration 6.73±0.16 nm. We note that the results obtained for diluted samples of ferrihydrite nanoparticles are close to the structural parameters determined previously by means of SAXS, HRTEM [11,13] and magnetogranulometry [9] methods on concentrated samples. This fact allows us to surmise that ultrasonic treatment does not produce any essential physico-chemical changes to the particles but facilitates to minimize the influence of aggregation on the scattering data. Higher q values contain more details regarding the molecular shape. Thus, in order to obtain more precise information about the characteristic structure of ferrihydrite, further investigations by SAXS experiments at higher angles are necessary.

References

[1] Schüler D 1999 *J. Molec. Microbiol. Biotechnol.*, **1** (1) 79-86
[2] Sumper M, Kröger N 2004 *J. Mater. Chem.*, **14** 2059-2065
[3] Shultz-Lam S, Harauz G, Beveridge T J 1992 *J. Bacteriol.*, **174** (24) 7971-7981
[4] Stolyar S V, Bayukov O A, et.al. 2006 *Inorganic Materials*, **42**(7) 763-768
[5] Ishchenko L A, Stolyar S V, Ladygina V P, et.al. 2010 *Physics Procedia*, **9** 279-282
[6] Stolyar S V, Bayukov O A, et.al. 2007 *Inorganic Materials* **43**(6) 638-641
[7] Stolyar S V, Bayukov O A, Ladygina V P, et al. 2011 *Phys. Solid State*, **53**(1) 100-104
[8] Raikher Yu L and Stepanov V I 2008 *J. Experim. Theor. Phys.*, **107**(3) 435-444
[9] Raikher Yu L, Stepanov V I, et.al. 2010 *Phys. Solid State*, **52**(2) 298-305
[10] Balasoiu M, Stolyar S V, et.al. 2010 *Rom. Journ. Phys.*, **55**(7-8) 782-789
[11] Balasoiu M, Ishchenko L I, Stolyar S V, et.al. 2010 *Optoelectronics and Advanced Materials – Rapid Communications* **4**(12) 2136-2139

[12] Ibrahim M, Alaam M, El-Haes H, et. al. 2006 *Ecl. Quim.* Sao Paulo **31**(3) 14-21
[13] Leone S, De Castro C, Parrillini M, et. al. 2007 *Eur. J. Org. Chem.* 5183-5189
[14] Baldi F, Marchetto D, Battistel D, et. al. 2009 *J. Appl. Microbiol.*, **107** 1241-1250
[15] Jao Y, Cody G, Harding A, et. al. 2010 *Appl. and Environ Microbiol.* **76**(9) 2916-2922
[16] Weidler P G 1997 *J. Porous Materials* **4** 165-169
[17] Svergun D I, et. al. 1992 *J. Appl. Cryst.* **25** 495-503
[18] Svergun D I, et. al. 2001 *Biophysical J.* **80** 2946-2953
[19] Glatter O, Kratky O 1982 *Small angle X-ray scattering* (Academic Press, London) 156-163

Investigation of DESO/LIPID membranes interaction by X-Ray scattering

Yu E Gorshkova[1], **O I Ivankov**[1], **A I Kuklin**[1,2], **V I Gordeliy**[1,2,3,4]

[1] Joint Institute for Nuclear Research, Dubna, Russia

[2] Bionanocenter, Moscow Institute of Physics and Technology, Dolgoprudny, Russia

[3] ICS-5, Forschungszentrum Juelich, Juelich, Germany

[4] Institute of Structural Biology J. P. Ebel, Grenoble, France

E-mail: valentin.gordeliy@ibs.fr

Abstract The influence of diethyl sulfoxide $(C_2H_5)_2SO$ (DESO) on the structure of the phospholipid DPPC (1,2-dipalmitoyl-sn-glycero-3-phosphocholine) membranes was investigated over a wide range of DESO molar concentrations $0.0 \leq X_{DESO} \leq 1.0$. The dependences of structural parameters - the repeat distance d of the multilamellar membranes and the thickness on the molar concentration X_{DESO} in the gel and liquid-crystalline phases were determined. The main phase transition temperature T_m for DPPC multilayers was identified at different molar concentrations of DESO. A comparative analysis of the data for the system DPPC/H_2O/DESO with the data previously published for the system DPPC/H_2O/DMSO was done.

1. Introduction

Cryoprotective agents are important tools of many biological, biophysical and medical studies. The mechanism of interaction of these cryoprotectants is different. Thus, the natural sugars provide protection without penetration into the cell, polyalcohols and glycerol enter the cell and reduce the likelihood of the formation of ice crystals. The greatest interest is the cryoprotectants, such as dimethyl sulfoxide, that can replace water molecules associated with cellular elements, and affect the structure of water. DMSO is widely used in cell biology, cryobiology, pharmacology, medicine and agriculture [1, 2]. However, the rapid development of molecular and cell biology poses a challenge to find new solvents and cryoprotectants that can maintain the integrity of the cells in the process of freezing - thawing. One of relatively new interesting and efficient cryoprotectants is diethyl sulfoxide (DESO).

First results obtained with DESO are quite encouraging. First of all, DESO shows strong interaction with water, even stronger than that of DMSO [3]. The investigation of the ability of DESO to act as an effective cryoprotectant on *E. coli* survival confirms that DESO, more than DMSO, is able to penetrate living tissues without causing significant damage [4]. It was shown that the temperature of thermal denaturation of *Human Serum Albumin* decreases in the presence of DESO. The hydrophobic interaction of sulfoxides with the protein surface is one of the probable reasons for this. The electron spectroscopy investigations indicate that for DESO this effect is observed at lower concentrations if compared with DMSO [5]. The fluorescent measurements testify that the formation of hydrophobic layer protects the protein against irreversible structural changes under low temperatures.

It is well known that a lipid bilayer is the structural basis of biological membranes. Consequently, the lipid bilayer influences or/and determines many properties of the membranes, in particular, proteins embedded in it [6]. Therefore it is not surprising to see intensive studies of the cryoprotectors, especially DMSO, on the structure and properties of lipid membranes [7 - 13]. The effect of these solvents on the structure of lipid membranes is different. For example, the main transition temperature T_m of the DPPC phospholipids dispersed in trehalose decreases from 42 to 24°C [14]. Glycerol significantly decreases T_m with increasing concentration [15]. In contrast, DMSO increases T_m of DPPC and DMPC multilayers [10, 12, 16].

DESO is a solvent which is somehow similar to well studied DMSO. The aim of the present study is to investigate the influence and interaction of diethyl sulfoxide with phospholipid DPPC *(1,2-dipalmitoyl-sn-glycero-3-phosphocholine)* membranes. The present paper presents the first results in the investigation of the membrane structure of DPPC multilayers in an x-ray diffraction experiment at a wide range of concentrations of DESO $0.0 \leq X_{DESO} \leq 1.0$. The study was carried out for all mole fractions X_{DESO} of DESO/water in the temperature interval from 20°C to 85°C.

2. Materials and methods

2.1. Materials

DPPC (over 99% pure) was obtained from SERVA (Germany).
DESO (over 99% pure) was produced by MERK (Darmstadt, Germany).
Water (18 MV/cm) was obtained with the help of Millipor.

2.2. Sample preparation

The given amount of DESO in a DESO/water solution was mixed with the lipid in 1:1 w/w solvent/lipid ratio and placed in an x-ray borosilicate capillary with a diameter of 1.5 mm and wall thickness of 0.01 mm (W. Muller, Berlin, Germany). The mixture was spun down by centrifugation. The capillary was hermetically sealed, and the specimens were held at a higher temperature than that of the phase transition for several hours.

2.3. Experimental technique

Wide-angle measurements were performed at the X-ray diffractometer DRON - 4 adapted for membrane studies. The X-ray diffractometer has a Cu-anode tube as a radiation source. Bragg-Brentano focusing and two slits are placed at a distance of 10 mm from each other, close to the output window of the tube for x-ray beam formation. An additional slit is placed just before the sample to suppress the scattering from the boundaries of the second slit. A single detector is used to detect the scattered x-rays. A θ-2θ scan is carried out in the horizontal plane (with the rotation of the detector around the vertical axis). Measurements were made at a Ni-filtered CuKα radiation x-ray source with a wavelength of 1.54 Å. The specimen container was kept at a constant temperature by using a liquid thermostat to the precision of $\pm 0.5\,°C$ and controlled by a thermocouple.

The repeat distance of the multilayer structure was determined from the position of diffraction peaks using the Bragg equation ($2d\sin(\theta) = \lambda$).

3. The results and discussion

3.1. The effect of DESO concentration on the lipid membranes in a gel phase

A typical diffraction pattern of the DPPC multilayers in DESO/water solution is presented in Figure 1. It corresponds to the DESO molar concentration $X_{DESO} = 0.2$ and the temperature $T = 20°C$.

The dependence of the repeat distance d of DPPC membranes on the DESO molar concentration at the temperature $T = 20°C$ is shown in Figure 2. The repeat period d drastically changes in the region of DESO concentrations $0 \leq X_{DESO} \leq 0.1$ in the $L_{\beta'}$ phase. The value d decreases from 64.2 Å for DPPC multilayers in pure water to 58.6 Å at $X_{DESO} = 0.1$ (see Table 1). A similar behavior in the change of the repeat distance takes place in a well investigated DPPC/DMSO/water system in the gel phase [12,

17]. An analogous pattern was observed in the study by SANS for the DMPC-d54 multilamellar membranes in the water/DMSO solution [18].

The second region corresponds to DESO concentrations $0.1 \leq X_{DESO} \leq 1.0$. In this case the repeat distance d decreases monotonously from 58.6 Å at X_{DESO} = 0.1 down to 57.1 Å at X_{DESO} = 1.0 as presented in Table 1. Thus we can assume that a further increase in the concentration of DESO does not result in structural changes in multilayer DPPC membranes in the gel $L_{\beta'}$ phase.

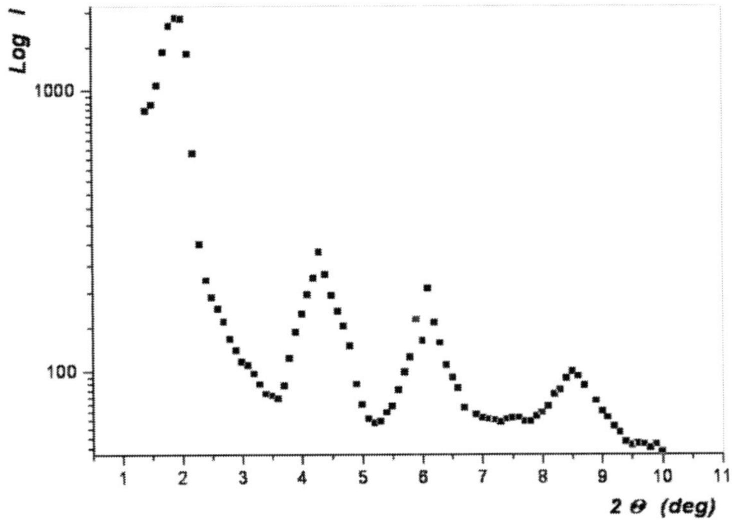

Figure1. X-ray diffraction pattern from DPPC/DESO/water mixture at the DESO concentration X_{DESO} = 0.2 and T = 20°C.

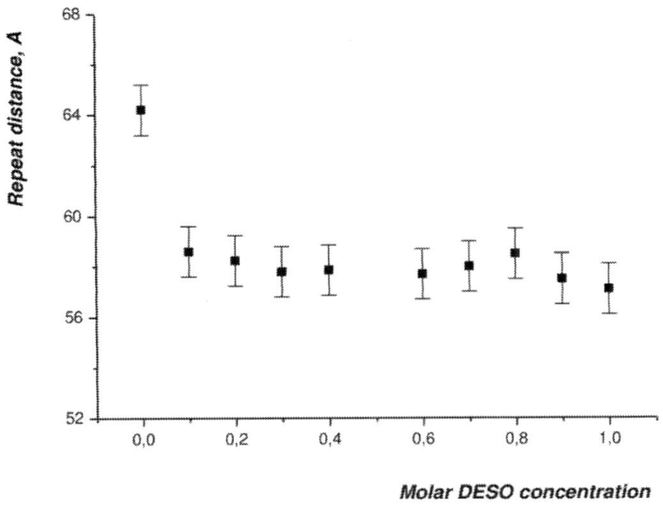

Figure2. The dependence of the repeat distance d of DPPC membranes on the DESO concentration at T = 20°C.

DESO as well as DMSO reduces the repeat distance of lipid membranes in contrast to other solvents (e.g. glycerol [19]). The influence of DESO on phospholipid membranes is similar to that of DMSO up to $X_{DESO} = 0.9$. However, in contrast to DMSO, it does not show a drastic reduction of the repeat distance (for DMSO it occurs at $X_{DMSO} = 0.9$) [12, 18]. A significant decrease of the repeat distance in the case of DMSO is associated with the transition of the lipid membranes to the interdigitated phase in which hydrocarbon chains of one monolayer of the bilayer penetrate to the other [20 - 22]. In case of DESO we did not observed any transition to the interdigitated $L_{\beta I}$ phase.

3.2. Effect of the DESO concentration on the lipid membranes in a liquid phase

The investigation of the DESO influence on the multilayer DPPC membranes was carried out at T= 70°C which is above the temperature of the main phase transition T_m for all molar concentrations X_{DESO} (see Figure 4 or Table 1). The repeat distance *d* decreases abruptly with increasing DESO fraction up to $X_{DESO} = 0.1$ as in the gel phase and then continues to decrease slowly.

Table 1. X-Ray data relative to the heating and cooling processes of DPPC multilayers in H_2O/DESO and H_2O/DMSO mixtures with increasing amounts of sulfoxides.

Molar concentration	Repeat distance (Å)				Temperature of main phase transition (°C)		
	Gel phase		Liquid phase				
X	d_{DESO} ±2Å	d_{DMSO} [a]	d_{DESO} ±2Å	d_{DMSO} [a]	T_{m_DESO}	T_{m_DMSO} [a]	T_{m_DMSO} [b]
0	64.2	64.0	57.5	58.0	41.0	40.7	40.7
0.1	58.6	57.8	46.0			44.0	44.0
0.2	58.2	58.0	41.1	50.0	48.0	49.0	49.0
0.3	57.8	57.6				50.0	50.0
0.4	57.9	57.0	40.8		56.0		
0.5		57.0		47.0		52.0	52.0
0.6	57.7	57.5			58.0	56.0	52.0
0.7	58.0	57.0					52.0
0.8	58.5	56.0	38.2	44.0	59.0	55.0	52.5
0.9	57.5	51.8				58.0	52.0
1	57.1	52.0		41.0	60.0	78.0	

[a] V. I. Gordeliy et al., Biophys.J. 75, 2343 (1998)

[b] M.A. Kiselev, JINR Preprint (2003)

In the first region $0 \leq X_{DESO} \leq 0.1$ the d value is changed by 11 Å. This is twice as much as the observed changes for the system DPPC/H_2O/DESO in the gel phase. With a further increase of the DESO concentration, d decreases monotonously from 46 Å at $X_{DESO} = 0.1$ down to 38 Å at $X_{DESO} =$

Journal of Physics: Conference Series **351** (2012) 012006 doi:10.1088/1742-6596/351/1/012006

0.8. The data for the DESO mole fraction dependence of the repeat period in the liquid L_α phase are included in Table 1.

It is clear that the influence of DESO and DMSO solvents on the multilayer DPPC membranes in the L_α phase is similar but, nevertheless, displays important differences.

3.3. Temperature dependence of membrane structural parameters.

The temperature dependence of the DPPC membrane repeat distance at the molar concentration of DESO $X_{DESO} = 0.2$ is presented in Figure 3. It is clear that structural changes of the system under these conditions take place at $T = 48$ °C. The repeat distance d = 58.24 Å is almost unchanged up to 45 °C. However, a sharp drop of the repeat distance occurs at $T_f = 48$ °C, which corresponds to the transition of the lipid membranes to the liquid L_α phase.

Temperature-dependent changes of the repeat period d at the DESO concentration X_{DESO} = 0, 0.2, 0.4, 0.6, 0.8, 1.0 have also been measured. In all cases we observed a similar qualitative change in the membrane structural parameters with the temperature variation.

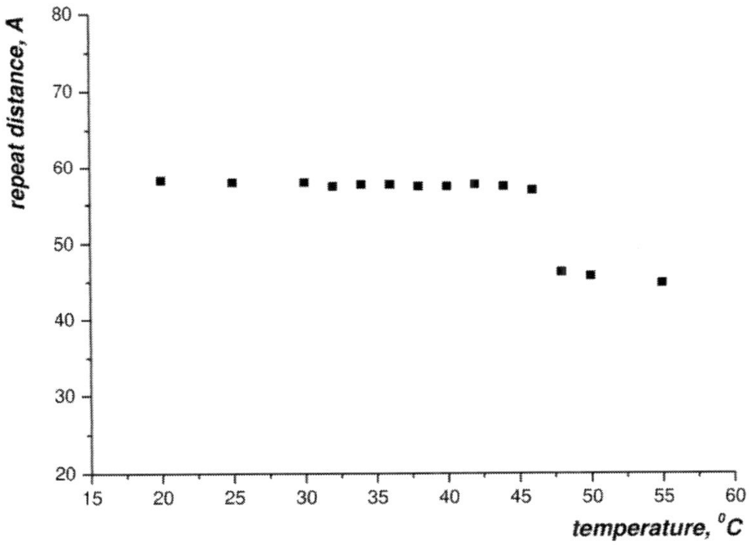

Figure 3. The temperature dependence of the DPPC membrane repeat distance at the DESO concentration $X_{DESO} = 0.2$.

The DESO concentration dependence of the phase transition temperature is presented in Figure 4. The temperature of the main phase transition T_{m_DESO} as T_{m_DMSO} (see Table 1) increased with increasing molar DESO concentration. The dependence of the phase transition temperature on the DESO concentration is exponential. One can see two characteristic behavior regions. In the first region at ratios $0 \leq X_{DESO} \leq 0.4$ one observes a fast, almost monotonic increase of the temperature of the main phase transition from 41°C for DPPC in pure water to 56 °C at $X_{DESO} = 0.4$ with increasing DESO concentration. In the second region ($0.4 \leq X_{DESO} \leq 1.0$) the T_{m_DESO} value changes slightly. We have not observed any transition to the interdigitated phase as shown by Gordeliy et al for the DPPC/DMSO/water system [12].

The fit of the experimental data shows that the dependence of the phase transition temperature on the DESO concentration has a non-linier (exponential) character.

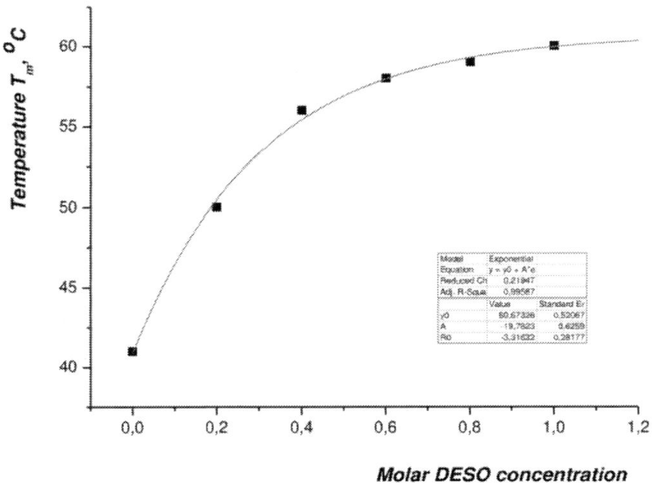

Figure 4. The dependence of the phase transition temperature vs. DESO concentration.
■ - experimental data
—— - fitting line

5. Conclusions

The repeat distance d in a gel phase decreases sharply (by 6 Å) with an increase in the mole fraction of DESO from 0 up to 0.1. With a further increase in the DESO concentration, d decreases monotonously from 58.6 Å at X_{DESO} = 0.1 down to 57.1 Å at X_{DESO} = 1.0. As in the gel phase, the repeat distance d in a liquid phase decreases abruptly with increasing DESO fraction up to X_{DESO} = 0.1 and then continues to decrease slowly.

The influence of both sulfoxides DMSO and DESO on the repeat distance of the DPPC phospholipid membranes is similar for $X_{DESO} < 0.9$. However, phospholipid membranes turn into an interdigitated phase with increasing amounts of dimethyl sulfoxide [10, 12, 18]. We have not observed a transition of the lipid membranes at X_{DESO} = 1.0 to the ripple phase at T = 20°C for the DPPC/H_2O/DESO system. This fact distinguishes diethyl sulfoxide from other well-known cryoprotectants. For instance, lipid chains interdigitate in the presence of the mixture of glycerol, ethylene glycol or methanol with water [16, 19].

The phase transition temperature changes from 41°C at X_{DESO} = 0 to 56°C at X_{DESO} = 0.4. The value of T_m changes slightly for the DESO concentration $0.4 \leq X_{DESO} \leq 1.0$. This region corresponds to a smooth decrease of the repeat distance. This means that diethyl sulfoxide has significant impact on the structure of phospholipid multilayers at the concentration ratio $0 \leq X_{DESO} \leq 0.4$. The temperature of the main phase transition is higher in case of DMSO at low concentrations than in DESO. A similar situation was observed in the DSC study on the effect of DMSO and DESO on phospholipid liposomes [23]. In the presence of low amounts of both sulfoxides (≤ 0.2) the main effect was an increase in the transition temperatures, which was more enhanced in the presence of DMSO. Our data support the hypothesis that diethyl sulfoxide should best behave as a cryoprotector at high concentrations. Our results show that at $X_{DESO} \geq 0.4$ the structure of the lipid multilayers does not change. This means that strong hydrophobic interactions take place between DPPC multilayers and diethyl sulfoxide. Perhaps this is the reason for the formation of a stable ice layer near the surface of the lipid surfaces and explains the effect of better protection of biological membranes during freezing as was shown earlier [4].

The comparison of the effect of the DESO and DMSO influence on the membrane structural parameters and phase transition shows that DESO changes properties of lipid bilayers similarly but in a more 'soft' way than DMSO. We assume that it is less stressful for biological membranes and this can be one of the reasons for higher survival of the cells when DESO is used for cryoprotection.

References

[1] Lovelock J E and Bishop M W H 1959 *Nature (London)* **183** 1394
[2] Ali J and Shelton J N 1993 *J. Reprod. Fertil.* **99** 471
[3] Markarian, S.A.; Zatikyan, A.L.; Bonora, S.; Fagnano, C. 2003 *J. Mol. Struct.*, **665** 285
[4] Markarian S A, Bonora S, Bagramyan K A, Arakelyan V B 2004 *Cryobiology* **49** 1
[5] Grigoryan K R Markarian S A Aznauryan M G 2009 *Problems of Cryobiology* **19** 3-9
[6] Gennis B R 1989 Biomembranes *Springer-Verlag (London)* **533**
[7] Crowe J H, Carpenter J F, Crowe L M, and Anchordoguy T J 1990 *Cryobiology* **27** 219–231
[8] Anchordoguy TJ, Rudolph AS, Carpenter JF, Crowe JH 1987 *Cryobiology* **24(4)** 324-31
[9] Anchordoguy T J, Carpenter J F, Crowe J H, Crowe L M 1992 *Biochim. Biophys. Acta* **1104** 117-122
[10] Tristram-Nagle S, Moore T, Petrache H, Nagle J F 1998 *Biochim. Biophys. Acta* **1369** 19-33
[11] Yu Z W, Quinn P J 1998 *Mol. Membr. Biol.* **15** 59-68
[12] Gordeliy V I, Kiselev M A, Lesieur P, Pole AV, Teixeira J 1998 *Biophys. J.* **75** 2343-51
[13] Smondyrev A M, Berkowitz M L 1999 *Biophys. J.* **76** 2472-78
[14] Crowe L M, Crowe J H 1988 *Biochim. Biophys. Acta* **946** 193- 201
[15] McDaniel R V, Simon S A, McIntosh T J 1983 *Biochim. Biophys. Acta* **731** 97 - 108
[16] Curatolo W 1985 *Biochim. Biophys. Acta* **817** 134- 138
[17] Yu Z W and Quinn P J 1995 *Biophys. J.* **69** 1456-63
[18] Gorshkova J E, Gordeliy V I 2007 *Crystallography Reports* **52(3)** 535–539
[19] McIntosh T J, Magid A D and Simon S 1989 *Biochemistry* **28** 7904–12
[20] Kim J T, Mattai J and Shipley G G 1987 *Biochemistry* **26** 6599–03
[21] Kirchner S and Cevc G 1993 *Europhys. Lett.* **23** 229–235
[22] Simon S A, McIntosh T J and Magid A D 1988 *J. Colloid Interface Sci.* **126** 74–83
[23] Bonora S, Markarian S A, Trinchero A, Grigorian K R 2005 *Thermochimica Acta* **433** 19–26

SANS spectra of the fractal supernucleosomal chromatin structure models

Andrey V Ilatovskiy[1,2], Dmitry V Lebedev[1], Michael V Filatov[1], Michael G Petukhov[1,2] and Vladimir V Isaev-Ivanov[1]

[1] Division of Molecular and Radiation Biophysics, Petersburg Nuclear Physics Institute, Russian Academy of Sciences, Gatchina, Russia

[2] Research and Education Center "Biophysics", PNPI RAS and St. Petersburg State Polytechnic University, St. Petersburg, Russia

E-mail: andreyi@omrb.pnpi.spb.ru

Abstract. The eukaryotic genome consists of chromatin—a nucleoprotein complex with hierarchical architecture based on nucleosomes, the organization of higher-order chromatin structures still remains unknown. Available experimental data, including SANS spectra we had obtained for whole nuclei, suggested fractal nature of chromatin. Previously we had built random-walk supernucleosomal models (up to 10^6 nucleosomes) to interpret our SANS spectra. Here we report a new method to build fractal supernucleosomal structure of a given fractal dimension or two different dimensions. Agreement between calculated and experimental SANS spectra was significantly improved, especially for model with two fractal dimensions—3 and 2.

1. Introduction

The eukaryotic genome is an highly packed nucleoprotein complex—chromatin, that has hierarchical structure based on nucleosomes [1]. The nucleosome was discovered in the mid-1970s [2, 3], its structure (Fig. 1) was solved at virtually full-atomic resolution 2.8 Å by X-ray diffraction method in 1997 (PDB [4] entry 1AOI) [5]. PDB database [4] contains more than 40 mononucleosome structures (PDB entry 1KX5 [6] has the best resolution of 1.9 Å) and one tetranucleosome structure (PDB entry 1ZBB [7], resolution of 9 Å).

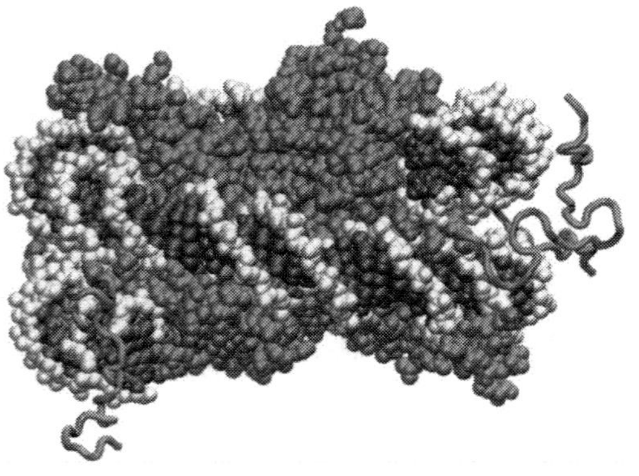

Figure 1. Mononucleosome structure (PDB entry 1KX5). Double-stranded DNA (shown by yellow (for sugar-phosphate backbone) and blue colors) forms ~1.7 turns of superhelix around protein core (shown by red color). Flexible protein tails (residues 1–37 of chains A and E and residues 1–22 of chain H) are shown in backbone representation.

30-nm chromatin fiber, observed with electron microscopy [3, 8], was proposed to be the next level of DNA compaction. It had been investigated with a number of experimental methods, including SANS [9, 10], however, obtained data were controversial and supported different models [1]. Growing evidence suggested irregular organization of 30-nm chromatin fiber based on 10-nm fiber (reviewed in [11]). Thus, structure of chromatin fiber is still an open question.

Recent studies of internal structure of the whole nucleus revealed its fractal nature [12]. New experimental Hi-C technique allowed to capture inter- and intra-chromosomal space interactions at megabase resolution [13]. Contact probability was calculated as a function of DNA strand distance a and it turned out that the probability was scaled as $1/a$ for the range of 0.5–7 Mb. This finding was consistent with fractal globule model [14], that had predicted fractal dimension of 3. Observations of diffusion behavior of fluorescent tracers provided another line of evidence to support fractal organization of chromatin [15]. Authors calculated fractal dimensions of euchromatin (2.6) and heterochromatin (2.2), that were close to estimates of fractal dimension of chromatin (2.3–2.4) made with tree fractal model of the second order [16].

We had investigated native nuclei of chicken erythrocytes [17] and other cells [18] with SANS. Obtained SANS spectra $I(s)$ had linear regions in double logarithmic scale for sizes from 15 nm to 1.5 μm, that indicated fractal properties of the nuclei [19]. Fractal dimensions of chromatin (calculated as negatives of slopes of linear regions of SANS spectra) were 2.4 for sizes below 420 nm and ~3 for larger sizes. Interestingly, fractal dimension of protein component of the nuclei (in 64% D_2O) was 2.5 for the whole range of sizes, while fractal dimensions of DNA component (in 40% D_2O) were 2.2 for sizes below 250 nm and ~3 for larger sizes.

Previously we have reported development of supernucleosomal chromatin structure model to interpret our SANS data [20]. As the first application of the method, SANS spectra were calculated for random-walk chains of 10^6 nucleosomes. It was shown that calculated and experimental spectra had both common and different features.

In this paper we describe general method to build fractal supernucleosomal structure models of a given fractal dimension (or two different fractal dimensions) and present results of SANS spectra calculations for these structures. New method of structure generation significantly improved agreement between calculated and experimental SANS spectra.

2. Methods

2.1. Nucleosome model

Nucleosome physical model was based on a mononucleosome structure from PDB database [4]. Scattering length density (SLD) parameterization was done at residue level. Scattering lengths in H_2O $\{L_i^H\}_{i=1}^{24}$ and in heavy water D_2O (with full substitution of exchangeable hydrogen atoms by deuterium atoms) $\{L_i^D\}_{i=1}^{24}$ and volumes $\{V_i\}_{i=1}^{24}$ for 24 residues (20 amino acids and 4 nucleotides) were taken from [21]. SLDs of H_2O (ρ_H) and D_2O (ρ_D) were taken from the same source.

Weight of a pair distance between two atoms of residues of types i and j (used to calculate SANS spectra) was defined as product of contrast values of the residues, concentration (volume fraction) of D_2O in solvent W_D and probability of H/D exchange $P_{H/D}$ were taken into account in this evaluation. First, SLD of solvent was calculated as

$$\rho_S(W_D) = p_H + W_D(\rho_D - \rho_H).$$

Next, SLD of a residue of type i was equal to

$$\rho_i(W_D, P_{H/D}) = \frac{1}{V_i}[L_i^H + W_D P_{H/D}(L_i^D - L_i^H)]$$

and its contrast was defined as $\Delta_i(W_D, P_{H/D}) = \rho_i - \rho_S$. Finally, weight of a pair distance was calculated as

$$w_{ij}(W_D, P_{H/D}) = \Delta_i \Delta_j \quad (1)$$

2.2. Structure generation

One of the methods to generate fractal is self-similarity transformation [22]. Let us consider a generation of a fractal of topological dimension $T = 1$. At the beginning fractal initiator (fractal of the 0th order) is straight interval of length L. At the first step, the initiator is replaced by generator that is broken line of n intervals of length L/f, thus producing fractal of the 1st order. Each interval of the generator is replaced by the whole generator at the next steps (Fig. 2). Fractal of the 2nd order consists of n^2 intervals of length $(L/f)^2$. For an ideal fractal obtained with this procedure (that consists of infinite number of self-similar subsets) fractal dimension d is defined as similarity dimension

$$d_s = \frac{\ln n}{\ln f}. \tag{2}$$

If generator is randomly changed (but n and f are kept constant) at each step of generation, than a stochastic fractal will be built (Fig. 2). Its structure is locally irregular, however, statistical characteristics of ideal and stochastic fractals are similar, especially for large structures.

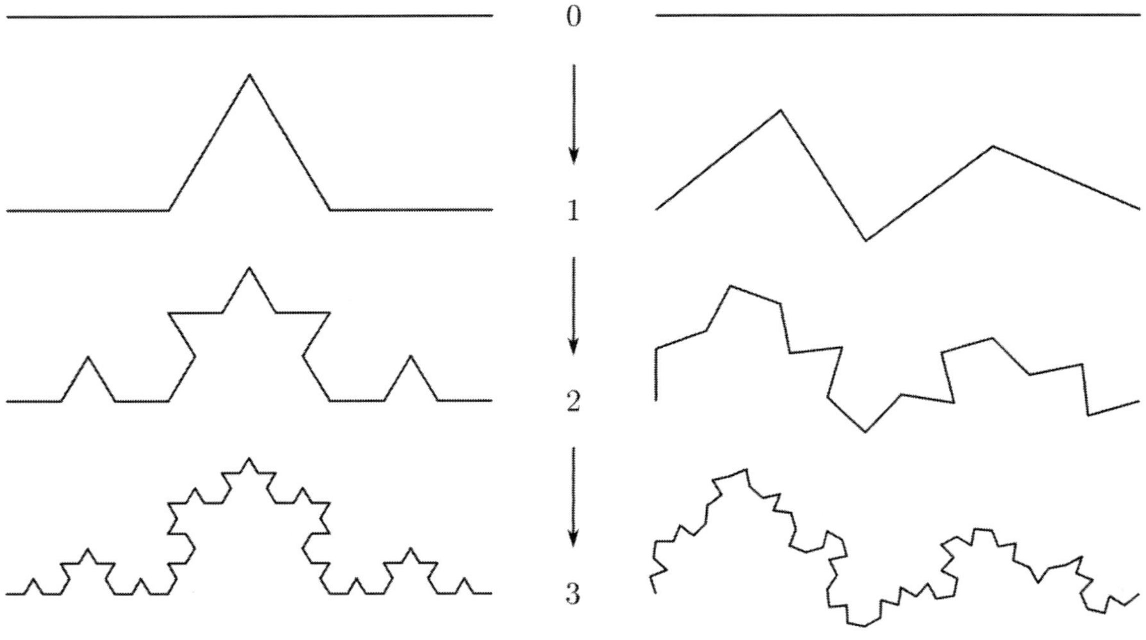

Figure 2. First three orders of generation of a fractal of topological dimension $T = 1$ (curve) by self-similarity transformation in space which Euclidean dimension is $E = 2$ (plane). Left panel shows growth of classic triadic Koch curve with $n = 4$ intervals of length reduced by a factor of $f = 3$, thus its fractal dimension is $d = \ln 4/\ln 3 \approx 1.2618 > T$. Right panel shows growth of stochastic fractal structure formed by random generators equal (in terms of n and f) to the generator of the Koch curve.

Self-similarity transformation in 3D space (Euclidean dimension $E = 3$) was used to generate fractal model of supernucleosomal chromatin structure of topological dimension $T = 1$. Initiator was straight interval of length L formed by two points that were the first and the last points of the final structure. Generator was broken line of n randomly oriented intervals of length $(L/f)^k$, where k—order of generation. Generation was started from the initiator (fractal of order $k = 0$) and was stopped at order $k = K$. Thus, final fractal structure consisted of $N = n^K$ points that were connected with intervals of length $\ell = (L/f)^K$.

N nucleosome structures were placed (in random orientation) to Cartesian coordinates of N points of fractal structure thus forming complete supernucleosomal chromatin structure model. To take into account finite size of the nucleosome structure self-intersections were not allowed during fractal structure generation: in case newly generated points were closer to previously generated points than effective size of the nucleosome structure $r_{\max}$, they were deleted and generation step was performed again using another random generator. In rare cases existing points left no space for new points and generation algorithm continued to the next step after definite number of unsuccessful tries (technically, generation algorithm was recursive tree-walk and there was no method to delete points once they were recorded into the structure). To eliminate self-intersections simple Monte Carlo dynamics was applied to the generated fractal structure. Since $r_{\max} \ll \ell$ this procedure had a slight impact on overall geometry of the structure.

Fractal dimension of the final structure could be estimated by Eq. (2) or obtained from expected dependence of mean radius of gyration r_g on number of points N_g

$$\langle r_g \rangle \propto [N_g]^{1/d_g}\,, \qquad (3)$$

where

$$r_g = \sqrt{\frac{1}{N_g}\sum_{i=1}^{N_g}(x_i - \langle x\rangle)^2 + (y_i - \langle y\rangle)^2 + (z_i - \langle z\rangle)^2},$$

$\{x_i, y_i, z_i\}$—Cartesian coordinates of N_g points of a continuous subset of the fractal structure. If dependence $\langle r_g(N_g)\rangle$ was linear in double logarithmic coordinates, fractal dimension d_g could be calculated as an inverse of slope of linear fit in accordance with Eq. (3). Despite generated fractal structures were not the ideal fractals, preliminary calculations shown good agreement between two estimates of fractal dimension made with Eqs. (2) and (3).

2.3. SANS spectra calculations

The method was described in full details in our previous work [20]. In brief, Cartesian coordinates of a fractal supernucleosomal structure and the physical nucleosome model were used to perform Monte Carlo simulation of unweighted histogram of distance distribution function (DDF). Each element of this histogram contained distance value and types of two residues, atoms of which were used to calculate the distance. The histogram was weighted by 0 and 1 to calculate simple geometrical DDF of DNA or protein parts of the whole structure. To obtain SANS spectra the histogram was weighted by products of contrast values of residues that formed particular distance (Eq. (1)), and Fourier-transformed into spectra.

3. Results and Discussion

3.1. Mononucleosome structure properties

Geometrical and SANS properties of 33 mononucleosome structures from PDB database [4] were analyzed in our previous work [20]. It was shown that all these structures were highly similar, an average nucleosome consisted of 12.1 thousand atoms, 146-bp double-stranded DNA formed superhelix with diameter of 84.0 Å and pitch of 25.8 Å.

Solvent match points (SMPs) of DNA and protein components of the average nucleosome structure varied in wide range of heavy water concentrations (SMP of protein was 0.308–0.391) depending on probability of H/D exchange $P_{H/D}$. For nucleosomal DNA $P_{H/D} \approx 1$ [23], while protein core had complex pattern of H/D exchange rate [24]. It was shown, that value of $P_{H/D}$ had significant impact on SANS spectra only for W_D values close to SMPs [20]. Thus for all further calculations $P_{H/D}$ was set to 1.

Among analyzed mononucleosome structures nucleosome 1KX5 [6] was selected for further calculations because this structure was solved at the best resolution of 1.9 Å and its properties were close to that of the average nucleosome.

Nucleosome 1KX5 had flexible protein tails (Fig. 1), that were located outside of core particle and poorly ordered—their particular conformation might have been affected by the crystal packing. DDF calculations shown that these tails increased maximal size of the nucleosome from 124.0 to 174.2 Å, while differences in corresponding SANS spectra were negligible [20]. Thus flexible protein tails were removed from the nucleosome structure and this model with effective size $r_{\max} = 120$ Å was used to build supernucleosomal chromatin structure.

3.2. SANS spectra of fractal supernucleosomal chromatin structure models

Fractal supernucleosomal chromatin structures of $\sim 10^5$ nucleosomes were generated for several values of n up to $n_{\max} = 10$. Value of $n = 2$ was found to be the most efficient in terms of computational time. Series of structures of $2^{20} \approx 10^6$ nucleosomes were generated for $n = 2$ and different values of internucleosomal distance ℓ and fractal dimension d_s.

"Soft" self-intersections (distance between i-th and j-th nucleosomes $r_{ij} \approx r_{\max}$) turned out to be relatively frequent and were observed for several percent of total number of nucleosomes. "Hard" self-intersections ($r_{ij} \ll r_{\max}$) were very rare (a few cases per the whole structure). Increase of internucleosomal distance ℓ decreased number of self-intersections, especially that of soft type. All self-intersections of both types were completely eliminated by simple Monte Carlo dynamics, having slight impact on overall geometry.

Dependence of radius of gyration $\langle r_g \rangle$ on size N_g of a continuous subset was calculated for the fractal structures. As expected, in all cases it had linear region in double logarithmic scale and fractal dimension d_g was evaluated by Eq. (3).

SANS spectra were calculated for all generated supernucleosomal structures with protocol, described in our previous work [20]. Key points of the protocol were summarized in the Methods.

As was noticed in the Introduction, fractal dimension d_I of the structure could be calculated from SANS spectrum assuming that

$$I \propto s^{d_I}, \tag{4}$$

i. e. the SANS spectrum had linear region in double logarithmic scale. We observed this behavior for all calculated SANS spectra of fractal supernucleosomal chromatin structure models and evaluated fractal dimension d_I by Eq. (4), in addition to known values of d_s and d_g.

Difference between continuous and infinite ideal fractals and our discrete and finite fractal structures led to dependence of fractal dimension of generated supernucleosomal structure on characteristic lengths $r_{\max}$ and ℓ. Value of $\ell = 300$ Å was found to provide good agreement of three estimates of fractal dimension d (Fig. 3).

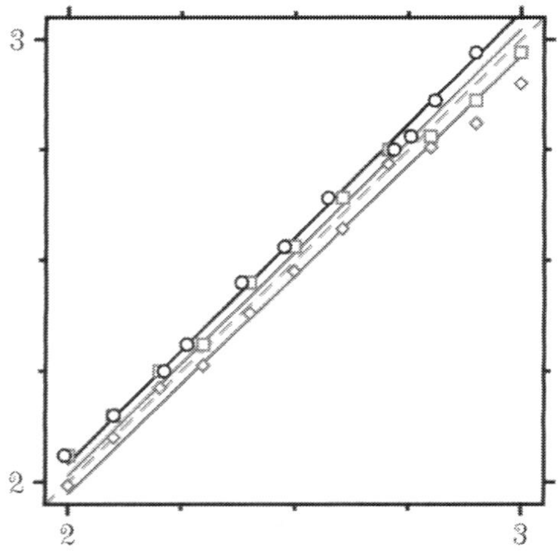

Figure 3. Correlation plot of three estimates of fractal dimension of generated supernucleosomal chromatin structures of 2^{20} nucleosomes ($n = 2$, $\ell = 300$ Å): d_s, d_g and d_I, calculated by Eqs. (2), (3) and (4), respectively. Symbols ◇, □ and ○ denote plots of d_g *vs.* d_s, d_I *vs.* d_s and d_I *vs.* d_g, respectively. Solid lines represent linear fits (without the intercept term, $y = \beta x$) of these correlations. Slopes (with error and coefficient of determination) were 0.987 ± 0.004 ($R^2 = 0.987$), 1.007 ± 0.005 ($R^2 = 0.984$) and 1.021 ± 0.002 ($R^2 = 0.997$), respectively. Dashed line with slope of 1.000 is shown for comparison.

We had to note that there was a difference between d_s (known by generation) and d_g and d_I (both based on the structure) at larger values ($d_s > 2.7$) (Fig. 3). Correlation between d_g and d_I at that region was significantly more robust than that between both of them and d_s, indicating some limitations of the generation method.

Fractal supernucleosomal structure of 2^{20} nucleosomes with fractal dimension $d_s = 2.4$ is shown on Fig. 4a. Initiator length was $L = 9.70 \times 10^4$ Å, while radius of gyration of the obtained structure was $R_g = 4.98 \times 10^4$ Å. 3D structure was irregular, with several clearly distinguished "focuses" of different sizes that were similar to the whole structure, illustrating scale invariance of the fractals.

As mentioned above, dependence of radius of gyration $\langle r_g(N_g)\rangle$ was calculated for the fractal structure and its plot in double logarithmic scale is shown on Fig. 4b. It had very short region ($N_g < 10$) with "slope" above 0.5 driven by effective repulsion of the nucleosomes. Then dependence $\langle r_g(N_g)\rangle$ was linear with slope of 0.42 that corresponded to $d_g = 2.38$ that was close to expected value $d_s = 2.4$.

SANS spectra of the structure were calculated for different values of volume fraction W_D of heavy water in solvent (shown on Fig. 4c) and had two distinct regions: supernucleosomal region ($s < 0.02$ Å^{-1}) and nucleosomal region. There was no dependence of SANS spectra on W_D value in supernucleosomal region. In contrast, in nucleosomal region protein ($W_D = 0.65$) and DNA ($W_D = 0.40$) parts of the structure had different SANS spectra with clearly seen patterns. Both observations were expected for supernucleosomal fractal structure as for discrete non-ideal fractal.

Nucleosomal part of the SANS spectra was resulted from locally-regular nucleosome packing (on a scale of several nucleosomes) that was a consequence of the generation algorithm. Observed SANS spectra patterns were more abundant in features in comparison with that of SANS spectra of random-walk structures [20]. These patterns are of particular interest in context of possible organization of 30-nm chromatin fibers. We have investigated a number of locally-regular supernucleosomal chromatin structures and are preparing data for publication.

Supernucleosomal part of the SANS spectra was linear in double logarithmic scale for $s < 0.01$ Å^{-1} with slope of -2.45, that corresponded to fractal dimension $d_I = 2.45$, while $d_s = 2.4$ and $d_g = 2.38$, thus there was a good agreement between different estimates of fractal dimension. Similar behavior was shown for supernucleosomal structures generated for several d_s values in range $[2, 3]$ (Fig. 3).

We had observed fractal dimensions d_I in range from 2.2 to ~3 in experimental SANS spectra for the whole nuclei [17]. Data obtained with other experimental methods also supported different estimates of fractal dimension of chromatin in the same range, from 2.2 (diffusion studies [15]) to 3 (Hi-C technique [13]). This motivated us to generate supernucleosomal structure with two different d_s values.

To introduce crossover point (in terms of d_s) in our supernucleosomal chromatin structure model, we modified original algorithm of structure generation described in the Methods. Generation started with fixed number of segments $n = 2$ and length ratio $f = f_1$ that provided fractal dimension d_{s1}. At definite order of generation $k = k_c$ length ratio was changed to $f = f_2$ that corresponded to d_{s2}.

Supernucleosomal structure model of 2^{20} nucleosomes with $\ell = 300$ Å, $n = 2$, $k_c = 14$ and fractal dimensions $d_{s1} = 3$ and $d_{s2} = 2$ is shown on Fig. 4d. It was similar to fractal structure with fractal dimension $d_s = 2.4$, but was more compact—radius of gyration was $R_g = 3.19 \times 10^4$ Å. Detailed analysis of dependence $\langle r_g(N_g)\rangle$ (Fig. 4e) revealed biphasic organization of the structure with crossover point 584 nucleosomes (2933 Å), while crossover point known by generation was 256 nucleosomes (1697 Å). This difference was slightly more than one order of fractal generation and may reflect transition from one dimension to another.

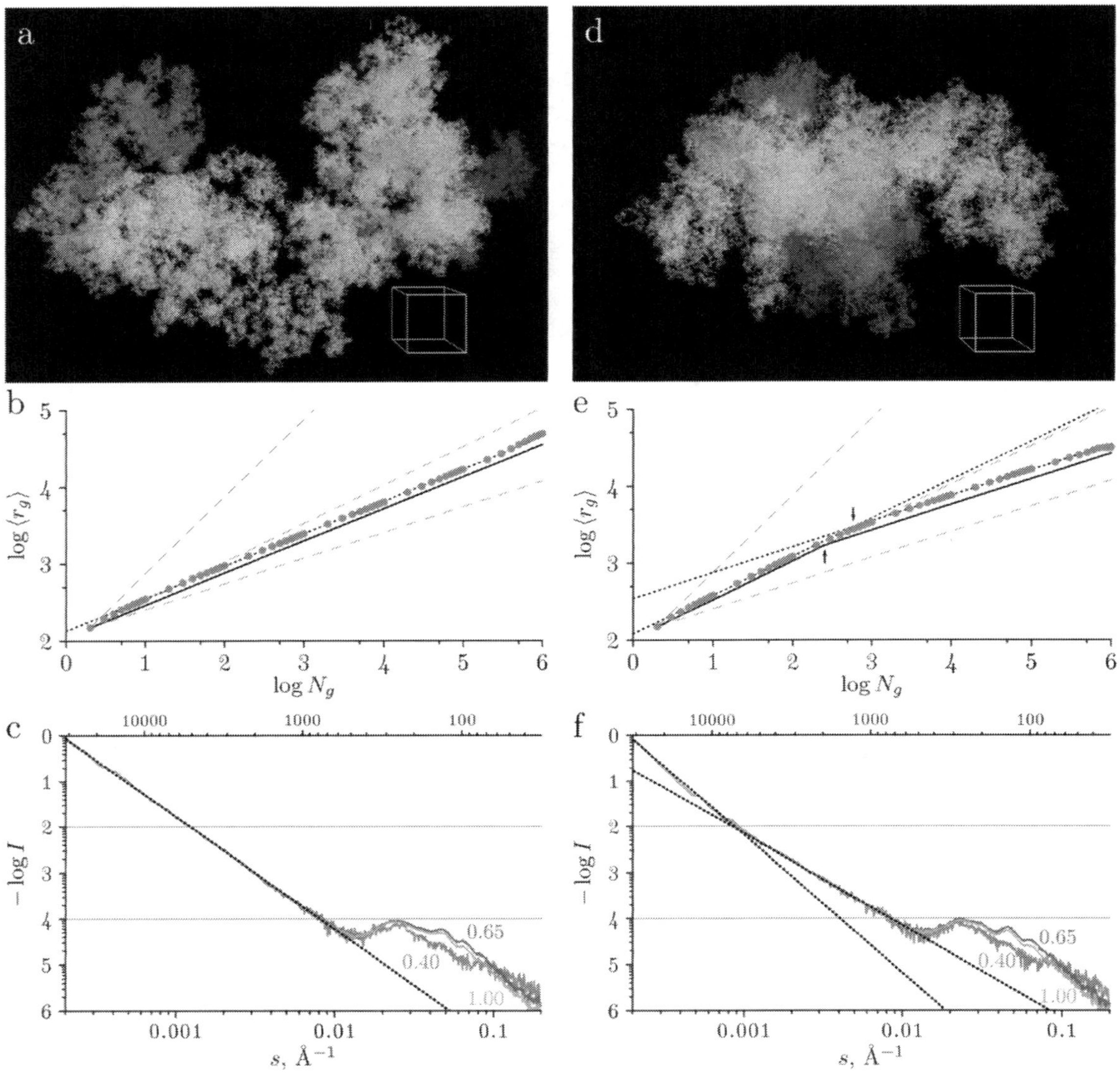

Figure 4. Two supernucleosomal chromatin structure models of 2^{20} nucleosomes, $\ell = 300$ Å. Fractal dimensions were (**a**–**c**) $d_s = 2.4$ and (**d**–**f**) $d_s = \{2, 3\}$. (**a**, **d**) 3D structures: each nucleosome is a sphere with diameter of $r_{\max}$, color is smoothly varying from blue (beginning of the chain) to red, cubes with edge of 2×10^4 Å are shown for comparison. (**b**, **e**) Dependence $\langle r_g(N_g) \rangle$: symbols • denote observed values, solid lines represent values predicted with d_s, dotted lines show linear fits, dashed lines for $d_g = \{1, 2, 3\}$ are shown for comparison. (**b**) Linear fit $y = \alpha + \beta x$: slope of 0.420 ± 0.001 ($R^2 = 0.999$) that corresponds to $d_g = 2.38$ shown on Fig. 3. (**e**) Crossover point (indicated by ↑) was 256 nucleosomes (1697 Å) by generation. Two linear fits $y = \alpha + x/d_s$ provided another crossover point (indicated by ↓)—584 nucleosomes (2933 Å), in both cases $R^2 = 0.999$. (**c**, **f**) SANS spectra: solid lines show spectra for specified W_D values, dotted lines represent linear fits, additional x-axis shows sizes in real space (in Å). (**c**) Linear fit $y = \alpha + \beta x$: slope -2.450 ± 0.002 ($R^2 = 0.999$) that corresponds to $d_I = 2.45$ shown on Fig. 3. (**f**) Linear fits $y = \alpha - 2x$ ($R^2 = 0.998$) and $y = \alpha - 3x$ ($R^2 = 0.991$).

SANS spectra of the structure with crossover (Fig. 4f) were similar to that of supernucleosomal structure with fractal dimension $d_s = 2.4$ and consisted of the supernucleosomal ($s < 0.01$ Å^{-1}) part with $I \propto s^{-d_I}$ and the nucleosomal part with characteristic pattern of locally-regular structure. Two different values of fractal dimension $d_I \approx d_s$ were observed, however, crossover point was less pronounced in comparison to $\langle r_g(N_g) \rangle$ plot and may be found in a range of 5–7 $\times 10^3$ Å.

Supernucleosomal part of calculated SANS spectra of fractal supernucleosomal chromatin structure models were in a good agreement with that we had observed experimentally for cell nuclei, especially spectra of the structure with two different values of fractal dimension—3 and 2 (for smaller sizes). Interestingly, diffusion studies had explored the nuclei at characteristic lengths up to 100 nm, and the obtained data supported low fractal dimension of 2.2–2.6 [15], while data acquired with Hi-C technique (resolution $\sim 5 \times 10^3$ nucleosomes) had been described by fractal globule model that predicted fractal dimension of 3 [13]. Thus, single chromatin structure model with crossover point fitted the data obtained with three different experimental methods.

In conclusion, analysis of available data suggested that supernucleosomal structures presented here are appropriate models of chromatin structure and are capable of reproducing its biphasic behavior. Future work is required to investigate dependence of observable parameters (radius of gyration, contact map captured with Hi-C technique, etc.) on fractal dimensions and crossover point of the supernucleosomal structure model.

Acknowledgments

This work was supported by the Russian Academy of Science (program "Physics of Elementary Particles", subprogram "Neutron physics", direction "Studies of structure, dynamics and non-ordinary properties of matter by neutron scattering") and by grants of the Ministry of Education and Science of the Russian Federation (RNP 2.2.1.1.4663, state contracts 02.740.11.5223 and 11.519.11.2002).

References

[1] Woodcock C L and Ghosh R P 2010 Chromatin higher-order structure and dynamics *Cold Spring Harb. Perspect. Biol.* **2** a000596
[2] Kornberg R D 1974 Chromatin structure: A repeating unit of histones and DNA *Science* **184** 868–871
[3] Oudet P, Gross-Bellard M and Chambon P 1975 Electron microscopic and biochemical evidence that chromatin structure is a repeating unit *Cell* **4** 281–300
[4] Berman H, Henrick K and Nakamura H 2003 Announcing the worldwide Protein Data Bank *Nat. Struct. Mol. Biol.* **10** 980
[5] Luger K, Mäder A W, Richmond R K, Sargent D F and Richmond T J 1997 Crystal structure of the nucleosome core particle at 2.8 Å resolution *Nature* **389** 251–260
[6] Davey C A, Sargent D F, Luger K, Maeder A W and Richmond T J 2002 Solvent mediated interactions in the structure of the nucleosome core particle at 1.9 Å resolution *J. Mol. Biol.* **319** 1097–1113
[7] Schalch T, Duda S, Sargent D F and Richmond T J 2005 X-ray structure of a tetranucleosome and its implications for the chromatin fibre *Nature* **436** 138–141
[8] Thoma F and Koller T 1977 Influence of histone H1 on chromatin structure *Cell* **12** 101–107
[9] Gerchman S E and Ramakrishnan V 1987 Chromatin higher-order structure studied by neutron scattering and scanning transmission electron microscopy *Proc. Natl. Acad. Sci. USA* **84** 7802–7806
[10] Hammermann M, Tóth K, Rodemer C, Waldeck W, May R P and Langowski J 2000 Salt-dependent compaction of di- and trinucleosomes studied by small-angle neutron scattering *Biophys. J.* **79** 584–594
[11] Fussner E, Ching R W and Bazett-Jones D P 2011 Living without 30 nm chromatin fibers *Trends Biochem. Sci.* **36** 1–6
[12] McNally J G and Mazza D 2010 Fractal geometry in the nucleus *EMBO J.* **29** 2–3
[13] Lieberman-Aiden E *et al* 2009 Comprehensive mapping of long-range interactions reveals folding principles of the human genome *Science* **326** 289–293
[14] Grosberg A, Rabin Y, Havlin S and Neer A 1993 Crumpled globule model of the three-dimensional structure of DNA *Europhys. Lett.* **23** 373–378

[15] Bancaud A, Huet S, Daigle N, Mozziconacci J, Beaudouin J and Ellenberg J 2009 Molecular crowding affects diffusion and binding of nuclear proteins in heterochromatin and reveals the fractal organization of chromatin *EMBO J.* **28** 3785–3798

[16] Takahashi M 1989 A fractal model of chromosomes and chromosomal DNA replication *J. Theor. Biol.* **141** 117–136

[17] Lebedev D V, Filatov M V, Kuklin A I, Islamov A K, Kentzinger E, Pantina R A, Toperverg B P and Isaev-Ivanov V V 2005 Fractal nature of chromatin organization in interphase chicken erythrocyte nuclei: DNA structure exhibits biphasic fractal properties *FEBS Lett.* **579** 1465–1468

[18] Isaev-Ivanov V V, Lebedev D V, Lauter H, Pantina R A, Kuklin A I, Islamov A K and Filatov M V 2010 Comparative analysis of the nucleosome structure of cell nuclei by small-angle neutron scattering *Phys. Solid State* **52** 1063–1073

[19] Schmidt P W 1989 Use of scattering to determine the fractal dimension *The Fractal Approach to Heterogeneous Chemistry: Surfaces, Colloids, Polymers* ed D Avnir (New York: John Wiley & Sons Ltd) chap 2.2, pp 67–79

[20] Ilatovskiy A V, Lebedev D V, Filatov M V, Grigoriev M, Petukhov M G and Isaev-Ivanov V V Modeling and SANS spectra of chromatin supernucleosomal structures at genome scale *J. Appl. Phys.* in press

[21] Harroun T A, Wignall G D and Katsaras J 2006 Neutron scattering for biology *Neutron Scattering in Biology* ed J Fitter, T Gutberlet and J Katsaras (Heidelberg: Springer) chap 1, pp 1–18

[22] Mandelbrot B B 1982 *The Fractal Geometry of Nature* (New York: W H Freeman and Co)

[23] Mitane Y, Nakanishi M, Tsuboi M, Kohwi-Shigematsu T, Enomoto T and Yamada M A 1980 Hydrogen-exchange study of a nucleosome *FEBS Lett.* **121** 130–132

[24] Black B E, Brock M A, Bédard S, Woods, Jr V L and Cleveland D W 2007 An epigenetic mark generated by the incorporation of CENP-A into centromeric nucleosomes *Proc. Natl. Acad. Sci. USA* **104** 5008–5013

Heat-induced structural transitions of alpha-crystallin studied by small-angle neutron scattering

A V Krivandin[1], A I Kuklin[2,3], K O Muranov[1], T N Murugova[2,3], S S Kozlov[1] and N K Genkina[1]

[1] Emanuel Institute of Biochemical Physics, Russian Academy of Sciences, 4 Kosygin Street, Moscow 119334, Russia
[2] Frank Laboratory of Neutron Physics, Joint Institute for Nuclear Research, 6 Joliot Curie Street, Dubna 141980, Moscow region, Russia
[3] Bionanocenter, Moscow Institute of Physics and Technology, Dolgoprudny, Moscow region, Russia

E-mail: a.krivandin@sky.chph.ras.ru

Abstract. Alpha-crystallin from the bovine eye lens was studied by small-angle neutron scattering (SANS) in 90% D_2O buffer solution at 20, 50, 60, 65, 75, 85 and 95°C. The temperature points for this study were specified on the basis of differential scanning calorimetric analysis of alpha-crystallin solutions which has shown two endothermic transitions with midpoints at 64.5 and 86°C. The SANS study revealed no significant alpha-crystallin quaternary structure alterations at 50°C as compared with 20°C. At 60-65°C the SANS data confirmed substantial alpha-crystallin quaternary structure rearrangements which resulted in the formation of alpha-crystallin oligomers with a similar shape but approximately twofold increased molecular weight as compared to the native state at 20°C. At higher temperatures (75, 85 and 95°C) the SANS patterns were very similar and were consistent with the scattering by rod-like particles with a cross-section radius of gyration ~55 Å. This transformation of alpha-crystallin to the rod-like particles was evidently irreversible as these particles remained in solution after cooling to 20°C. *Ab initio* shape models of the native and high-temperature alpha-crystallin were retrieved with DAMMIN and DAMAVER software. Schematic model of alpha-crystallin heat-induced quaternary structure transitions was considered.

1. Introduction

Alpha-crystallin is one of the main proteins of the vertebrate eye lens. It is a polydisperse oligomeric protein with an average molecular weight about 700-800 kDa which consists of polypeptide subunits with molecular weight of about 20 kDa each [1, 2].

Alpha-crystallin belongs to the small heat shock protein family and has a property to form complexes with destabilized proteins and suppress a nonspecific aggregation of such proteins (chaperone-like activity of alpha-crystallin) [2–6]. It is considered that this property of alpha-crystallin promotes the maintenance of the eye lens transparency through the whole life [2, 4].

The spatial structures of alpha-crystallin oligomers and subunits are still unknown. Computer modeling of the tertiary structure of alpha-crystallin subunits was performed [7], and several distinct models of its quaternary structure were proposed (please see [7–10] and reviews [1, 2]).

Heat-induced structural transitions of alpha-crystallin were studied earlier by various experimental techniques [5, 9–17]. One of the distinct features of these transitions is the rearrangement of alpha-crystallin quaternary structure at temperatures above approximately 50°C which results in the substantial increase of its size and molecular weight [5, 9, 11, 14–17]. In spite of the numerous studies the molecular basis of this rearrangement remains to be poorly understood.

In order to get some more insight in heat-induced structural transitions of alpha-crystallin we performed in this work the small-angle neutron scattering (SANS) study of alpha-crystallin quaternary structure in the wide temperature range. The temperature points for this study were specified on the basis of differential scanning calorimetric analysis of alpha-crystallin.

2. Materials and methods

Alpha-crystallin was separated from the bovine eye lenses in the aqueous buffer solution (20 mM Na_2HPO_4, 20 mM NaH_2PO_4, 100 mM NaCl, 3 mM NaN_3, 1 mM EDTA, pH 6.8) by gel permeation chromatography and concentrated by ultrafiltration as described in detail elsewhere [18]. In the SANS study the buffer solution of the same composition but with 90% D_2O was used. Concentration of alpha-crystallin solutions was determined on a UV-1700 spectrophotometer (Shimadzu, Japan) by absorption at 280 nm using specific absorption coefficient $A_{280}(1\%, 1\text{ cm}) = 8.2$ as in [19]. Differential scanning calorimetry analysis of alpha-crystallin solutions was done using a DASM-4 high-sensitivity differential scanning microcalorimeter (Institute of Biological Instrument Engineering, Pushchino, Russia) at a scanning rate 1°C/min, constant pressure 2.2 atm and concentration of alpha-crystallin from 1 to 8 mg/ml.

SANS study was carried out with the YuMO spectrometer (IBR-2, JINR, Russia) using alpha-crystallin solution with protein concentration 7.2 mg/ml successively at 20, 50, 60, 65, 75, 85 and 95°C. At each temperature two successive SANS measurements with 30 min exposition were done (thereafter these two successive measurements will be referred as exposition No1 and exposition No2). Alpha-crystallin solution after the SANS study was turbid indicating that large protein aggregates have appeared in solution at high temperatures. Experimental SANS spectra were converted into the dependence of the coherent differential scattering cross-section (intensity) I on the scattering vector magnitude $S = (4\pi\sin\theta)/\lambda$ (2θ is scattering angle, λ is radiation wavelength) taking into account sample transmission and the incoherent component of scattering, and then normalized using a vanadium scatterer by the standard procedure with SAS software [20]. Then SANS data were analyzed with PRIMUS small-angle scattering data analysis software [21] and indirect Fourier transformation method implemented in the program GNOM [22]. *Ab initio* shape models of alpha-crystallin were retrieved with DAMMIN software [23] and averaged with DAMAVER program suite [24]. The basics of the SANS theory and data treatment can be found elsewhere [25, 26]. Some details of SANS application to alpha-crystallin structure studies were considered in [19].

3. Results and discussion

Differential scanning calorimetry thermogram of alpha-crystallin solution at concentration 1.2 mg/ml shows two broad endothermic peaks (figure 1) which indicate two heat-induced structural transitions with midpoints at 64.5 and 86°C. Thermograms obtained for alpha-crystallin solutions at concentrations 2.5 and 8 mg/ml have shown essentially the same endothermic peaks (results are not presented in this article). Repeated differential scanning calorimetry scans performed after cooling of the protein solutions revealed that the first transition at 64.5°C is mainly irreversible whereas the second transition at 86°C is almost totally reversible (results are not presented in this article).

Similar endothermic peak indicative of the first transition but at slightly lower temperature (60–63°C) was observed earlier [10, 12], but a heat-induced transition at 86°C as far as we know was detected in our study for the first time.

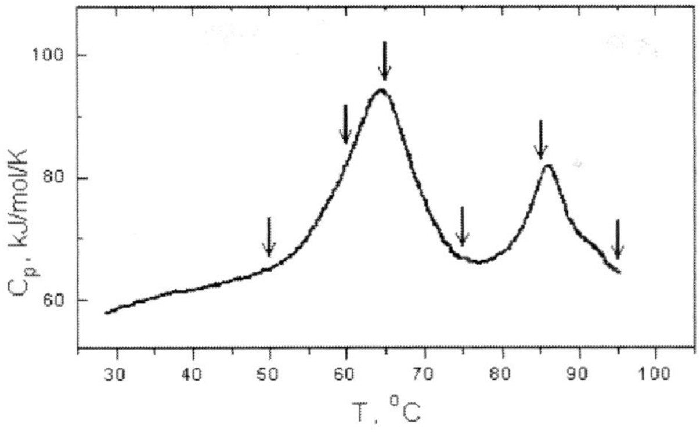

Figure 1. Differential scanning calorimetry thermogram of alpha-crystallin at concentration 1.2 mg/ml. Arrows indicate temperatures at which the SANS study was done.

On the basis of differential scanning calorimetric data the temperatures for the SANS study were selected as 20, 50, 60, 65, 75, 85 and 95°C. For correlation with calorimetric data these temperatures are marked in figure 1 with arrows. Initial parts of experimental SANS curves recorded at these temperatures are shown in figures 2a, 2b and 2c. For each temperature (except 20°C before heating) two successive SANS measurements are depicted in these figures.

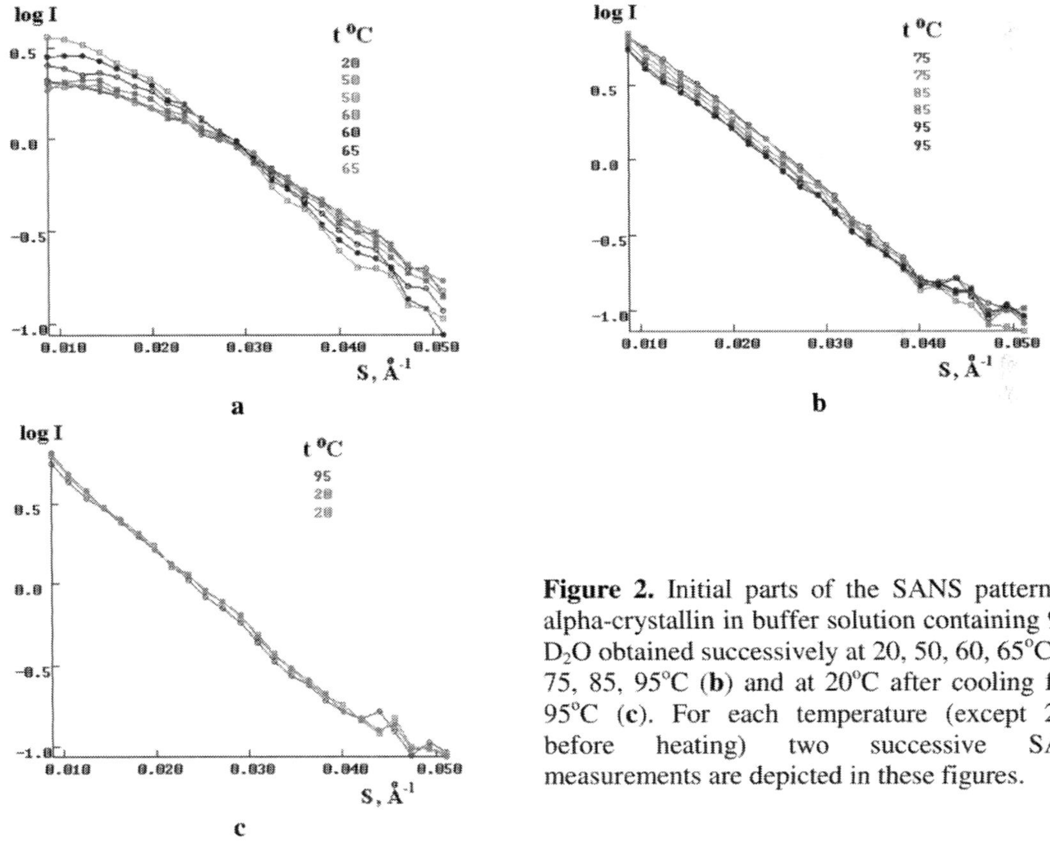

Figure 2. Initial parts of the SANS patterns of alpha-crystallin in buffer solution containing 90% D_2O obtained successively at 20, 50, 60, 65°C (**a**), 75, 85, 95°C (**b**) and at 20°C after cooling from 95°C (**c**). For each temperature (except 20°C before heating) two successive SANS measurements are depicted in these figures.

The SANS patterns measured at 20 and 50°C were mainly the same (figure 2a) indicating that no essential alpha-crystallin quaternary structure perturbations have occurred at 50°C.

At temperatures 60 and 65°C the SANS intensity at S<0.03 Å^{-1} increased and intensity decline grew progressively with the increase of temperature and elapsed time at each temperature (figure 2a). This denotes progressive quaternary structure alterations of alpha-crystallin at 60–65°C.

At 75°C another type of SANS curve was observed (figure 2b) as compared with the lower temperatures. To successive expositions at 75°C (figure 2b) showed almost one and the same intensity curve indicating that structural equilibrium was achieved at this temperature.

At 85 and 95°C the SANS curves were similar to the curve observed at 75°C (figure 2b). All SANS curves obtained at 75, 85, 95°C had practically one and the same slope, yet the overall intensity progressively lowered with temperature increase from 75 to 95°C. Successive measurements at constant temperatures at 85 and 95°C, as in the case of 75°C, provided at each of these temperatures nearly one and the same intensity curve indicating structural equilibrium at each of these temperatures.

No essential differences in SANS patterns taken at 75, 85 and 95°C were found which could be correlated with endothermic alpha-crystallin transition detected by differential scanning calorimetry at 86°C.

Only slight variations of intensity curve were observed after cooling the protein solution from 95 to 20°C as compared with 95°C (figure 2c) indicating irreversibility of alpha-crystallin quaternary structure transitions at high temperatures.

All SANS patterns taken for alpha-crystallin at 20, 50, 60 and 65°C have shown at small S values well defined linear regions in the plots of ln(I) vs S^2. For example such plots drawn with PRIMUS software are shown in figures 3a and 3b for alpha-crystallin SANS data obtained at 20°C before heating (native state) and at 65°C.

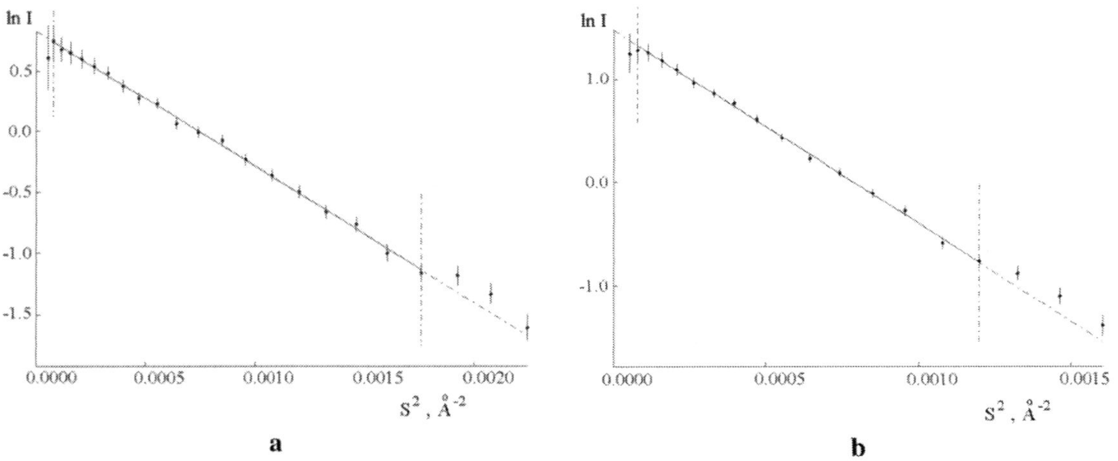

Figure 3. Plots of ln(I) vs S^2 for alpha-crystallin SANS data obtained at 20°C before heating (**a**) and at 65°C (**b**, exposition No 2).

Radii of gyration R_g and intensities at the zero angle $I(0)$ determined for alpha-crystallin at different temperatures by approximation of linear regions in the plots of ln(I) vs S^2 with Guinier formula are given in table 1.

It follows from table 1 that radius of gyration R_g and zero angle intensity $I(0)$ found for alpha-crystallin at 50°C within experimental error do not differ from those values for the native alpha-crystallin at 20°C. This means that according to the SANS data no essential quaternary structure rearrangements of alpha-crystallin have occurred at 50°C as compared with the native state.

The heating of the alpha-crystallin solution up to 60 and 65°C resulted in the increase of R_g up to ~66 Å and ~75 Å respectively (table 1) indicating substantial alpha-crystallin quaternary structure alterations at these temperatures. These quaternary structure alterations of alpha-crystallin correlate with the endothermic transition which has midpoint at 64.5°C shown by differential scanning calorimetry (figure 1).

Table 1. Radii of gyration R_g and SANS intensities at the zero angle $I(0)$ determined for alpha-crystallin at different temperatures by Guinier approximation with PRIMUS software.

t, °C	Exposition No 1		Exposition No 2	
	R_g, Å	I(0), rel. units	R_g, Å	I(0), rel. units
20	57.7±0.8	2.28±0.06	–	–
50	57.8±0.8	2.34±0.07	58.0±0.9	2.29±0.07
60	61.5±0.8	2.64±0.08	65.9±0.8	3.18±0.09
65	69.9±0.7	3.70±0.10	75.3±0.9	4.40±0.12

According to the values of intensity scattered at the zero angle $I(0)$ (table 1) it may be concluded that molecular weight of alpha-crystallin oligomers increased approximately twofold at 65°C as compared with the native state at 20°C. This conclusion is in agreement with the previous results obtained by small-angle X-ray scattering [5, 16].

It is worth to note that SANS curves obtained for alpha-crystallin at 20, 50, 60 and 65°C intersect at one and the same point, namely at S≈0.031 Å^{-1} (figure 2a). Similar isoscattering point for alpha-crystallin at various temperatures was observed also by small-angle X-ray scattering [15, 16]. The existence of such isoscatterng point indicates that alpha-crystallin heat-transition at 60–65°C may be associated with a two-state process, i.e. at such transition the native alpha-crystallin most probably transforms into the high-temperature form without intermediates. Taking into account that a high-temperature form of alpha-crystallin has a molecular weight approximately twice higher than a native form, a two-state quaternary structure transition at 60–65°C may be assumed as merging of two oligomers of alpha-crystallin. Additional studies are required to answer unequivocally the question whether this is a simple dimerization of two native oligomers of alpha-crystallin or rearrangement of subunits of two merging oligomers into a new oligomeric structure.

At higher temperatures (75, 85, 95°C) and after cooling from 95 to 20°C the SANS patterns were consistent with the scattering by rod-like particles. An example of the linear plot ln(*IS*) vs S^2 indicative of such particles is given in figure 4. Radius of gyration of the cross-section of the rod-like particles for temperatures 75, 85 and 95°C was found from such plots to be ~55 Å. Probably these rod-like particles were formed by aggregation of alpha-crystallin oligomers rearranged at 60-65°C.

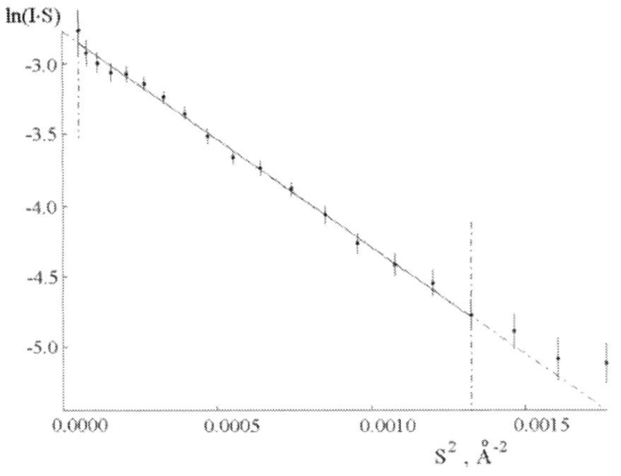

Figure 4. Plots of ln(*IS*) vs S^2 for alpha-crystallin SANS pattern taken at 85°C (exposition No 2).

Existence of alpha-crystallin rod-like particles detected at high temperatures by the SANS method in this work is consistent with chain-like structures observed by transmission electron microscopy for alpha-crystallin after incubation at temperatures higher than 60°C [14].

The SANS data analysis by indirect Fourier transformation method with GNOM software confirmed results outlined above. Essentially the same R_g and $I(0)$ values as in table 1 were obtained by this method for alpha-crystallin at 20, 50 60 and 65°C. Examples of the SANS curves fitting and retrieved distance distribution functions $P(R)$ with GNOM software are shown for alpha-crystallin at 20 and 65°C in figures 5 and 6.

The distance at which $P(R)$ reaches zero gives the maximal size of the scattering particle. So we obtain the maximal size of alpha-crystallin to be ~170 Å at 20°C and ~230 Å at 65°C (figures 5 and 6).

The $P(R)$ functions for alpha-crystallin at 20 and 65°C (figures 5 and 6) have a very similar form. This means that in spite of the fact that alpha-crystallin essentially increases its size and molecular

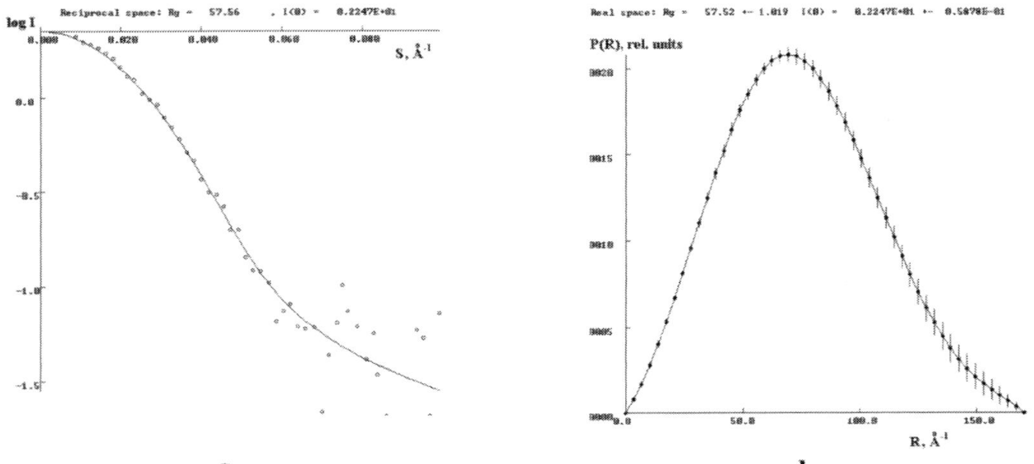

Figure 5. Approximation of the SANS curve for alpha-crystallin at 20°C by indirect Fourier transformation method with GNOM software (**a**) and retrieved distance distribution function (**b**).

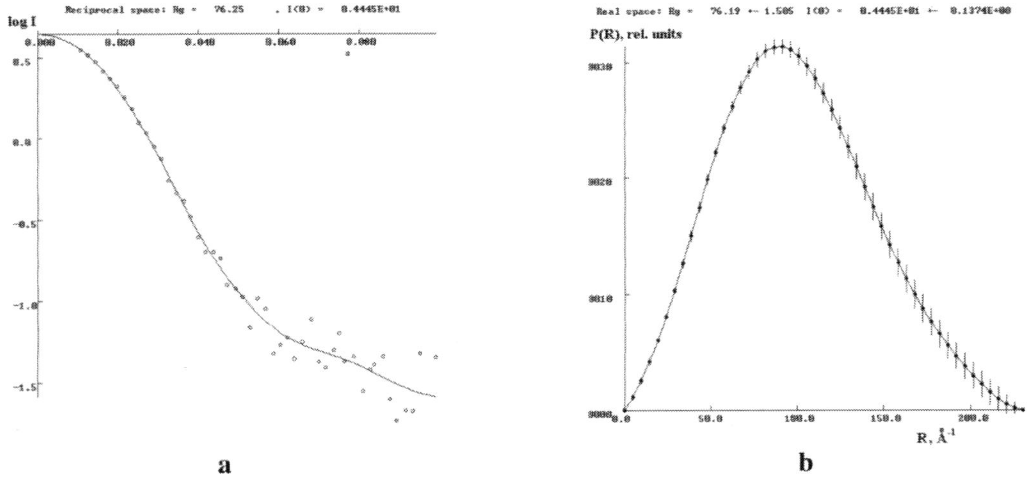

Figure 6. Approximation of the SANS curve for alpha-crystallin at 65°C (exposition No 2) by indirect Fourier transformation method with GNOM software (**a**) and retrieved distance distribution function (**b**).

weight at 65°C its overall shape after this structural rearrangement is similar to the native shape at 20°C.

Results of the SANS data processing with GNOM software for alpha-crystallin high-temperature form at 85°C assuming a rod-like shape of the scattering particles is shown in figure 7.

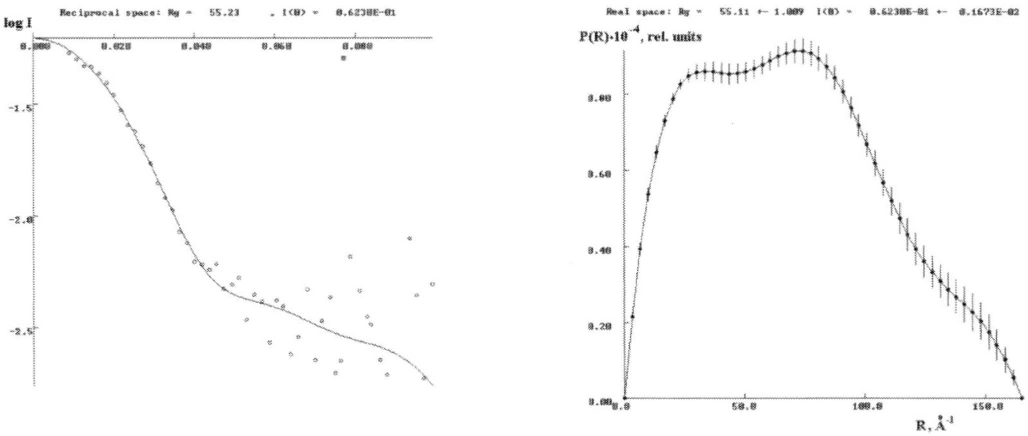

a **b**

Figure 7. Processing of the SANS data for alpha-crystallin at 85°C (exposition No 2) by indirect Fourier transformation method with GNOM software assuming a rod-like shape of the scattering particles: **a** – experimental intensity fitting, **b** – retrieved distance distribution function of a rod-like particle cross-section.

Again as in the case of a rod-like particle approximation with PRIMUS software (figure 4) we obtain by indirect Fourier transformation method the radius of gyration of the cross-section of the rod-like particle to be ~55 Å (figure 7). The distance distribution function $P(R)$ of the cross-section of the rod-like particle reaches zero at R = 165 Å (figure 7b). Thus the maximal size of the cross-section of such high-temperature rod-like particles of alpha-crystallin is about 165 Å. This size is approximately equal to the maximal size of the native alpha-crystallin.

Ab initio shape models for the native (20°C) and high-temperature (65 and 95°C) alpha-crystallin obtained with DAMMIN and DAMAVER software are shown in figure 8. A model depicted for the high-temperature (95°C) alpha-crystallin (figure 8c) should be considered as a fragment of a rod-like particle.

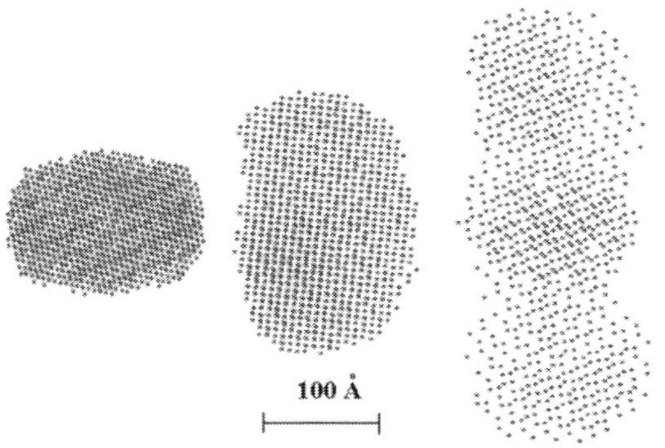

a b c

Figure 8. *Ab initio* shape models of alpha-crystallin at 20°C (**a**), 65°C (**b**) and 95°C (**c**, fragment of a rod-like particle). Each model results from the averaging with DAMAVER program suite of at least 10 models retrieved with DAMMIN software.

According to the sizes of alpha-crystallin models (figure 8) and twofold increase of alpha-crystallin molecular weight at 60–65°C it can be assumed that the high-temperature form of alpha-crystallin which appears at 60–65°C results from the side to side dimerization of native alpha-crystallin oligomers and then at higher temperatures (75–95°C) these alpha-crystallin dimers aggregate and form the rod-like particles. Additional experimental proof of this schematic model of alpha-crystallin heat-induced quaternary structure transitions is required.

Acknowledgements
We gratefully acknowledge the Russian Foundation for Basic Research for financial support of this work (grant No 08-04-01787-a) and Dr. A. Kh. Islamov for assistance in the accomplishment of SANS experiment. A. I. Kuklin is grateful for support to the Seventh Framework Programme [FR7/2007-2013] under grant agreement №211800.

References

[1] Groenen P J T A, Merck K B, De Jong W W, Bloemendal H 1994 *Eur. J. Biochem.* **225** 1
[2] Horwitz J 2003 *Exp. Eye Res.* **76** 145
[3] Horwitz J 1992 *Proc. Natl. Acad. Sci. USA.* **89** 10449
[4] Rao P V, Huang Q, Horwitz J, Zigler J S 1995 *Biochim. Biophys. Acta.* **1245** 439
[5] Krivandin A V, Muranov K O and Ostrovsky M A 2004 *Mol. Biol. (Moscow)* **38** 447
[6] Krivandin A V, Muranov K O, Poturaeva I D, Poliansky N B and Ostrovsky M A 2006 *Dokl. AN* **409** 550
[7] Groth-Vasselli B, Kumosinski T F and Farnsworth P N 1995 *Exp. Eye Res.* **61** 249
[8] Peschek ., Braun N, Franzmann T ., Georgalis Y, Haslbeck M, Weinkauf S and Buchner J 2009 *Proc. Natl. Acad. Sci. USA* **106** 13272
[9] Tardieu A, Laporte D, Lichino P, Krop B and Delae M 1986 *J. Mol. Biol.* **192** 711
[10] Walsh M T, Sen A C and Chakrabarti B 1991 *J. Biol. Chem.* **266** 20079
[11] Siezen R J, Bindels J G and Hoenders H J 1980 *Eur. J. Biochem.* **111** 435
[12] Surewicz W K and Olesen P R 1995 *Biochemistry* **34** 9655
[13] Vanhoudt J V, Abgar S, Aerts T and Clauwaert J 2000 *Biochemistry* **39** 4483
[14] Burgio M R, Bennett P M and Koretz J F 2001 *Molecular Vision* **7** 228
[15] Regini J W, Grossmann J G, Burgio M R, Malik N S, Koretz J F, Hodson S A and Elliott G F 2004 *J. Mol. Biol.* **336** 1185
[16] Scouri-Panet F, Quevillon-Cheruel S, Michiel M, Tardieu A and Finet S 2006 *Biochim. Biophys. Acta.* **1764** 372
[17] Regini J W, Grossmann J G, Timmins P, Harding J J, Quantock A J, Hodson S A and Elliott G F 2007 *Invest. Ophthalm. Vis. Sci.* **48** 2695
[18] Krivandin A V, Muranov K O, Yakovlev F Yu, Poliansky N B, Wasserman L A and Ostrovsky M A 2009 *Biochemistry (Moscow)* **74** 633
[19] Krivandin A V, Murugova T N, Kuklin A I, Muranov K O, Poliansky N B, Aksenov V L and Ostrovsky M A 2010 *Biochemistry (Moscow)* **75** 1324
[20] Solovyev A G, Solovyeva T M, Stadnik A V, Islamov A Kh and Kuklin A I 2003 *JINR Communication P10-2003-86* (JINR, Dubna)
[21] Konarev P V, Volkov V V, Sokolova A V, Koch M H J and Svergun D I 2003 J Appl Cryst. **36** 1277
[22] Svergun D I 1992 *J. Appl. Cryst.* **25** 495
[23] Svergun D I 1999 *Biophys J.* **76** 2879
[24] Volkov V V and Svergun D I 2003 J. Appl. Cryst. **36** 860
[25] Feigin L A and Svergun D I 1987 *Structure Analysis by Small-Angle X-Ray and Newtron Scattering* (New York: Plenum Press)
[26] Ostanevich Yu M and Serdyuk I I 1982 *Uspekhi Fiz. Nauk* **137** 85

Comparative study on low resolution structures of apoferritin via SANS and SAXS

A I Kuklin[1,2], T N Murugova[1,2], O I Ivankov[1,3], A V Rogachev[1,2,4], D V Soloviov[1,3], Yu S Kovalev[1,2], A V Ishchenko[5,6], A Zhigunov[7], T S Kurkin[8], V I Gordeliy[1,2,5,9]

[1] Joint Institute for Nuclear Research, Dubna, Russia
[2] Bionanocenter of Moscow Institute for Physics and Technology, Dolgoprudny, Russia
[3] Taras Shevchenko National University of Kyiv, Kyiv, Ukraine
[4] Skobeltsyn Institute of Nuclear Physics of Moscow State University, Moscow, Russia
[5] Institute of Complex Systems (ICS), ICS-5: Molecular Biophysics, Research Centre Juelich, 52425 Juelich, Germany
[6] Institute of Crystallography, University of Aachen (RWTH), Jaegerstrasse 17-19, 52056 Aachen, Germany
[7] Institute of Macromolecular Chemistry of the Czech Republic Academy of Sciences, Prague, Czech Republic
[8] Institute of Synthetic Polymeric Materials of the Russian Academy of Sciences, Moscow, Russia
[9] Institute of Structural Biology J.P.Ebel, Grenoble, France

E-mail: kuklin@nf.jinr.ru; valentin.gordeliy@ibs.fr

Abstract. The results of small angle scattering investigation of protein apoferritin are presented. The sizes and shapes, including those determined by indirect Fourier transform method, are calculated. The pair-distance distribution function for both small angle neutron (SANS) and X-ray small angle scattering (SAXS) are obtained. It is shown that SANS and SAXS methods give similar shape (spherical shell with holes) of apoferritin. At the same time, fits of experimental data for SAXS and SANS curves give a little different sizes and volumes of the molecule. The reasons for these differences are discussed.

1. Introduction

The protein complex ferritin plays a key role in iron metabolism. Ferritin is a protein which resolves the oxide radicals and iron storage problems. Apoferritin protein or its analogues have been found in all living organisms [1-3].The technical applications of ferritin with modified central part are also discussed. As it is well known, ferritin is a large protein (about 120A in diameter), consists of 24 ferritin polypeptides and comprises central part containing iron based molecules. The apoferritin is a protein complex which is a ferritin shell (without iron core). Together with mineral nuclear apoferritin forms the ferritin. The name of "ferritin" was given by V.Laufberger due to protein containing over 20% (wt/wt) of iron according to [5, 6].

First investigations of iron metabolism were started in 1940-th by Granick [7], followed by studies on magnetic properties of ferrittin and ferric compounds [8] and the mechanisms of iron metabolism [9-11]. First X-ray diffraction experiments with the ferritin were performed in the 40-th of previous century [12]. Nevertheless, at present, there is still quite big activity in the field. Recently, the structure of protein-protein complexes in native solution has received much attention [4]. Despite the fact that a lot of structures of these complexes are presented in PDB bank and the history of their studies is quite long there are still several open questions. In particular, recent structural studies are focused on interparticle correlations [13], and the protein complex appeared to be very useful as a test sample for SAS instruments [14].

The aim of this work is to apply recent SAS approaches to low resolution studies of the mentioned above protein complexes with a particular goal - to compare obtained in such way SAXS and SANS structures.

2. Materials and methods

Horse spleen apoferritin was purchased from Sigma-Aldrich Chemie GmbH (Product number A3660). Storage buffer (50% glycerol and 0.075 M NaCl) was replaced for buffer containing 150 mM NaCl and 5 mM Na2HPO4, pH=7.3. Buffer for neutron experiment was based on heavy water form best contrast and decreasing incoherent background. The buffer for x-ray experiments was based on light water. The buffer replacement was fulfilled by centrifugal ultrafiltration with filters Centricon Y-100 (Millipore corporation). The final concentration of protein was 25 mg/ml for SANS experiments and 50 mg/ml for SAXS.

The neutron experiments were performed at YuMO spectrometer with two detectors system mode [16, 17]. The beam was collimated to a diameter of 14 mm on the sample. The obtained spectra were normalized and absolute values of the neutron scattering intensity obtained using a vanadium scatterer by the standard procedure with SAS software [18]. The investigations by small angle X-ray (SAXS) method were carried out on Brucker Nanostar instrument which is available at the Institute of Synthetic Polymer Materials RAS (Moscow, Russia) and in Prague, the experiments were performed using a pinhole camera of Molecular Metrology SAXS System at the Institute of Macromolecular Chemistry CAS (Prague, Czech republic).

A pinhole camera Molecular Metrology SAXS System is attached to a microfocused X-ray beam generator (Osmic MicroMax 002) that operates at 45 kV and 0.66 mA (30 W). The camera was equipped with a multiwire, gas-filled area detector with an active area diameter of 20 cm (Gabriel design). Two experimental setups were used to cover the q range of 0.0045 – 1.1 Å ($q=(4\pi/\lambda)\sin\Theta$, where λ is the wavelength and 2Θ is the scattering angle). The scattering intensities were put on absolute scale using a glassy carbon standard.

3. Results and discussion

3.1. The pair-distance distribution function

The pair-distance distribution function P(r) was obtained using indirect Fourier transformation of experimental small angle neutron scattering data treatments for apoferritin. The function P(r) was reconstructed using program GNOM [19] of ATSAS program package [20]. The pair-distance distribution function (P(r)) obtained by this procedure and its approximation of experimental curves are shown in Figure.1. Functions are bell-shaped, the maxima of this function are shifted from the center position, which indicates the displacement of the scattering density from the center (unlike for spherical shell) to the periphery of the molecule. The maximum of the P(r) function corresponds to the most probable distance between two points in a protein molecule. This distance is about 85 Å. For the obtained pair-distance distribution functions P (r) the values of the radii of gyration, according to the [21], are 51.3 and 51.73 Å for the SANS and SAXS, respectively. The values of the radius of gyration obtained from the indirect transform are equal to those obtained previously from fitting the experimental curves by the model of spherical shell scattering function. From the indirect transform, we can also estimate the maximum size of protein molecule, which is about 120 Å.

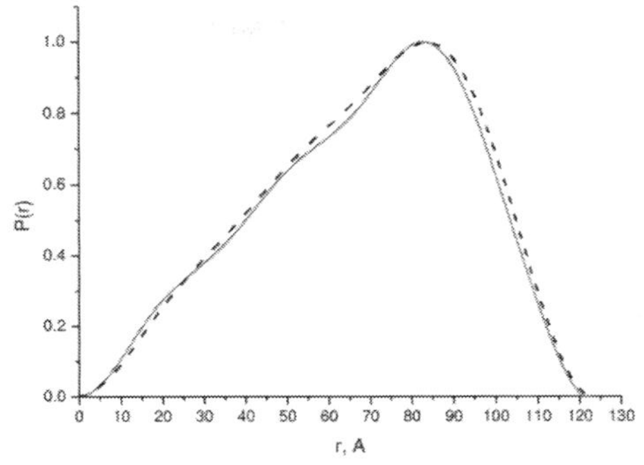

A

Figure 1. The result of the pair-distance distribution function reconstruction for the apoferritin molecule for small-angle neutron scattering (SANS) and X-ray (SAXS)

(A) Pair-distance distribution function P(r) SANS (continuous line) and X-ray (SAXS) (dash line).

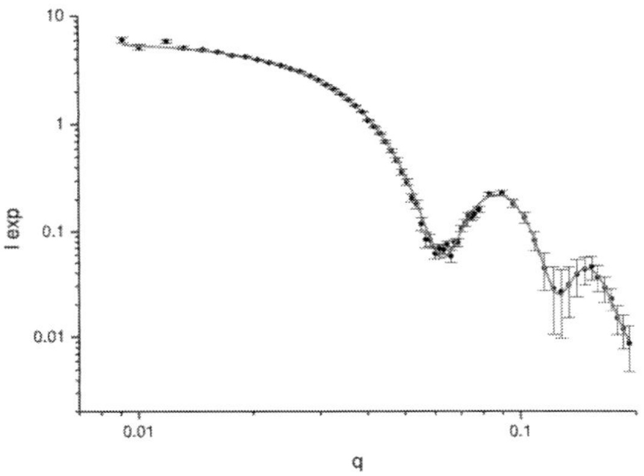

(B) Fit of SAXS experimental data by the scattering curve calculated from P(r)

B

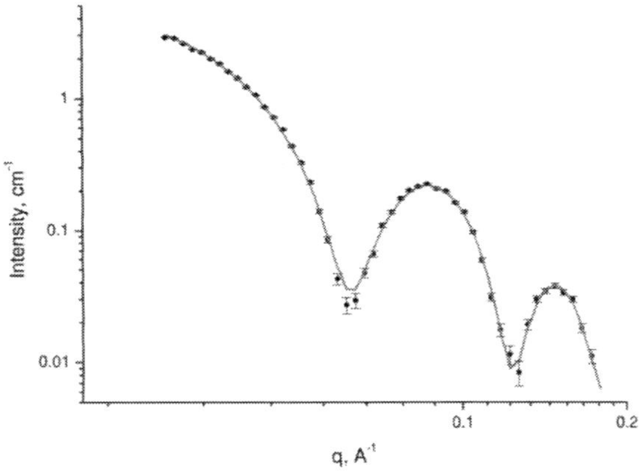

(C) Fit of SANS experimental data by the scattering curve calculated from P(r)

C

3.2. Determination of the volume of the molecules by Porod approximation

The Porod invariant [22] characterizes the integral scattering from object and it is calculated from Eq.(1):

Journal of Physics: Conference Series **351** (2012) 012009 doi:10.1088/1742-6596/351/1/012009

$$Q = \int_0^\infty q^2 I(q)dq \tag{1}$$

The volume of the object is calculated from Eq.(2).

$$Q = 2\pi^2\gamma(0) = 2\pi^2 \int_V \rho^2(r)dr \tag{2}$$

In particular for homogenous particle the volume is related to Porod invariant by:

$$V = 2\pi^2 I(0)/Q = 2\pi^2/Q_0 \tag{3}$$

Where

$$Q_0 = \int_0^\infty i(q)q^2 dq$$

and (4)

$$i(q) = \frac{I(q)}{I(0)}$$

The apoferritin molecule volume was estimated from Porod invariant Q. It is equal to the area under q-dependence of the curve $I(q)q^2$. The calculation was performed by the Primus program [23] from the ATSAS software package [20]. The $I(q)q^2$ vs q is shown in Figure. 2.

The volume values are 560 x10^3 and 600 x10^3 Å^3 for SANS and SAXS, respectively.

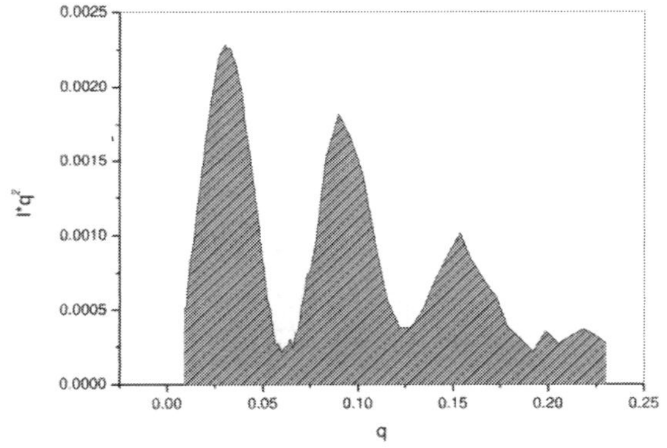

A

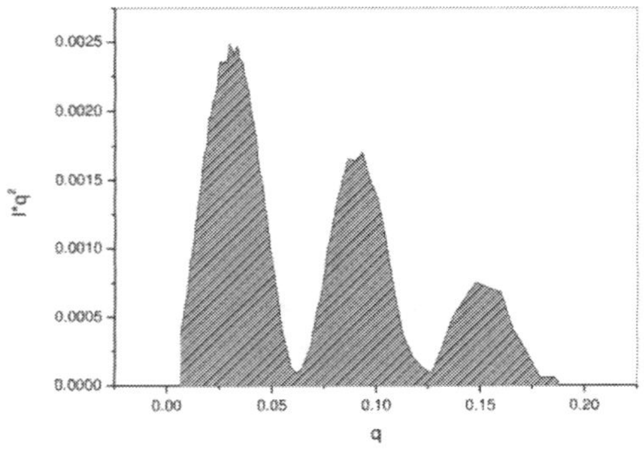

B

Figure 2. The Porod dependence Iq^2 vs q obtained from SANS **(A)** and SAXS **(B)** data of apoferritin. The hatched area corresponds to the Porod invariant Q (1).

3.3. Three-dimensional model of apoferritin molecule, constructed by Monte-Carlo method

Often, three-dimensional representation of the object's shape can be obtained by fitting the experimental scattering curve by theoretical one, which corresponds to the selected model. The model search without any *a priori* information about the object structure could be done using small-angle scattering data, for example, by an algorithm, implemented in the program DAMMIF [24]. This algorithm uses the Monte Carlo method and represents a model of an object as a set of small spheres (dummy-atoms). The resulting model is a low resolution mode, because SANS method is a representative of low-resolution techniques, which give structural information in the range 10-1000 Å. The scattering intensity for the dummy-model is calculated as follows:

$$I(q) = 2\pi 2 \sum_{l=0}^{\infty} \sum_{m=-l}^{l} |A_{lm}(q)|^2, \tag{4}$$

where A_{lm} amplitude is calculated as:

$$A_{lm}(q) = i^l (2/\pi)^{1/2} v_a \sum_{\substack{j=1 \\ X(j)=1}}^{M} j_l(qr_j) Y^*_{lm}(\omega_j) \tag{5}$$

(r_j, ω_j) – polar coordinates, $v_a = \frac{4\pi r_0^3}{3} \frac{1}{0.74}$ – dummy-atom volume, $Y^*_{lm}(\omega_j)$ – corresponding spherical harmonic, $j_l(qr_j)$ – Bessel function. For the optimal model searching the following function f (X) should be minimized:

$$f(X) = R^2(I, X) + \sum_k \alpha_k P_k(X), \tag{6}$$

The first term is the model intensity deviation from the experiment, the second term is the sum of penalties imposed on the solution, with their weights. The simulation result is a compact model of the balls that approximates the experimental data of small-angle scattering in the best way according to (7).

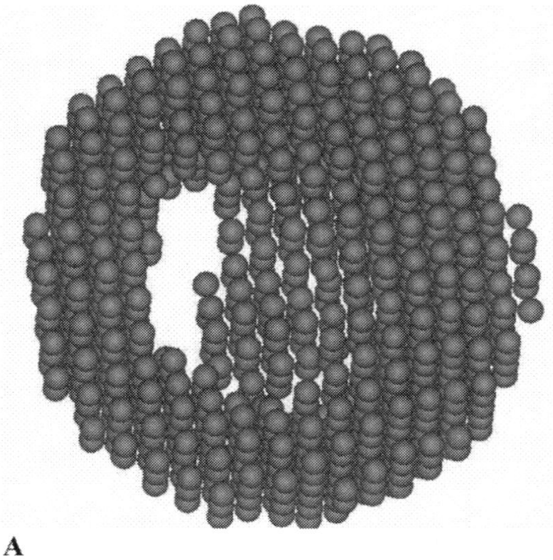

A

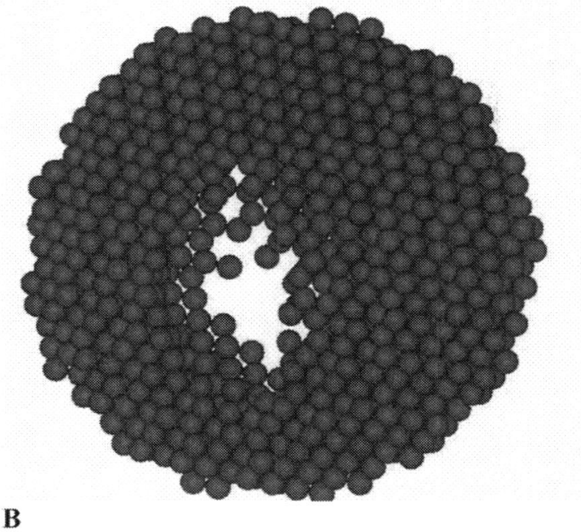

B

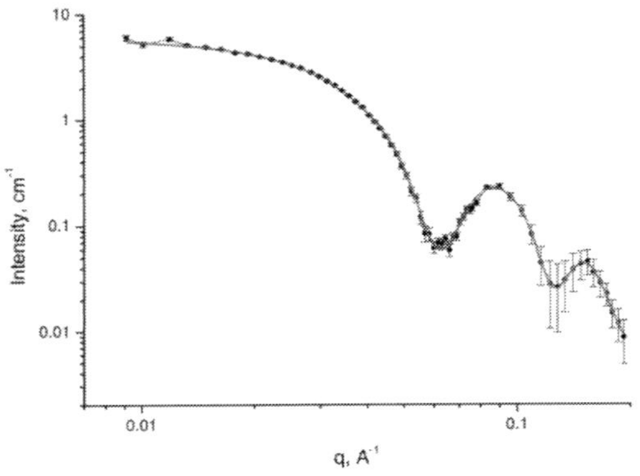

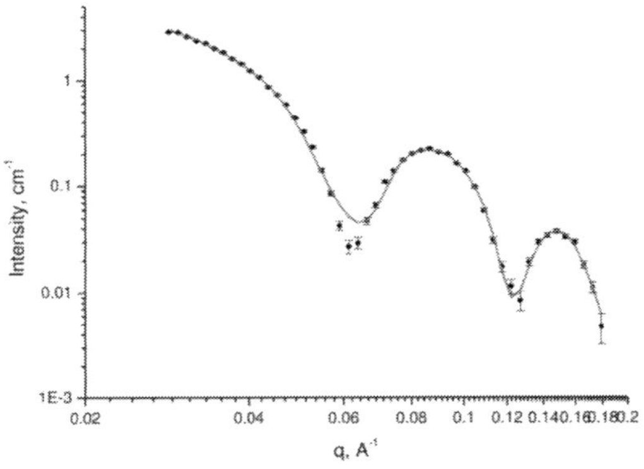

C **D**

Figure 3. The projection of apoferritin 3D model obtained by DAMMIF program for SANS **(A)** and for SAXS **(B)**. Approximations of experimental curves for each 3D model for SANS(**C**) and SAXS (**D**) correspondently.

SANS and SAXS data for the apoferritin solution were treated by the DAMMIF program. The projections of the obtained structures are shown in Figure.3. The model is a spherical shell with outer and inner diameters of about 120 Å and 75 Å respectively. The volume of this structure is 510 $\times 10^3$Å^3 for SANS and 640 $\times 10^3$Å^3 for SAXS. There are "holes" in the shell, which could appear due to the problem of instability solution due to the small thickness of the protein shell (20 Å) or by the presence of regions in the protein shell with the same scattering density as for the solvent. Last assumption requires further analysis.

3.4. The scattering curves approximation by spherical shell function

According to the crystallographic data (Protein Data Bank, PDB) and data obtained by authors using DAMMIF program the apoferritin molecule could be defined by a spherical shell. The neutron and x-ray scattering curves for apoferritin were approximated by the scattering function for a spherical shell (8):

$$I_{sph}(q) = \left[V_1\Phi(qR_1) - V_2\Phi(qR_2)\right]^2, \tag{7}$$

where $V_1 = \frac{4}{3}\pi R_1^{\,3}$, $V_2 = \frac{4}{3}\pi R_2^{\,3}$, $\Phi(t) = 3\frac{\sin t - t\cos t}{t^3}$, R_1 and R_2 – outer and inner spherical shell radius correspondingly.

For a spherical shell the radius of gyration, inner and outer radii are related as:

$$R_g = \frac{3}{5}\frac{R_1^{\,5} - R_2^{\,5}}{R_1^{\,3} - R_2^{\,3}} \tag{8}$$

Experimental curves approximation was carried out by program Fitter 2.1.2 [25] and presented in Figure. 4. The maxima size of molecule is $2R_1$ and the volume of the spherical shell is:

$$V_{shell} = \frac{4}{3}\pi(R_1^{\,3} - R_2^{\,3}) \tag{9}$$

Table 1. The main parameters of fitting obtained for spherical shell model. R1 and R2 values were calculated by program Fitter, other parameters were calculated according to Eqs.(9,10).

№	Parameter	SANS	SAXS
1	Outer radius, R_1, Å	59.1±0.3	62.0 ±0.1
2	Inner radius R_2, Å	40.1±0.4	37.5 ±0.1

3	Radius of Giration R_g,Å	51.1±0.3	52.2 ±0.1
4	Volume of Shell, Vp, $x10^3$ $Å^3$	594.2±23	776 ±18
5	Volume (from Porod approximation) Vp $x10^3$ $Å^3$	560	600
6	Maximum size, Å	118.2±0.3	124±0.1

The parameters for the spherical shell, obtained by Fitter [25] and formula (9,10) are given in Table 1. These parameters (inner and outer radii, radius of gyration) are self-consistent. Nevertheless, there are small systematic differences in the parameters obtained by SANS and SAXS. The calculated volumes of the molecules for the models calculated by program DAMMIF is $510x10^3Å^3$ and $640x10^3Å^3$ for SANS and SAXS data respectively. In comparison, the difference between SANS and SAXS volumes obtained from Porod approximation (560 $x10^3Å^3$ and 600 $x10^3$ $Å^3$) is much smaller. This can be explained by the following: (1) the shape of apoferritin is not exactly spherical; (2) the scattering density distribution of molecules are slightly different; (3) the values, obtained by Porod approximation are quite sensitive to the background level; (4) the difference between volumes and inner and outer radius obtained by SANS and SAXS could be also be explained by H-D exchange of protein and solvent in SANS curves case.

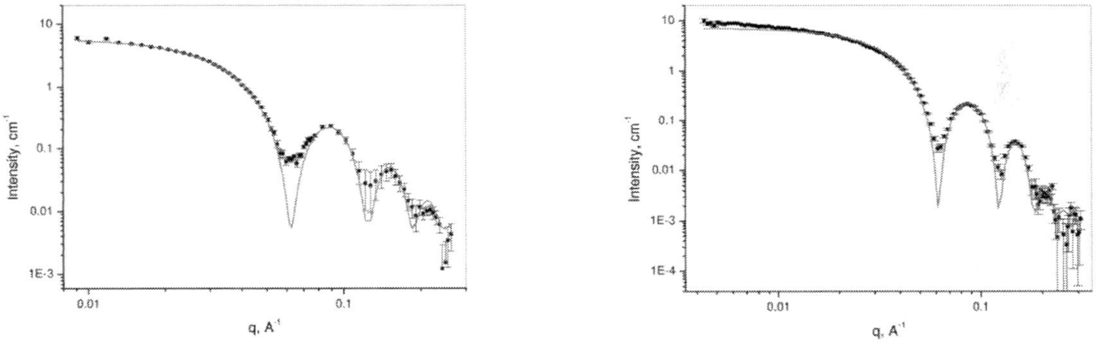

A **B**

Figure 4. SANS (A) and SAXS (B) curves for apoferritin solution. Experimental data shown by points and spherical shell model by line.

4. Conclusion

Obtained SANS and SAXS data for the apoferritin in solution were treated by the DAMMIF program. The structural models correspond to spherical shell with outer and inner diameter of about 120 Å and 75 Å respectively. The shape and 3D model obtained by the low resolution method is consistent with the known results (PDB bank data). The calculated volumes for the models of the molecules is $510x10^3Å^3$ and $640x10^3Å^3$ for SANS and SAXS data respectively, however, the apoferritin molecule volumes determined from Porod approximation are 560 $x10^3Å^3$ and 600 $x10^3Å^3$ for SANS and SAXS, respectively. We have observed a slight systematic difference of structural parameters obtained for SAXS and SANS data. We suggest that this should be taken into account in the interpretation of the results of low resolution studies of proteins.

Acknowledgements

We are grateful to Prof. G. Bueldt and his Laboratory at MIPT for the support of this work. We greatly acknowledge financial support from the NIU program and the ONEXIM group. The work was supported by Federal Target Program "Scientific and Pedagogical Staff of an Innovative Russia" (state contracts P-1160 and 02.740.11.0542). We are also grateful to Seventh Framework Programme [FR7/2007-2013] under grant agreement №211800.

VG acknowledges support from ANR Chair of Excellence and the BMBF for PhoNa project. We are grateful to Dr.S.Sergeenkov for fruitful discussion.

Journal of Physics: Conference Series **351** (2012) 012009 doi:10.1088/1742-6596/351/1/012009

References

[1] Andrews S C 1998 Iron Storage in Bacteria *Advances in Microbial Physiology* **40** 281-351
[2] Crichton R R and Ward R J 1998 Iron homeostasis *Met Ions Biol Syst* **35** 633-65
[3] Massover W H 1993 Ultrastructure of ferritin and apoferritin: A review *Micron* **24** 389-437
[4] Pons C, D'Abramo M, Svergun D I, Orozco M, Bernado P and Fernandez-Recio J 2010 Structural characterization of protein-protein complexes by integrating computational docking with small-angle scattering data *J Mol Biol* **403** 217-30
[5] Kresge N, Simoni R D and Hill R L 2004 The Characterization of Ferritin and Apoferritin by Leonor Michaelis and Sam Granick *Journal of Biological Chemistry* **279** e9-e11
[6] Laufberger V 1937 Sur la cristallisation de la ferritine *Bulletin de la Société de chimie biologique* **19** 1575-82
[7] Granick S 1943 Ferritin. Occurrence and immunological properties of ferritin *Journal of Biological Chemistry* **149** 157-67
[8] Michaelis L, Coryell C D and Granick S 1943 Ferritin. The magnetic properties of ferritin and some other colloidal ferric compounds *Journal of Biological Chemistry* **148** 463-80
[9] Lawson D M, Artymiuk P J, Yewdall S J, Smith J M A, Livingstone J C, Treffry A, Luzzago A, Levi S, Arosio P, Cesareni G, Thomas C D, Shaw W V and Harrison P M 1991 Solving the structure of human H ferritin by genetically engineering intermolecular crystal contacts *Nature* **349** 541-4
[10] Massover W H 1991 Iron Biominerals, ed F B Frankel and R P Blakemore (New York: Plenum Press) 349-58
[11] Massover W H 1992 Pathways for penetration of iron and negative stain across the protein shell of ferritin ultrastructural perspectieves. In: *50-th Annual Meeting Electron Microcs. Soc. Am.*, ed G W Barley, *et al.* (San Francisco: San Francisco Press) pp 1012-3
[12] Fankuchen I 1943 Ferritin. V. X-Ray diffraction data on ferritin and apoferritin *J. Biol. Chem.* **150** 57-9
[13] Häußler W, Wilk A, Gapinski J and Patkowski A 2002 Interparticle correlations due to electrostatic interactions: A small angle x-ray and dynamic light scattering study. I. Apoferritin *Journal of Chemical Physics* **117**: AIP)
[14] Toyokawa H, Fujisawa T, Inoko Y, Nagayoshi T, Nishi Y, Nishikawa Y, Ochi A, Suzuki M and Tanimori T 2001 Performance of a micro-strip gas chamber in solution X-ray scattering Nuclear Instruments and Methods in Physics Research Section A: Accelerators, Spectrometers, Detectors and Associated Equipment *Nucl Instrum Methods Phys* **467-468,** Part 2 1144-7
[15] Harrison P M 1963 The structure of apoferritin: molecular size, shape and symmetry from x-ray data *J Mol Biol* **6** 404-22
[16] Kuklin, A I, Islamov, A Kh and Gordeliy V I 2005 'Scientific Reviews: Two-Detector System for Small-Angle Neutron Scattering Instrument' *Neutron News* **16:3**, 16—18.
[17] Kuklin A I, Islamov A K, Kovalev Y S, Utrobin P and Gordelii V I 2006 Optimization two-detector system small-angle neutron spectrometer YuMO for nanoobject investigation *Journal of Surface Investigation: X-ray, Synchrotron and Neutron Techniques* **6** 74-83
[18] Soloviev A G, Solovieva T M,Stadnik AV, Islamov A Kh and Kuklin A I 2003 *Communication of JINR* **P10-2003-86**
[19] Svergun D 1992 Determination of the regularization parameter in indirect-transform methods using perceptual criteria *Journal of Applied Crystallography* **25** 495-503
[20] Data analysis software ATSAS 2.3.
[21] Svergun D 1991 Mathematical methods in small-angle scattering data analysis *Journal of Applied Crystallography* **24** 485-92
[22] Porod G 1952 The x-ray small-angle scattering of close-packed colloid systems II *Kolloid-Zeitschrift* **125** 51-7
[23] Konarev P V, Volkov V V, Sokolova A V, Koch M H J and Svergun D I 2003 PRIMUS: a Windows PC-based system for small-angle scattering data analysis *Journal of Applied Crystallography* **36** 1277-82

[24] Franke D and Svergun D I 2009 DAMMIF, a program for rapid ab-initio shape determination in small-angle scattering *Journal of Applied Crystallography* **42** 342-6

[25] Soloviev A G, Stadnik A V, Islamov A K and Kuklin A I 2008 Fitter. The package for fitting a chosen theoretical multi-parameter function through a set of data points. Application to experimental data of the YuMO spectrometer. *JINR Communication* **E10-2008-2**

Ripple Phase Behavior in Mixtures of DPPC/POPC lipids: SAXS and SANS Studies

D V Soloviov[1,2], Yu E Gorshkova[1], O I Ivankov[1,2], A N Zhigunov[3], L A Bulavin[2], V I Gordeliy[4,5,6*], A I Kuklin[1,6*]

[1] Joint Institute for Nuclear Research, Dubna, Russia
[2] Taras Shevchenko National University of Kyiv, Kyiv, Ukraine
[3] Institute of Macromolecular Chemistry of the Academy of Sciences of the Czech Republic, Prague, Czech Republic
[4]Institute of Structural Biology J.P.Ebel, Grenoble, France
[5] Institute of Complex System–5, Juelich, Germany
[6] Bionanocenter of Moscow Institute for Physics and Technology, Dolgoprudny, Russia

E-mail: kuklin@nf.jinr.ru; valentin.gordeliy@ibs.fr

Abstract. Multilayer membranes of DPPC (1,2–dipalmitoyl–sn–glycero–3–phosphatidylcholine) and mixture DPPC/POPC (1–palmitoyl–2–oleoyl–sn–glycero–3–phosphatidylcholine) in excess of water were investigated by small angle X–ray scattering and neutron small angle scattering. Structural parameters of lipid bilayers were determined in a wide temperature range. It was shown that addition of the POPC to the DPPC/H_2O does not change the temperature of the DPPC main phase transition T_m. The observed phase separation occurs for the system DPPC/POPC/H_2O up to the pre–transition temperature T_r (gel phase – ripple phase) of the DPPC multilayers.

1. Introduction

Studies of the physical and physics–chemical aspects of phase transitions in membranes (real biological and model lipid) have been going on for many years [1-3]. This interest is due both to the biological importance of the membranes and their promising applications (such as sensors, transducers, filters), as well as the fundamental physical issues related to phase transitions, fluctuations, and dynamics of complex systems.

Despite of the large number of papers about phase transitions, including those in biological systems, the biological role of phase transitions is still unclear. For example, some researchers believe that having a phase transition in membranes determines how temperature of warm–blooded organisms stabilize and cooperative processes of cell [4].

Intermembrane interactions are determined by the balance of forces between uncharged membranes. The main contribution is made by a strong short–range repulsion (often called the "hydration" force), long–range van der Waals attraction, and long–range Helfrich's undulation repulsion.

In the work [5] new approaches of intermembrane interactions studies are developed and the universal constant, based on a study of the temperature dependence of intermembrane interactions with the complementary use of small–angle scattering of thermal neutrons and high resolution diffraction at the synchrotron source, is determined. In particular, a value of the coupling constant, is coincides with the theoretically predicted. In addition, in [5] it was first shown what the true value of undulation forces are and they really make a significant contribution to the balance of the intermembrane interactions and, moreover, these forces become dominant at distances greater than 20Å.

Back in 1975 a theoretical model of the lipid layer [5] was proposed. The extensive surface area and the conjugate surface pressure were introduced into a cooperatively interacting hydrocarbon chain model. The exact statistical mechanical method of solution is extended to accommodate the new variables. In [6], these authors also offer a refined model of the lipid bilayer, which takes into account the weak interactions between the heads of the lipids. It is shown that even weak interactions between the lipid heads are similar to the bilayer surface pressure which influences the temperature and pressure at which the phase transition occurs. The phase diagram of lipid was calculated and the value of surface pressure was determined.

Using wide–angle diffraction, the authors [7] found two orientation distribution of lipid molecules at low temperature for mixtures of two lipids with cholesterol. This orientation ordering is observed in a wide temperature range. Is this a true microphase separation of the lipids or just the effect of "forced" changes in the concentration, still remains a question.

Numerous publications are devoted to studying the effect of various additives on the points of lipid membranes phase transitions. For example, in [8] the authors proposed a model to explain the phase transition of cylindrical micelles – bilayer vesicles in solutions of phospholipids and surfactants. The model predicts that the first order transition and transition temperature depends of the relative concentrations of two components. The phase transition boundary is calculated as a function of the area of phospholipids and surfactants. The proposed model, the authors believe, is in qualitative agreement with experimental data. In [9] the effect of electrolytes on solutions of phospholipids at a temperature below the main phase transition was studied by small angle neutron and X–ray scattering and by differential scanning calorimetry. Effect of cholesterol, anesthetics, as well as polypeptides on the main phase transition temperature of phospholipids was investigated by nuclear magnetic resonance [10] and by double electron–electron resonance. In the case of the double electron–electron resonance to the ends of the hydrocarbon chains electronic tags were attached. Using the distance between these tags the thickness of the phospholipid bilayer was determined [11].

Using X–ray diffraction and electron microscopy the effect of dichlorphenol on the main lipid phase transition in DPPC was shown [12]. It was established that the structural changes depend on the concentration of dichlorphenol.

During phase transition investigations it is interesting to change some external parameters, like temperature, pressure, pH level etc. For lipid membranes pressure is a particularly interesting option, as the natural feature of the life of some living organisms, as well as an external parameter. In the professor Winter's papers devoted to high pressure studies [10, 13-15] different research methods have been used: small–angle neutron and X–ray scattering, X–ray diffraction, nuclear magnetic resonance method, and PVT studies. The article presents the phase diagrams for the various lipids, the influence of proteins, polypeptides and some surfactants on the phase transition point. The kinetics of phase transitions and influence of phospholipid hydrocarbon tails type on the bilayer packing in the different phases was investigated [16].

Earlier we investigated the aqueous solution of DMPC with PVT method and showed the critical behavior of the isothermal compressibility at the main phase transition [17, 18].

Despite on the considerable list of publications about the effect of high hydrostatic pressure on the phase transitions, the key experiments have not yet been delivered. In our view, this is due to two factors: the technical implementation of the experiment and the difficulty in preparing the samples.

In this work multilayer membrane DPPC and DPPC/POPC (3:1) in excess water are investigated by small–angle X–ray and neutron scattering. These two lipids have similar length of hydrocarbon chains but different temperatures of the main phase transition.

The aim of our work is evaluation the impact on the structure of the lipid close to the temperature shift of the phase transition point at atmospheric pressure. Though, the effect of certain additives (such as alcohols, hydrocarbon chains, cholesterol) on the phase transition is well–studied and there are several models of this semiphenomenological influence, however, there is still no clarity in the main phase transition. Additional difficulties in the study of phase transitions of mixtures are due to the fact that the components of the system at a sufficient degree of homogenization may be in a nonequilibrium state. Arising in this state are connected primarily with the effects of hydration–dehydration, which essentially determines the morphology of the membranes.

Lipid mixtures investigations have great interest for understanding of the structure and the important functions of membranes. Biological membranes are multicomponent lipid systems. Many biomembranes have about 100 different lipids. It is also of considerable interest from the standpoint of understanding the influence of the lipid bilayer on the function of membrane proteins. In addition, these studies are important for understanding the crystallization of membrane proteins in lipid membranes.

Our task is investigation of the structural changes with temperature in DPPC and POPC multilayer membranes in the case of two–component mixture.

2. Material and methods

Synthetic 1,2–dipalmitoyl–sn–glycero–3–phosphatidylcholine (DPPC) and 1–palmitoyl–2–oleoyl–sn–glycero–3– phosphatidylcholine (POPC) was purchased from Avanti (Birmingham, Al) and was used without further purification. Water (18 MV/cm) was obtained with the help of Millipore (USA).

Multilamellar vesicles (MLVs) were prepared in the following way: DPPC or DPPC/POPC (3:1) was added to the water. For mixing the system a shaker was used. Homogeneous dispersion was obtained by freezing – heating method about ±20°C below and above the main phase transition temperature. For SAXS experiments the mixture was placed into borosilicate capillary with 1.5 mm diameter and 0.01 mm wall thickness (W. Muller, Berlin, Germany). The capillary was hermetically sealed, and the specimens were held at the temperature higher than that of phase transition for several hours. For SANS experiments the mixture was placed in quartz cell.

The final concentration of the lipid/water was 10 mg/ml. We have used the same way for preparation of all samples. The samples prepared for neutron measurements had final concentration 2% (lipids/heavy water wt/wt).

The investigations by small angle X–ray (SAXS) method were performed on a pinhole camera Molecular Metrology SAXS System at the Institute of Macromolecular Chemistry CAS (Prague, Czech Republic).

A pinhole camera Molecular Metrology SAXS System was attached to a microfocused X–ray beam generator (Osmic MicroMax 002) operating at 45 kV and 0.66 mA (30 W). The camera was equipped with a multiwire, gas–filled area detector with an active area diameter of 20 cm (Gabriel design). Two experimental setups were used to cover the q range of 0.0045 – 1.1 Å ($q=(4\ \pi/\lambda)\sin\theta$, where λ is the wavelength and 2Θ is the scattering angle).

The neutron experiments were performed at YuMO spectrometer using two–detector set up with ring wire detectors [19, 20]. The beam was collimated to a diameter of 14 mm on the sample. The data treatment was performed by SAS program with smoothing mode [21, 22].

The data acquisition was performed using the SONIX Software Package [23].

3. Results and discussions

3.1. SAXS results

In Figure 1 the small angle X–ray scattering curves of pure DPPC (a) and mixture DPPC/POPC (3:1 wt/wt) (b) near phase transition temperature points are shown. Above 34.5°C we observe the shift of the diffraction peak, which corresponds to the change repeat distances of lipid bilayers, and thus manifests lipid phase transition.

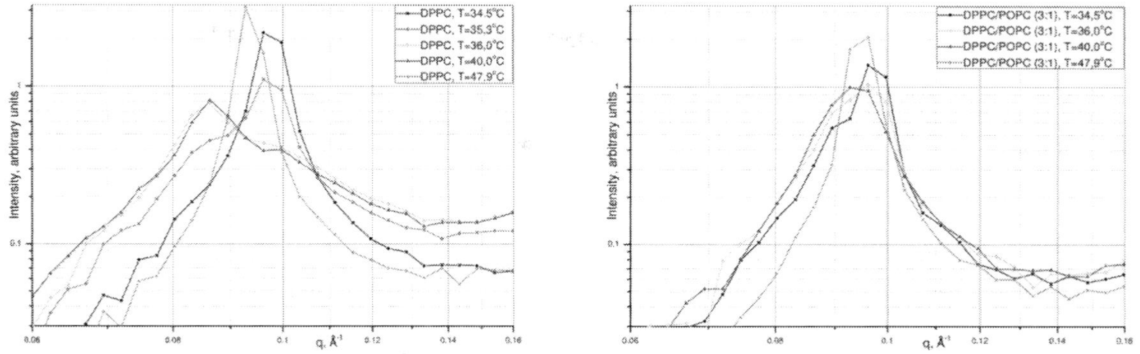

a **b**

Figure 1. SAXS patterns comparison for DPPC (**a**) and DPPC/POPC (3/1) lipid mixture (**b**) in water excess.

As expected, from 20°C to 34.5°C DPPC membranes are in the gel phase and at the temperature range from 35,3°C to 36°C they are in the ripple phase, and at temperature range from 40°C to 47,9°C they are in the liquid phase. This is the test of the experimental accuracy.

The repeat distance of multilamellar bilayers was calculated by the equation (1):

$$D = \frac{2\pi}{q_{peak}}, \tag{1}$$

where D is repeat distances of the multilammelar bilayer, q_{peak} corresponds to the maximum of the diffraction peak.

In Figure 2 we present the results of the calculation of the periodicity of multilayers of the DPPC and DPPC/POPC (3:1) membranes vs temperature. Due to sufficiently high resolution we observed second peak on scattering curves for two lipids DPPC/POPC mixture.

As follow from the data, presented in Figure 2, the adding of POPC to DPPC (1:3) reduce the lattice constant of the ripple phase in comparison with pure DPPC solution.

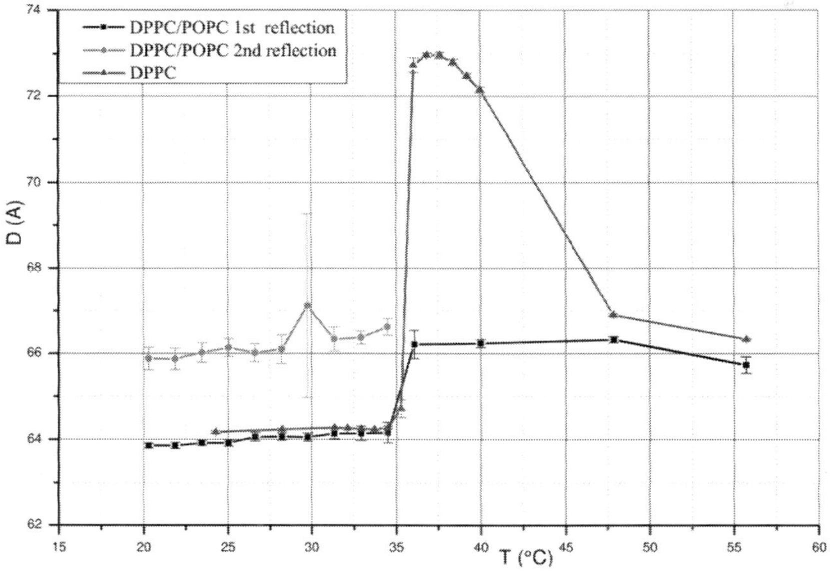

Figure 2. Repeat distance of multilayers of DPPC and DPPC/POPC membranes vs temperature (SAXS data). Triangles corresponds to the repeat distance calculated from scattering curves of DPPC sample, rectangle – the repeat distance calculated from scattering curves used first peak and circles corresponds to the data calculated from scattering curves used second peak DPPC/POPC (3:1) sample.

3.2. SANS results

The measurements at YuMO small angle neutron scattering spectrometer were done under the following conditions: the size of the neutron beam is 14 mm instead of 0.3 mm for the X–ray beam, the concentration of lipid/heavy water is 2 % (wt/wt) instead of 10% lipid/water (wt/wt), and volume of sample is 1cm^3 instead of 0.5 mm^3.

In Figure 3 the typical small angle neutron scattering curves of DPPC (a) and DPPC/POPC (b) in heavy water solution (2% wt/wt) for 11°C, 37°C, 41°C, 61°C and pressure 1 atm are shown.

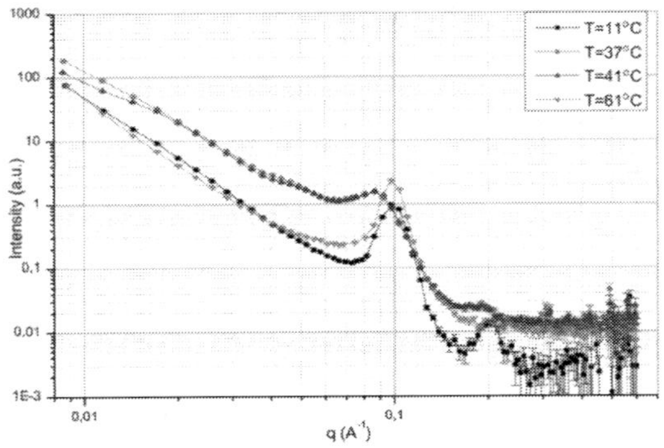

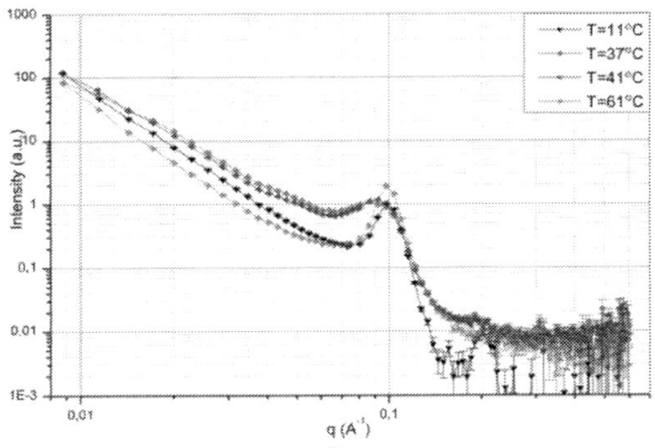

Figure 3. Small–angle neutron scattering curves of DPPC (**a**) and DPPC/POPC membranes (**b**) in heavy water (2% wt/wt) for 11°C, 37°C, 41°C, 61°C and pressure 1 atm.

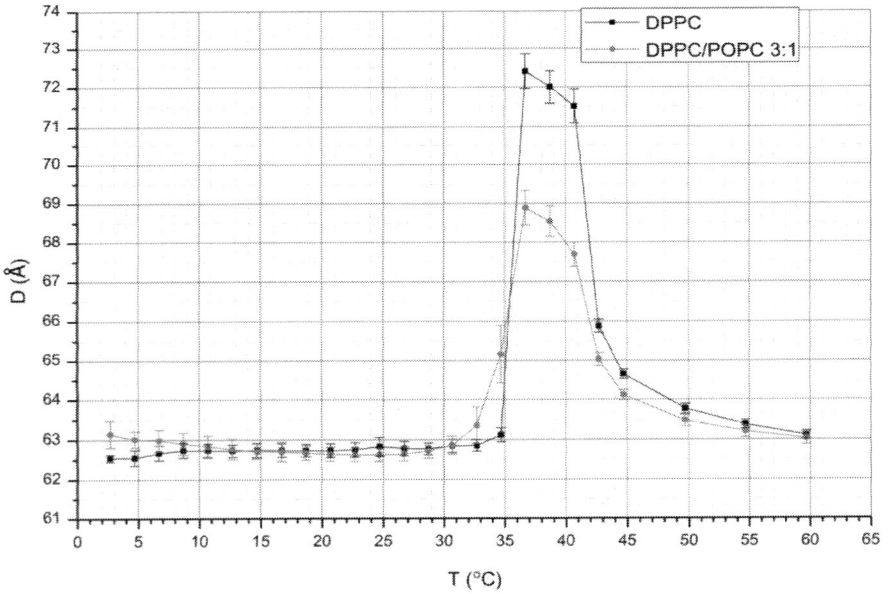

Figure 4. The repeat distance of the multilayers DPPC and DPPC/POPC vs temperature obtaining at the YuMO spectrometer. Rectangle corresponds to DPPC, circle to DPPC/POPC (3:1) membranes.

In Figure 4 the repeat distance of multilamellar structures of DPPC and DPPC/POPC (3:1) mixture in heavy water are presented. Behavior of the scattering curves is different in ripple phase, namely in the former case the periodicity is 72 Å and for the mixture with POPC it is equal to 69 Å.

Using two different methods (SAXS and SANS) for two concentration and different volumes of water (heavy water) for the mixture of POPC/DPPC we found that by adding of POPC (1:3) the periodicity of the ripple phase of DPPC is suppressed in comparison with pure DPPC solution.

4. Acknowledgement

The work was supported by the program "Chaires d'excellence" edition 2008 of ANR France, CEA(IBS) – HGF(FZJ) STC 5.1 specific agreement and the MC grant for training and career development of researchers (Marie Curie, FP7–PEOPLE–2007–1–1– ITN, project SBMPs) and by an EC FP7 grant for the EDICT consortium (HEALTH–201924). The work was done with the support of Russian State Contracts Nos. 02.740.11.0299, No. 02.740.11.5010, No. P974 of activity 1.2.2 and No. state contracts P–1160, P211 of activity 1.3.2 of the Federal Target Program "Scientific and academic research staff of innovative Russia" for 2009–2013 years. Part of this work was supported by the German Federal Ministry for Education and Research (PhoNa – Photonic Nanomaterials). We are also grateful to A.Kh. Islamov and R.V. Erhan for fruitfull discussions.

References

[1] Biltonen R L 1990 A statistical-thermodynamic view of cooperative structural changes in phospholipid bilayer membranes: their potential role in biological function *The Journal of Chemical Thermodynamics* **22** 1-19

[2] Nagle J F and Tristram-Nagle S 2000 Structure of lipid bilayers *Biochimica et Biophysica Acta (BBA) - Reviews on Biomembranes* **1469** 159-95

[3] Tristram-Nagle S and Nagle J F 2004 Lipid bilayers: thermodynamics, structure, fluctuations, and interactions *Chemistry and Physics of Lipids* **127** 3-14

[4] Kharakoz D P 2001 Uspehi biologicheskoy khimii (in Russian) **41** 333-64

[5] Nagle J F 1975 Chain model theory of lipid monolayer transitions *The Journal of Chemical Physics* **63** 1255-61

[6] Nagle J F 1976 Theory of lipid monolayer and bilayer phase transitions: Effect of headgroup interactions *Journal of Membrane Biology* **27** 233-50

[7] Mills T T, Tristram-Nagle S, Heberle F A, Morales N F, Zhao J, Wu J, Toombes G E, Nagle J F and Feigenson G W 2008 Liquid-liquid domains in bilayers detected by wide angle X-ray scattering *Biophys J* **95** 682-90

[8] Andelman D and et al. 1994 Phase Transitions between Vesicles and Micelles Driven by Competing Curvatures *EPL (Europhysics Letters)* **25** 231

[9] Fukada K and Miki N 2009 Phase Separation of Phosphatidylcholine–Water Systems below the Main Transition Temperature Induced by Monovalent Ions *Bulletin of the Chemical Society of Japan* **82** 439-45

[10] Winter R 2003 *Annual Reports on NMR Spectroscopy,* ed W G. A: Academic Press) pp 163-200

[11] Chiang Y-W, Costa-Filho A J and Freed J H 2007 Dynamic Molecular Structure and Phase Diagram of DPPC–Cholesterol Binary Mixtures: A 2D-ELDOR Study *The Journal of Physical Chemistry B* **111** 11260-70

[12] Csiszár Á, Klumpp E, Bóta A and Szegedi K 2003 Effect of 2,4-dichlorophenol on DPPC/water liposomes studied by X-ray and freeze-fracture electron microscopy *Chemistry and Physics of Lipids* **126** 155-66

[13] Reis O, Winter R and Zerda T W 1996 The effect of high external pressure on DPPC-cholesterol multilamellar vesicles: a pressure-tuning Fourier transform infrared spectroscopy study *Biochimica et Biophysica Acta (BBA) - Biomembranes* **1279** 5-16

[14] Winter R 2001 Effects of hydrostatic pressure on lipid and surfactant phases *Current Opinion in Colloid & Interface Science* **6** 303-12

[15] Winter R 2002 Synchrotron X-ray and neutron small-angle scattering of lyotropic lipid mesophases, model biomembranes and proteins in solution at high pressure *Biochimica et Biophysica Acta (BBA) - Protein Structure and Molecular Enzymology* **1595** 160-84

[16] Winter R and Pilgrim W C 1989 A SANS Study of High Pressure Phase Transitions in Model Biomembranes *Berichte der Bunsengesellschaft für physikalische Chemie* **93** 708-17

[17] Kuklin A I, Soloviov D V, Rogachev A V, Utrobin P K, Kovalev Y S, Balasoiu M, Ivankov O I, Sirotin A P, Murugova T N, Petukhova T B, Gorshkova Y E, Erhan R V, Kutuzov S A, Soloviev A G and Gordeliy V I 2011 New opportunities provided by modernized small-angle neutron scattering two-detector system instrument (YuMO) *Journal of Physics: Conference Series* **291** 012013

[18] Solov'ev D, Kuklin A, Utrobin P, Ozerin A, Kurkin T, Ivan'kov O, Bulavin L and Gordelii V 2011 X-ray scattering and volumetric P-V-T studies of the dimyristoylphosphatidylcholine-water system *Journal of Surface Investigation: X-ray, Synchrotron and Neutron Techniques* **5** 7-10

[19] Kuklin A I, Islamov A K and Gordeliy V I 2005 Two-detector System for Small-Angle Neutron Scattering Instrument *Neutron News* **16** 16-8

[20] Kuklin A I, Islamov A K, Kovalev Y S, Utrobin P and Gordely V I 2006 Optimization two-detector system small-angle neutron spectrometer YuMO for nanoobject investigation *Journal of Surface Investigation: X-ray, Synchrotron and Neutron Techniques* **6** 74-83

[21] Soloviev A G, Litvinenko E I, Ososkov G A, Islamov A K and Kuklin A I 2003 Application of wavelet analysis to data treatment for small-angle neutron scattering *Nuclear Instruments and Methods in Physics Research Section A: Accelerators, Spectrometers, Detectors and Associated Equipment* **502** 500-2

[22] Soloviev A G, Sioloviova T M, Stadnik A V, Islamov A K and Kuklin A I 2003 The Package for Small-Angle Neutron Scattering Data Treatment *JINR Communication* **P10-2003-86**

[23] Kirilov A S, Litvinenko E I, Astakhova N V, Murashkevich S M, Petukhova T B, Yudin V E, Gordelii V I, Islamov A K and Kuklin A I 2004 Evolution of the SONIX Software Package for the YuMO Spectrometer at the IBR-2 Reactor, Instruments and Experimental Techniques *Pribory i tekhnika eksperimenta* **47** 334-6

Lipid bilayer – DNA interaction mediated by divalent metal cations: SANS and SAXD study

D Uhríková[1], N Kučerka[1,2], A Lengyel[1,3], P Pullmannová[1], J Teixeira[4], T Murugova[5], S S Funari[6], P Balgavý[1]

[1]Faculty of Pharmacy, Comenius University, Odbojárov 10, 832 32 Bratislava, Slovakia
[2]Canadian Neutron Beam Centre, National Research Council, Chalk River, Ontario, Canada
[3]Max-Planck-Institut für Biophysik, Abteilung für Strukturbiologie, Max-von-Laue-Straße 3, 60438 Frankfurt am Main, Germany
[4]Laboratoire Leon Brillouin, CEA, Saclay, 91191 Gif sur Yvette Cedex, France
[5]Frank Laboratory of Neutron Physics, JINR, Dubna, Russia
[6]Hasylab at DESY, Notkestr. 85, D-22607 Hamburg, Germany

E-mail: uhrikova@fpharm.uniba.sk

Abstract. The structure of aggregates formed due to DNA interaction with dipalmitoylphosphatidylcholine (DPPC) in presence of Ca^{2+} and Zn^{2+} is examined using small-angle synchrotron X-ray diffraction (SAXD) and small-angle neutron scattering (SANS). SAXD detected two structures: L^C - condensed lamellar phase and L^X - lamellar phase with DNA strands intercalated between the adjacent lipid bilayers, without regular packing at low cation concentration (~1 mM). The high concentration of Zn^{2+} induces a macroscopic phase separation in mixtures. The SANS curves of DPPC+ions^{2+} vesicles evaluated using the strip function model have shown different modes of cations binding to the DPPC bilayers.

1. Introduction

More than three decades ago Budker at al. [1] reported the ability of divalent metal cations to mediate the interaction between DNA and phosphatidylcholine vesicles. Electron freeze fracture micrographs of DNA – phosphatidylcholine - Ca^{2+} mixtures suggested structures with long-range organization [2,3]. The small angle X-ray diffraction confirmed the presence of condensed lamellar phase (L^C) with DNA strands regularly packed between phospholipid bilayers in DNA – dipalmitoylphosphatidylcholine (DPPC) - Mg^{2+} and Ca^{2+} system [4,5]. In addition to biologically relevant divalent cations, calcium and magnesium, also Fe^{2+}, Co^{2+} and Mn^{2+} have shown the ability to mediate DNA – neutral phospholipid binding [6]. The DNA interaction with neutral phospholipid bilayer in the presence of divalent metal cations (Ca^{2+}, Mg^{2+}, Co^{2+}, Ni^{2+}, Zn^{2+}) was thoroughly studied in our laboratory [7,8]. All studied cations facilitate regular packing of DNA strands between DPPC bilayers forming the condensed lamellar phase (L_β^C) in the temperature range corresponding to the gel state of DPPC. However, the DNA regular packing is disrupted when the lipid goes into a liquid-crystalline state, as shown in [9]. Aggregates phospholipid – DNA – cations have shown high structural heterogeneity as a function of temperature and cations concentration.

In this contribution we try to shed more light onto observed structural variety in DNA+phospholipid+ion^{2+} system, particularly in aggregates with zinc. Zinc plays a fundamental role in several critical cellular functions such as protein metabolism, gene expression, structural and functional integrity of biomembranes, and in metabolic processes [10]. Compared with other micronutrients, zinc is present in biological systems in high concentrations, particularly in biomembranes. According to Williams [11], the concentration of zinc in animal systems range from $<10^{-9}$ M in cytoplasm to $>10^{-3}$ M in some membrane vesicles.

Generally, in the system consisiting of DNA, neutral phospholipid and divalent cations, in addition to a formation of triple complex (DNA + phospholipid + ion^{2+}), one must consider also the interaction DNA + ion^{2+} and phospholipid + ion^{2+}. Alkaline earth metal cations (Ca^{2+}, Mg^{2+}) preferentially interact with the phosphate groups of DNA, thereby reducing the charge repulsion between the opposing strands of double-helix [12] and stabilizing the polynucleotide molecule. Transition metal cations interact more extensively with DNA bases causing disruption of base pairing and destabilization of DNA molecule [12, 13]. According to Duguid et al. [14], the affinity of divalent cations to DNA bases decreases in the order $Hg^{2+}>Cu^{2+}>Pb^{2+}>Cd^{2+}>Zn^{2+}>Mn^{2+}>Ni^{2+}$, $Co^{2+}>Fe^{2+}>Ca^{2+}>Mg^{2+}$, Ba^{2+}. Our fluorescence and UV-VIS experiments have confirmed the ability of zinc to condense DNA in presence of lipid bilayer, and to protect it against thermal denaturation up to the level comparable with calcium [8].

In this contribution we demonstrate the structural differences in aggregates prepared due to DNA interaction with DPPC bilayer in the presence of calcium and zinc observed in our SAXD experiments. We have studied recently polymorphic behaviour of the DNA-DPPC in a solution of divalent transition metals and Zn^{2+} in large concentration range [7, 15]. In aggregates with phosphatidylcholines (DPPC, DMPC, DOPC), we have identified structures: L^C - condensed lamellar phase; L^X - lamellar phase with DNA strands intercalated between the adjacent lipid bilayers, without regular packing; coexistence of a lamellar phase L^{PC} of pure lipid with L^X or L^C phase. High concentrations of transition metal ions (Co^{2+}, Ni^{2+}) and Zn^{2+} induce a macroscopic phase separation. In addition to the condensed lamellar phase a partially ordered lamellar phase L^V was identified, showing the periodicity ~ 20 - 8 nm, formed due to screening of the phospholipid positive surface charge at high concentration of ions in the solution.

With the aim to understand the observed polymorphic behaviour we have studied the effect of two cations, calcium and zinc, on the DPPC bilayer itself employing small-angle neutron scattering (SANS). The study of DPPC+Ca^{2+} system has revealed structural changes in DPPC bilayer due to calcium binding [16]. The analysis of SANS curves of DPPC+Zn^{2+} system in the concentration range 1 – 60 mM of $ZnCl_2$ indicates different mode of ions binding to DPPC bilayer that offers an explanation for the macroscopic phase separation observed in DNA+DPPC+Zn^{2+} mixtures.

2. Material and Methods

2.1. Sample preparation

SAXD experiment: Highly polymerized calf thymus DNA (Sigma Chemicals Co., USA) was dissolved in 0.5 mM NaCl, pH~7, at concentration ~2 mg/ml. The precise value of concentration was determined by measuring the absorbance A_λ at λ=260 nm. The purity of DNA was checked by measuring the absorbance A_λ at λ=260 nm and 280 nm, we have obtained A_{260}/A_{280}=1.8. Dipalmitoylphosphatidylcholine (DPPC) was purchased from Avanti Polar Lipids, USA. Solutions of $CaCl_2$ and $ZnCl_2$ (Merck, Germany) were prepared in 0.5 mM NaCl, pH~7. Redistilled water was used for the preparation of all solutions. Samples were prepared by the hydration of a thin lipid film (~5 mg of lipid per sample) with a solution of DNA and divalent metal cations to obtain the required molar ratio lipid : DNA base (3:1) and concentration of ions. The samples were vortexed for a short time; a few minutes after the preparation a sediment formed in the sample. The supernatant was gently removed by a Pasteur pipette and the sediment was placed between two Kapton foils (Dupont, France), which constitute the windows of the sample holder for X-ray diffraction. Each sample was let at rest at least 30 min before to be transferred to the sample holder. Multilamellar liposomes prepared from fully hydrated DPPC were used as a reference sample.

SANS experiment: Heavy water of isotopic purity 99.9% D_2O was purchased from Merck (Germany). Solutions of $CaCl_2$ and $ZnCl_2$ were prepared in D_2O (5 mM NaCl, pH~7). Lipid was dissolved in methanol and portioned (10 mg per sample) into plastic tubes. The solvent was gently evaporated under a stream of gaseous nitrogen to create a thin lipid film. Traces of solvent were removed by an oil vacuum pump. The dry lipid was hydrated by adding 1 ml of the XCl_2 solution (X = Ca^{2+}, Zn^{2+}) at wished concentration. The dispersion was vortexed and homogenized in an ultrasound bath (at 60 °C) and by at least tenfold freezing-thawing process to obtain a homogeneous distribution of calcium ions between lipid multilayers. After this procedure, the samples showed a slight opalescence, typical for dispersions of lipid vesicles. The reference sample of unilamellar vesicles from DPPC was prepared by extrusion of DPPC multilamellar dispersion through a polycarbonate filter (Nuclepore, Plesanton, USA) with pores of 50 nm diameter, using the LiposoFast Basic extruder (Avestin, Ottawa, Canada) fitted with two gas-tight Hamilton syringes (Hamilton, Reno, USA) as described by MacDonald et al. [17]. The samples were subjected to 51 passes through the filter at a temperature above the main phase transition temperature of pure phospholipid. The prepared samples were placed in 2-mm quartz cells (Hellma, Müllheim, Germany), closed and kept at room temperature before measurement.

2.2. *Experimental methods*

Small-angle X-ray diffraction: SAXD experiments were performed at the soft-condensed matter beam line A2 at HASYLAB at the Deutsches Elektronen Synchrotron (DESY) in Hamburg (Germany), using a monochromatic radiation of wavelength λ=0.15 nm. The evacuated double-focusing camera was equipped with two linear delay line readout detectors. The sample was equilibrated at selected temperature for 5 min before exposure to radiation. The raw data were normalized against the incident beam intensity using the signal of the ionization chamber. The SAXD detector was calibrated using rattail collagen [18]. Each diffraction peak of SAXD region was fitted with a Lorentzian above a linear background.

Small-angle neutron scattering: SANS experiments were performed on the *PAXE* spectrometer located at the G5 cold neutron guide of the Orphée reactor (Laboratoire Léon Brillouin, Saclay, France). The sample to detector distance was 2.75 m and the neutron wavelength was λ=6 Å ($\Delta\lambda/\lambda$=10%) covering the scattering vector range 0.016-0.224 Å^{-1}. The temperature of the sample was set to 20 °C, controlled electronically within an accuracy of ±0.1°C. The acquisition time for one sample was 30 minutes. The normalized SANS intensity $I(q)$ as a function of the scattering vector modulus q was obtained as described previously [19].

3. Results and Discussion

3.1. *SAXD of DNA – DPPC – ion^{2+}*

Fully hydrated DPPC at 20 °C forms a lamellar phase (Figure 1a), we determined the repeat distance d = $2\pi/(q_2-q_1)$ = 6.36 ±0.01 nm, where q_i is the position of i-th peak's maximum in reciprocal space units. Typical diffractograms observed in the DNA-DPPC-ion^{2+} (Ca^{2+}, Zn^{2+}) mixture at 20 °C are shown in Figure 1b-h. The diffractogram of the aggregate DPPC:DNA=3:1 mol/base prepared at 1 mM Ca^{2+} (Figure 1b) is interpreted as a superposition of two one-dimensional periodic structures.

Journal of Physics: Conference Series **351** (2012) 012011 doi:10.1088/1742-6596/351/1/012011

The reflections $L(1)^x$ and $L(2)^x$ correspond to the lamellar phase (L^x). We determined its repeat distance d_{Lx} = 8.04±0.01 nm at 20 °C. Supposing the DPPC bilayer thickness d_L ~ 5.3 nm [16], the water layer thickness $d_W=d-d_L$ provides enough space to accommodate DNA strands with diameter ~2.5 nm [20]. Because the packing of DNA strands in this phase is not obvious, we denote this phase as L^x. The reflections $L(1)^{DPPC}$ and $L(2)^{DPPC}$ were identified as reflections due to the DPPC bilayer stacking (lamellar phase L^{DPPC}). Such coexistence of two phases in one aggregate was observed and discussed earlier in [5,6,21]. An increase in Ca^{2+} concentration improved the organization of DNA strands inside of L^x phase, although the DNA peak is broad, indicating a poor organization of the DNA lattice. Diffractograms in Figure 1c, d, and f are typical diffractograms of a condensed lamellar phase with regularly packed DNA strands. Structural parameters are given in Table 1. Aggregates prepared at low concentration of zinc (c_{ZnCl2} ≤ 20 mM) show similar microstructure (Figure 1e and f). The diffraction pattern of DNA+DPPC in 40 mM of $ZnCl_2$ (Figure 1g) is different in comparison to the previous one. A deconvolution of peaks revealed the superposition of two phases: the condensed lamellar phase with repeat distances d=8.54 nm and d_{DNA}=6.14 nm, and a lamellar phase L^V with periodicity d=13.5 nm. The periodicity ~ 13 nm (with the water gap d_W ~8 nm) is too big in comparison to structural parameters of L^C phase, also to accommodate DNA strands or to be formed from neutral lipid itself. This indicates a destruction of the lamellar structure yielding its swelling into excess of water. The periodicity of the L^V phase is modulated with increasing concentration of zinc

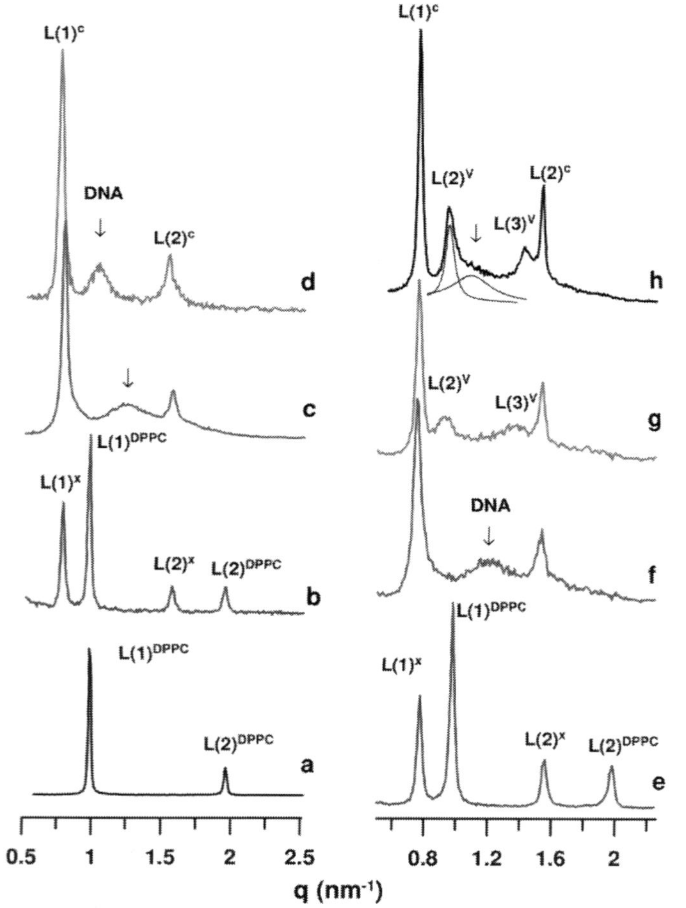

Figure 1.
SAXD diffractograms (at 20 °C) of:
a) DPPC
DNA+DPPC in:
b) 1 mM $CaCl_2$
c) 20 mM $CaCl_2$
d) 50 mM $CaCl_2$
e) 1 mM $ZnCl_2$
f) 20 mM $ZnCl_2$ (*Is*=65 mM)
g) 40 mM $ZnCl_2$
h) 20 mM $ZnCl_2$ (*Is*=122 mM)*

The peak related to the DNA-DNA packing is marked by arrow.
Is is ionic strength of solution
* adapted from [15].

(Table 1) that was studied in [15]. We believe that only a fraction of lipid is bound by DNA and Zn^{2+} forming L^C phase, the rest of DPPC forms L^V phase, which shows the periodicity decreasing with increasing concentration of ions in solution, and which is macroscopically separated from L^C.

The driving force for mutual condensation of DNA by cationic vesicles to form an ordered, composite phase is the gain in electrostatic free energy. The electrostatic free energy depends on the surface charge densities of the separated macro-ions, the structure and composition of the condensed phases, and the salt concentration in solution [22]. A detailed analysis of the phase behaviour of DNA, cationic amphiphiles and neutral phospholipids mixtures can be found in [22-24]. If salt is added to the system, the mobile salt ions screen electrostatic interactions between fixed charges along DNA and the P^--N^+ dipole of phospholipid headgroups. High ionic strength can screen the DNA – lipid interaction up to level the aggregate dissolution [4]. Calcium has shown an ability to destroy and to rebuild lamellar organization of DPPC bilayers as a function of concentration [25-27]. Yamada et al. [26] detected so-called "unbound state" of DPPC bilayers (e.g. unilamellar vesicles) up to ~ 50 mM of $CaCl_2$ at 20 wt% of DPPC in the liquid-crystalline phase, and this phenomenon was dependent on the DPPC concentration. In the gel phase, the DPPC + Ca^{2+} mixture forms a lamellar phase with decreasing periodicity ~ 33 – 6.5 nm in the concentration range 4 – 400 mM of $CaCl_2$. The phenomenon was discussed through electrostatic cation – dipole interaction [28]. Figure 1h displays the diffractogram of DNA+DPPC in 20 mM of $ZnCl_2$, where the ionic strength of solution was modulated by NaCl (total ionic strength I_S=122 mM). The structural parameters (Table 1) are close to the system DNA+DPPC in 40 mM of $ZnCl_2$. Neither the repeat distance nor the lipid bilayer thickness of neutral phospholipid changed in 1–500 mM NaCl [29]. Evidently, the electrostatic screening of Zn^{2+} charge due to ions accumulation and formation of a diffuse double layer are responsible for the observed structural changes. It should be stressed that 0. 144 M of NaCl was found to destroy the L^C phase in DNA+DPPC+Mg^{2+} mixture [4].

Table 1. Structural parameters of DNA+DPPC+ion^{2+} aggregates

Cation	Concentration (mM)	d (nm)	d_{DNA} (nm)	d_{DPPC} (nm)	d^V (nm)
-	0	6.36±0.01	-	-	-
Ca^{2+}	1	8.04±0.01	-	6.49±0.01	-
Ca^{2+}	20	8.19±0.01	4.87±0.05	-	-
Ca^{2+}	50	8.33±0.01	6.08±0.01	-	-
Zn^{2+}	1	8.27±0.01	-	6.53±0.01	-
Zn^{2+}	20	8.48±0.01	5.25±0.01	-	-
Zn^{2+}	40	8.54±0.01	6.14±0.02	-	13.47±0.03
Zn^{2+}	50	8.54±0.01	6.38±0.01	-	12.19±0.01
Zn^{2+}	20*	8.18±0.01	5.70±0.02	-	13.12±0.03

* ionic strength I_S=122 mM, data from [15]

3.2. SANS on DPPC – ion^{2+}

We employed SANS to examine the effect of cations (Ca^{2+} and Zn^{2+}) on DPPC bilayer. Figure 2 shows typical scatttering curves. For dispersions of monodisperse centrosymmetric particles, the SANS intensity is given by

$$I(q) = N_P \cdot |F(q)|^2 \cdot S(q) \tag{1}$$

where q is the scattering vector q=4πsinθ/λ, 2θ the scattering angle, N_P the number density of particles, $F(q)$ their form factor and $S(q)$ the interparticle structure factor. The interparticle structure factor $S(q)$ is approximately equal to 1 for dilute and weakly interacting particles, what is a good approximation for unilamellar vesicles at PC concentrations <2 wt. % [30, 31]. The model of weakly interacting particles describes well unilamellar vesicles prepared by extrusion [32]. A typical SANS curve of unilamellar vesicles prepared by extrusion is shown in Figure 2a. Unilamellar vesicles are also formed spontaneously when the surface charge density is higher than 1-2 μC/cm^2 [33], e.g. multilamellar DPPC dispersion in 1 mM $CaCl_2$ completely converts into unilamellar vesicles [25, 34]. Figure 2b-e shows SANS curves of unilamellar DPPC vesicles formed spontaneously in solutions of studied ions.

The SANS curves were analyzed using a strip-function model of the phospholipid bilayer [35]. In this model, the bilayer is divided into three strips corresponding to two polar head group regions, each of thickness d_H, (one on each side of the bilayer) and the bilayer centre spanning the hydrocarbon region (with thickness d_C). The form factor $F(q)$ is the Fourier transform of the contrast $\Delta\rho(r)$ between the coherent neutron-scattering length density (SLD) of the bilayer and the solvent. For unilamellar vesicles bilayer, the strip model gives:

$$F(q) = 4\pi \sum_{i=1}^{3} \int_{R_{i-1}}^{R_i} \Delta\rho_i \frac{\sin(qr)}{qr} r^2 dr \qquad (2)$$

where $\Delta\rho_i(r)$ is the SLD contrast and $\Delta d_i = R_i - R_{i-1}$ is the thickness of the i-th strip. The total lipid bilayer thickness d_L is then given by $d_L = 2d_H + d_C$. In addition to d_L, the model includes the number of water molecules n_W per lipid located in the head group region of lipid, and the surface area per 1 lipid molecule on water lipid interface A_L. The obtained structural parameters of the DPPC bilayer and effect of both cations as a function of their concentration (at 20 °C) is summarized in Figure 2. With increasing concentration of zinc (Figure 2, empty symbols) we observe the increase in the DPPC bilayer thickness d_L, accompanied by the decrease in A_L and in the hydration of the polar head group region. For comparison, full symbols in Figure 2 represent data with calcium thoroughly discussed in our previous study [16]. Evidently, the effect of both ions on the DPPC bilayer is different. The observed changes follow from the nature of the ion itself. The electronic structure of zinc is different

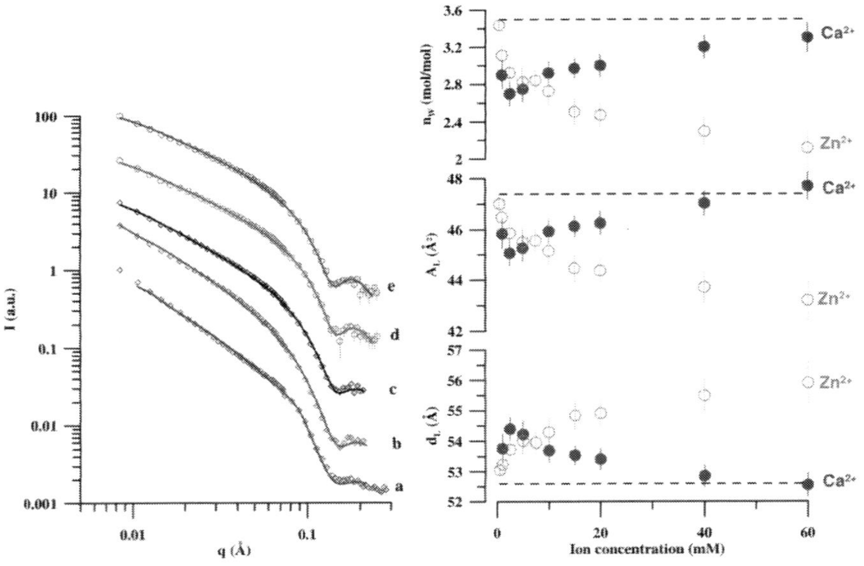

Figure 2. SANS curves of DPPC unilamellar vesicles (a) and DPPC vesicles prepared in 20 mM of $CaCl_2$ (b); and in $ZnCl_2$ at 0.5 mM (c), 5 mM (d) and 20 mM (e). Structural parameters of DPPC bilayer as a function of ion concentration: $ZnCl_2$ (empty symbols) and $CaCl_2$ [16] (full symbols).

from that of divalent alkali earth metal ions. Zinc cation possesses a higher affinity to electronegative groups [36], and therefore also other electronegative moieties such as ester oxygens and/or carbonyl groups of the lipid headgroup can be directly involved in complex formation. According to [37], Zn^{2+} bridges neighboring zwitterionic lipid molecules, forming a complex lipid : Zn^{2+} = 2 : 1, reaching the ratio 1 : 1 at saturation. The interaction of phospholipid headgroups with Zn^{2+} is conveniently described by a hydrated zinc-phosphate complex, which key energy contribution has a more covalent than electrostatic nature. Our analysis has shown the decrease in both, the area per lipid A_L and hydration of polar headgroup (n_W) of DPPC what supports the finding mentioned above. The dependence of d_L *vs. ion concentration* shows a course of "a classic adsorption", the most probably indicating a formation of a diffuse layer of the zinc and counter-ions at the DPPC surface.

The minima in electrostatic free energy of DNA – cationic liposomes complexes occur when the fixed negative charges on DNA surface are balanced by the same number of positive charges on the bilayer surface, i.e., at the isoelectric point. Because of the high mobility of metal cations, the evaluation of the isoelectric point in our system is not a trivial task. Our experiments do not allow us to determine the composition of the mixture at the isoelectric point. However, a short summarization can provide a picture "how compactly are the aggregates packed". We found the repeat distance of L_β^C phase $d = 8.51 \pm 0.03$ nm in the concentration range 20 – 50 mM of $ZnCl_2$. The steric lipid bilayer thickness of DPPC in 20 – 60 mM of $ZnCl_2$ at 20°C is $d_L = 5.54 \pm 0.05$ nm. Thus, for the interbilayer distance with the DNA accommodated, we get $d_W = d - d_L = 2.97 \pm 0.08$ nm. The diameter of DNA strands is 2 nm. The minimal interbilayer distance 2.6 nm is usually treated in theoretical works [22, 24], representing the short-range repulsive forces arising from hydration, protrusion, and other excluded volume interactions [38]. The packing of DNA in aggregates is thus rather "soft". The obtained d_W offers enough room for a layer of ions between DNA and the DPPC bilayer surfaces. It has been proved that millimolar concentrations of zinc cause the sedimentation of DNA [39]. Thus at higher zinc concentrations, binding sites of both, DNA and DPPC are saturated, and zinc does not mediate the binding. This saturation is responsible for the macroscopic phase separation observed in the studied mixtures, and the temperature instability of DNA + DPPC + Zn^{2+} phase particularly in aggregates with short fragmented DNA as shown in [15, 40].

4. Conclusion

As follows from our experiments, both ions mediate DNA binding to DPPC bilayer forming the condensed lamellar phase. With alkaline earth metals, in a large concentration range of ions, either the condensed lamellar phase with regularly packed DNA strands or coexistence of two phases have been detected at 20 °C [5, 6, 9, 21]. In aggregates with zinc, at low concentration ($\leq$ 20 mM $ZnCl_2$) we detected the condensed lamellar phase. High concentrations of zinc induce a macroscopic phase separation in the mixture, and in the addition to L_β^C phase, some volume fraction of lipid forms a partially ordered lamellar phase with periodicities significantly dependent on the ionic strength of solution. The observed structural polymorphism results from the nature of cations themselves, their affinity and way of binding to electronegative groups. Our experiments revealed that zinc at very low concentrations (of the order of millimoles) mediates the DNA-DPPC binding. These concentrations can be physiologically relevant, meaning that our experimental results could contribute to understanding of the role of zinc in biological processes.

Acknowledgements
Financial support provided by the European Community's Seventh Framework Program (FP7/2007-2013) under grant agreement no 226716 (HASYLAB project II-20100372 EC), by the JINR project 07-4-1069-09/2011, and by grants VEGA 1/0292/09 and APVV-0212-10 is gratefully acknowledged.

References

[1] Budker V G, Kazatchkov Y A and Naumova L P 1978 *FEBS Letters* **95** 143
[2] Tarahovsky Y S, Khusainova R S, Gorelov A V, Dawson K A and Ivanitskij G R 1996 *Dokl. Akad. Nauk* **350** 411
[3] Khusainova R S, Dawson K A, Rochev I, Gorelov A V and Ivanitskii G R 1999 *Dokl. Akad. Nauk* **367** 553
[4] Uhríková D, Rapp G and Balgavý P 2001 *Challenges for Coordination Chemistry in the new Century* ed Melník M and Sirota A (Slovak Technical University Press, Bratislava) pp. 219-224
[5] McManus J, Radler J O and Dawson K A 2003 *Langmuir* **19** 9630
[6] Francescangeli O, Stanic V, Gobbi L, Bruni P, Iacussi M, Tosi G and Bernstorff S 2003 *Physical Review E* **67** 011904
[7] Lengyel A 2010 *Interaction of DNA with phospholipid liposomes in presence of divalent cations* PhD Thesis (Faculty of Pharmacy, Comenius University, Bratislava, Slovakia)
[8] Lengyel A, Uhríková D, Klacsová M and Balgavý P 2011 *Colloids Surf. B* **86** 212
[9] Uhríková D, Lengyel A, Hanulová M, Funari S S and Balgavý P 2007 *Eur. Biophys. J.* **36** 363
[10] Christianson D W 1991 *Adv. Protein. Chem.* **42** 281
[11] Williams R J P 1988 *Zinc in human biology* (Springer-Verlag, London, UK) pp. 15-31
[12] Luck G and Zimmer C 1972 *Eur. J. Biochem.* **29** 528
[13] Eichhorn G L and Shin Y A 1968 *Journ. Am. Chem. Soc.* **90** 7323
[14] Duguid J, Bloomfield V A, Benevides J and Thomas G J, Jr. 1993 *Biophys. J.* **65** 1916
[15] Uhríková D, Pullmannová P, Bastos M, Funari S S and Teixeira J 2009 *Gen. Physiol. Biophys.* **28** 146
[16] Uhríková D, Kučerka N, Teixeira J, Gordeliy V and Balgavý P 2008 *Chem. Phys. Lipids* **155** 80
[17] MacDonald R C, MacDonald R I, Menco B Ph M, Takeshita K, Subbarao N K and Hu L R 1991 *Biochim. Biophys. Acta* **1061** 297
[18] Roveri N, Bigi A, Castellani P P, Foresti E, Marchini M and Strocchi R 1980 *Boll. Soc. Ital. Biol. Sper.* **56** 953
[19] Kučerka N, Uhríková D, Teixeira J and Balgavý P 2003 *Acta Fac. Pharm. Univ. Comenianae* **50** 78
[20] Radler J O, Koltover I, Salditt T and Safinya C R 1997 *Science* **275** 810
[21] Uhríková D, Hanulová M, Funari S S, Khusainova R S, Šeršeň F and Balgavý P 2005 *Biochim. Biophys. Acta* **1713** 15
[22] May S, Harries D and Ben Shaul A 2000 *Biophys. J.* **78** 1681
[23] Bruinsma R and Mashl J 1998 *Europhys. Lett.* **41** 165
[24] Harries D, May S, Gelbart W M and Ben Shaul A 1998 *Biophys. J.* **75** 159
[25] Inoko Y, Yamaguchi T, Furuya K and Mitsui T 1975 *Biochim. Biophys. Acta* **413** 24
[26] Yamada N L, Seto H, Takeda T, Naga M, Kawabata Y and Inoue K 2005 *J. Phys. Soc. Japan* **74** 2853
[27] Gorshkova J, Kuklin AI, Gordeliy V 2011 The influence of the Ca2+ ions concentration on the unbinding of the phospholipid membranes *5th European Conference on Neutron Scattering (Prague, Czech Republic, July 17-22 2011)* Abstracts, p. 89
[28] Izumitani Y 1994 *J. Coll Interface Sci.* **166** 143
[29] Pabst G, Hodzic A, Štrancar J, Danner S, Rappolt M and Laggner P 2007 *Biophys J.* **93** 2688
[30] Knoll W, Haas J, Sturhrmann H B, Fuldner H H, Vogel H and Sackmann E 1981 *J. Appl. Cryst.* **14** 191
[31] Nawroth T, Conrad H and Dose K 1989 *Physica B* **156&157** 477
[32] Balgavý P, Dubničková M, Uhríková D, Yaradaikin S, Kiselev M and Gordeliy V 1998 *Acta Physica Slovaca* **48** 509

[33] Hauser H 1993 *Phospholipid Handbook* ed Cevc G (Marcel Dekker, Inc., New York) pp. 603-637.
[34] Uhríková D, Teixeira J, Lengyel A, Almasy L and Balgavý P 2007 *Spectroscopy- Int. J.* **21** 43
[35] Kučerka N, Nagle J F, Feller S E and Balgavý P 2004 *Phys. Rev. E* **69** 051903
[36] Gresh N and Šponer J 1999 *J. Phys. Chem. B* **103** 11415
[37] Binder H, Arnold K, Ulrich A S and Zschornig O 2001 *Biophys. Chemistry* **90** 57
[38] Israelachvili J N and Wennerstrom H 1990 *Langmuir* **6** 873
[39] Kejnovsky E and Kypr J 1998 Nucleic Acid Res. **26** 5295
[40] Uhríková D, Pullmannová P, Kučerka N, Funari S S, Teixeira J and Balgavý P 2009 *Eur. Phys. J. Special Topics* **167** 191

Journal of Physics: Conference Series **351** (2012) 012012 doi:10.1088/1742-6596/351/1/012012

SANS contrast variation method applied in experiments on ferrofluids at MURN instrument of IBR-2 reactor

Maria Balasoiu[1,2], **Alexander Kuklin**[2]
[1]Horia Hulubei National Institute of Physics and Nuclear Engineering, Bucharest, Romania
[2]Joint Institute of Nuclear Research, Dubna, Russia

E-mail: balasoiumaria@yahoo.com; balas@nf.jinr.ru

Abstract. Separate determination of the nuclear and magnetic contributions to the scattering intensity by means of a contrast variation method applied in a small angle neutron scattering experiment of nonpolarized neutrons in ferrofluids in early 90's at the MURN instrument is reviewed. The nuclear scattering contribution gives the features of the colloidal particle dimensions, surfactant shell structure and the solvent degree penetration to the macromolecular layer. The magnetic scattering part is compatible to the models where is supposed that the particle surface has a nonmagnetic layer. Details on experimental "Grabcev method" in obtaining separate nuclear and magnetic contributions to the small angle neutron scattering intensity of unpolarized neutrons are emphasized for the case of a high quality ultrastabile benzene-based ferrofluid with magnetite nanoparticles.

1. Introduction

In the last more than four decades, studies of ferrofluids have attracted an increased interest (e.g. references [1-46]) due to their intrinsic physical properties of these materials as well as to their remarkable technological importance. Among some other experimental techniques (magnetization and hydrodynamic measurements, electron microscopy (EM), high resolution transmission electron microscopy (HRTEM), photon correlation spectroscopy, light scattering, X-ray small angle scattering, Mossbauer spectroscopy, inelastic scattering, neutron depolarization, muon spectroscopy, synchrotron radiation) the small angle neutron scattering (SANS) has been efficiently used for the determination of particle structure (see e.g. references [10-19]), structural and magnetic correlations (see e.g. references [10-35]), aggregation (see e.g. references [33-40]), dynamics (see e.g. references [20,41-43]), particle size distribution and polydispersity effects (see e.g. references [17,33-40,44,45]) in ferrofluids, of new stabilization compounds and methods [9,32,35, 46].

In the present paper, SANS experiments with the use of contrast variation method (for general information see references [47-49], studying the structure of magnetic colloidal particles, performed in 90's on the small angle scattering facility MURN [50,51] (named YuMO lately [51-54]) installed at the IBR-2 high pulsed reactor of the Joint Institute of Nuclear Research, Dubna, Russia, are reviewed.

Resulting from these experiments, it was established that within the Guinier [55] scattering range where the magnetic form factor, *f*, of the magnetite is approximately 1 with an error less than 1% [56], using the contrast variation method - when nuclear contrast changes the sign inside the D/H

Published under licence by IOP Publishing Ltd

concentration variation range, the contribution of nuclear and magnetic components of the experimental intensity can be obtained separately even in the case of the unpolarized neutrons [14-18]. This method now is known as the "Grabcev method" upon the name of the coordinator of the experiments.

2. Experimental

The magnetite ferrofluids were prepared [57] in D- and H-benzene with oleic acid surfactant. The size distribution obtained by EM analyzes showed log-normal form with a standard deviation of $\ln R$ $s = 0.3$ and the most probable radius $R_0 = 5.8$ nm. By dilution with measured quantities of D-benzene and/or H-benzene volume fraction $f_v = 0.6\%$ was obtained. The D-benzene volume fraction in the carrier x was produced to take six equidistant values in the range of 0.0 to 1.0.

The SANS experiments consisted in measuring, for each value of x, $I_f(Q)$, the intensity as a function of Q, the momentum transfer, $Q = (4\pi/\lambda)\sin(\theta/2)$, where λ is the wavelength of the neutron while θ is the scattering angle. Runs were also performed on all the compositions of the carrier liquids, free of particles, measuring in such way $I_c(Q)$. The determination from the experimental data of $d\Sigma_f(Q)/d\Omega$ and respectively $d\Sigma_c(Q)/d\Omega$, the differential elastic coherent scattering cross-sections per unit solid angle and unit sample volume, of the ferrofluids (subscript f) and, respectively, of the carrier liquid (subscript c), was performed according to the procedure described in [51]. The difference,

$$\frac{d\Sigma}{d\Omega}(Q) = \frac{d\Sigma_f}{d\Omega}(Q) - \frac{d\Sigma_c}{d\Omega}(Q),$$

represents the excess scattering power of colloidal particles.

3. Results and discussions

Neglecting interparticle correlations and size polydispersity, on one hand, and presuming approximately spherical particles, on the other, for values of Q small enough to ensure the Guinier's law to be valid and magnetite magnetic form factor $f \sim 1$, the differential scattering cross-section for unpolarized neutrons may be written as

$$\frac{d\Sigma}{d\Omega}(Q) = \frac{d\Sigma_n}{d\Omega}(Q) + \frac{d\Sigma_m}{d\Omega}(Q) \tag{1}$$

or, approximately,

$$\frac{d\Sigma}{d\Omega}(Q) \approx \frac{d\Sigma_n}{d\Omega}(0)\exp(-R_{gn}^2 Q^2/3) + \frac{d\Sigma_m}{d\Omega}(0)\exp(-R_{gm}^2 Q^2/3) \tag{2}$$

The first term in the right side of the above equations represents the nuclear component of the cross-section while the second, the magnetic one. The quantities R_{gn} and R_{gm} are corresponding gyration radii while

$$\frac{d\Sigma_n}{d\Omega}(0) = N_p \rho^{*2} V_d^2 \tag{3}$$

$$\frac{d\Sigma_m}{d\Omega}(0) = (2/3) N_p \rho_{mm}^2 V_m^2 \tag{4}$$

and, $N_p = 3 \cdot 10^{15}$ particles/cm^2 is the particle concentration; V_d - represents the volume of the magnetite particle with the adsorbed surfactant macromolecules on its surface; V_m - represents the magnetic volume; it is considered having a constant magnetization value.

ρ^* - represents the contrast [47-49]; to calculate its expression, following conditions are considered:

The colloidal particle is central-symmetrical and composed from a magnetite core surrounded by a surfactant shell ;

The surfactant shell has a nonuniform structure; partially, it can be penetrated by the carrier and empty spaces;

The scattering length densities inside the volumes occupied by magnetite particles, surfactant and carrier, are uniform.

and,

$$\rho^* = (1/V_d)\int_v [\rho_n(\vec{r}) - \rho_c] d\vec{r} =$$

$$= (4\pi/V_d)[\int_0^{R_0}(\rho_{mn} - \rho_c)r^2 dr + \phi\int_{R_0}^{R_t}(\rho_s - \rho_c)r^2 dr + \gamma\int_{R_0}^{R_t}(\rho_{mn} - \rho_c)r^2 dr +$$

$$+(1-\phi-\gamma)\int_{R_0}^{R_t}(0-\rho_c)r^2 dr] = 1/V_d\{(\rho_{mn} - \rho_c)V_0 + [\phi\rho_s - (1-\gamma)\rho_c]V_s\} \quad (5)$$

ρ_c is the scattering length density of the carrier:

$$\rho_c = x\rho_{C_6D_6} + (1-x)\rho_{C_6H_6}$$

where $\rho_{C_6D_6}$ and $\rho_{C_6H_6}$ are the scattering length densities of deuterated and, respectively, hydrogenated benzene; ρ_{mm} is the magnetic scattering length density of the magnetite core; V_p is the total volume of a colloidal particle, including the surfactant layer; V_m is the magnetic active volume, i.e. the volume with a nonvanishing magnetization density; the numerical ratio $2/3$ in equation (4) is the polarization factor for unpolarized neutrons.

In Figure 1, the radial distribution of ρ_n (density of nuclear coherent scattering length), inside as well as outside a colloidal particle, is shown. The notations ρ_{mn} and ρ_s are the nuclear scattering length densities of magnetite and, respectively, oleic acid surfactant. The ultimate values of carrier scattering length density are shown to illustrate the basic concept of the contrast variation method. The radial distribution of ρ_{mm}, the magnetic coherent scattering length density for the Guinier region where the magnetic form factor $f \sim 1$ with an error less than 1%, is plotted in Figure 2. Here R_m, the magnetic radius of the magnetite core may be smaller than R_0. At the same time, there has been assumed that for $r < R_m$ the magnetization density acquires the same value as in the bulk state. For the calculation of ρ_{mm}, a mass density $d_m = 5.2$ g/cm^2 and a magnetic moment $\mu = 4.2$ Bohr magnetons/molecule [58] are considered for Fe_3O_4.

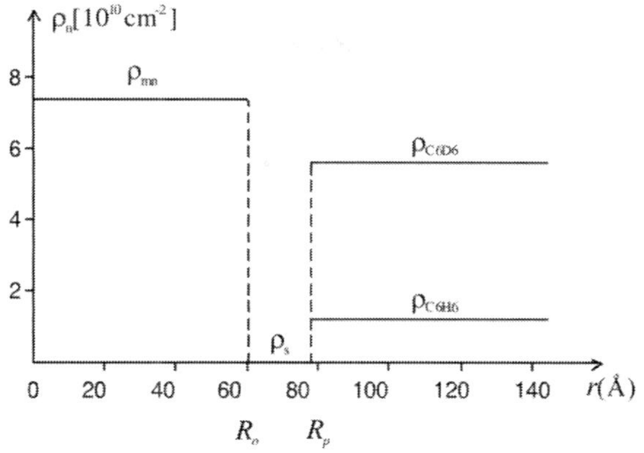

Figure 1. Method of contrast variation: ρ_n - is the radial distribution of the density of nuclear scattering of the particle; ρ_s - is the radial distribution of the nuclear scattering density of the surfactant; ρ_{C6D6} and ρ_{C6H6} - are the radial distributions of the nuclear scattering density of the carrier - deuterate benzene, respectively of hydrogenate benzene.

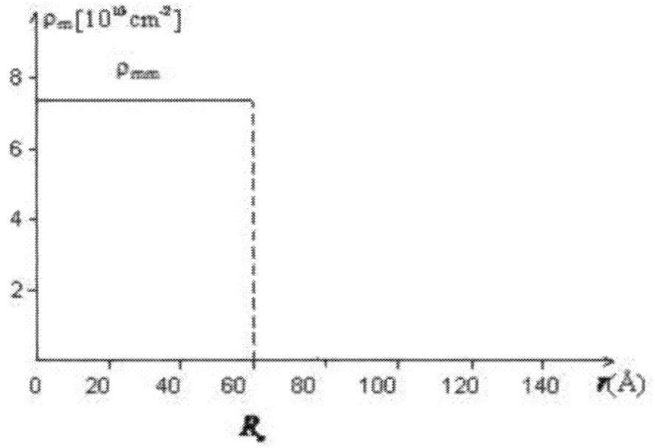

Figure 2. Radial distribution of the magnetic scattering density of the particle - ρ_m for the momentum transfer Q = 0, where the magnetic form factor has a maximal value.

The numerical values of scattering length densities (Figure 1) are:
$\rho_{mn} = 6.977 \cdot 10^{10}\text{cm}^{-2}$; $\rho_s = 0.077 \cdot 10^{10}\text{cm}^{-2}$; $\rho_{C_6H_6} = 1.182 \cdot 10^{10}\text{cm}^{-2}$; $\rho_{C_6D_6} = 5.437 \cdot 10^{10}$ cm^{-2} and $\rho_{mm} = 3.068 \cdot 10^{10}\text{cm}^{-2}$.

In Figure 3, the $d\Sigma(0)/d\Omega$ obtained by extrapolation of experimental data, is plotted against x. A minimum appears here, in the vicinity of $x = 0.6$, one of the values used in the experiment for C_6D_6 volume fraction in the carrier. An inspection of equations (1)-(4) shows that only the first term, namely that corresponding to nuclear scattering, depends of x. One may conclude hence that at $x = x_0$ the particle contrast, ρ^* as a linear function of x, changes the sign.

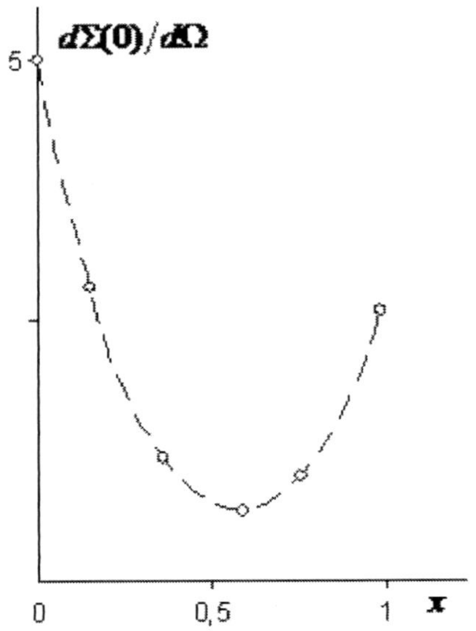

Figure 3. Dependence of $d\Sigma(0)/d\Omega$ on x, the D-benzene value fraction in the carrier. There appears a minimum at $x = 0.6$. The dashed line is just an approximate guide for the eye.

However, although theoretically it is not possible to equal rigorously to zero the contrast value of an inhomogeneous particle embodied in a homogenous medium, the basic hypothesis of the following model data treatment is that at $x = x_0$ the nuclear contrast, ρ^*, is negligibly small as compared with the value of the magnetic scattering length density, ρ_{mm}, so that at this point the residual contribution of nuclear scattering cross section may be disregarded, or in other words

$$\left.\frac{d\Sigma}{d\Omega}(0)\right|_{x=x_0} = \frac{d\Sigma_m}{d\Omega}(0) \qquad (6)$$

And, according to equations (2) and (3) and taking into account the independence of the magnetic scattering cross-section upon x, one obtains

$$\frac{d\Sigma_m}{d\Omega}(Q) = \left.\frac{d\Sigma}{d\Omega}(Q)\right|_{x=x_0} \qquad (7)$$

Consequently, in virtue of equations (1)-(3), within the Guinier's scattering range, one may consider that for any value of x

$$\frac{d\Sigma_n}{d\Omega}(Q) = \frac{d\Sigma}{d\Omega}(Q) - \left.\frac{d\Sigma}{d\Omega}(Q)\right|_{x=x_0} \qquad (8)$$

Equation (8) allows separating from experimental data the nuclear component of the scattering cross-section. Further, by plotting $\ln[\frac{d\Sigma_n}{d\Omega}(Q)]$ versus Q^2, one can establish the limits of the Guinier's range for every value of x (Table 1).

Table 1. The limits of the Guinier's range for every value of x

X	$Q^2_{min}(nm^{-2})$	$Q^2_{max}(nm^{-2})$
0.0	0.02272	0.2218
0.2	0.02272	0.2218
0.4	0.02272	0.2218
0.6	0.02272	0.2218
0.8	0.02272	0.1247
1.0	0.02272	0.1571

Extrapolation of the Guinier's plots to $Q = 0$ gives $\frac{d\Sigma_n}{d\Omega}(0)$, and the square root of this, $[\frac{d\Sigma_n}{d\Omega}(0)]^{1/2}$, is drawn in Figure 4 as a function of x. The straight line representing the linear fit obtained by least squares method, has the following equation:

$$[\frac{d\Sigma_n}{d\Omega}(0)]^{1/2} = 2.20(\pm 0.08) - 3.60(\pm 0.14)x$$

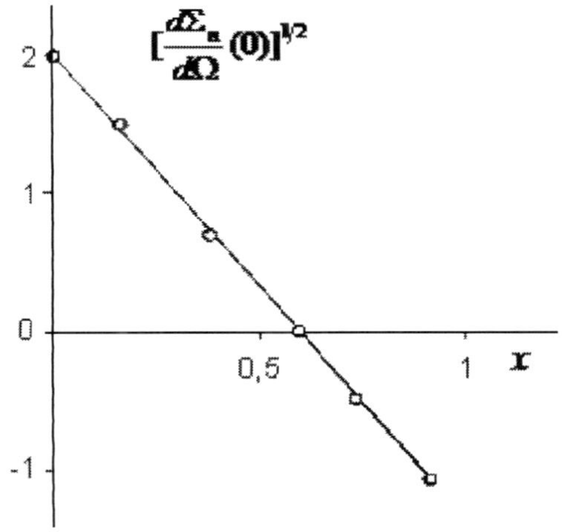

Figure 4. Representation of $[\frac{d\Sigma_n}{d\Omega}(0)]^{1/2}$ as a function of x, together with the best straight line fit. The change of sign follows from the equation (3); actually, the quantity $N_p^{1/2}\rho^* V_d$ is shown in the figure. Its value vanishes at $x = x_0 = 0.6$. The errors are smaller than the dimensions of the dots (circles).

The linear dependence on x of $[\frac{d\Sigma_n}{d\Omega}(0)]^{1/2}$, on one hand, and the vanishing value of this quantity at $x = x_0 = 0.6$, on the other, allows confirming the validity of the experimental data interpretation method.

For the interpretation of experimental data, a central-symmetrical colloidal particle with an impenetrable core of magnetite surrounded by a surfactant layer is considered. The volume of the layer is $V_s = (4\pi/3)[(R_0+\delta)^3 - R_0^3] = 1.19\cdot10^{10}\text{cm}^3$; $\delta = 20$ Å is the length of an oleic acid molecule (Figure 5).

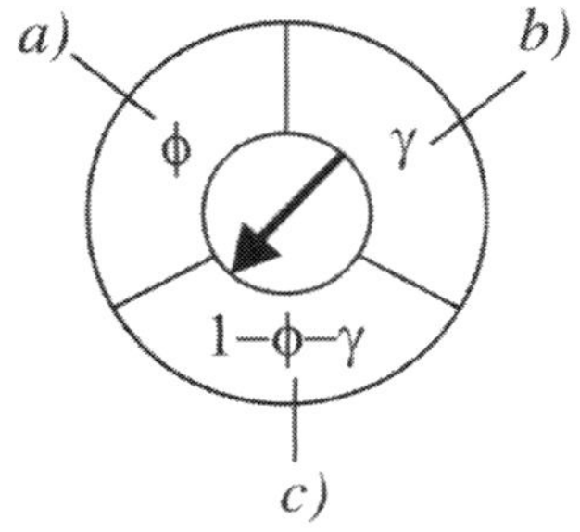

Figure 5. Diagram of the volume fraction distribution of the components of the outer shell on the surface of the particle; (a) ϕ, the volume fraction of the surfactant; (b) γ, the volume fraction of the carrier penetrated inside the surfactant shell; (c) $(1-\phi-\gamma)$, the empty volume fraction inside the surfactant shell.

At the same time, is assumed a nonuniform structure of the surfactant shell. So, it may be penetrated by the carrier and/or partially unoccupied. Assuming uniform densities inside the volume occupied by magnetite, surfactant and carrier and introducing ϕ, the volume fraction (referred to V_s) of the surfactant and γ, the volume fraction of the carrier, taking into account the definition of ρ^* from equation (5) and the value of ρ^* for $x = x_0 = 0.6$, $\rho^* = 0$, the following expression can be written:

$$(\rho_{mn} - \rho_c)V_0 + [\phi\rho_s - (1-\gamma)\rho_c]V_s = 0$$

Replacing with the given values ρ_{mn}, V_0 the particle volume calculated from EM analysis data and V_s and $\rho_c(x_0) = 3.777\cdot10^{10}\text{cm}^{-2}$ for $x_0 = 0.6$, and considering $\phi\rho_s$ negligible small, γ, the volume fraction of the carrier inside the surfactant shell is obtained:

$$\gamma = 0.25 \pm 0.1$$

Further the volume and radius of the colloidal particle extracted from the carrier V_d and R_d are obtained:

$$V_d = V_0 + (1-\gamma)V_s$$

$$R_d = (3V_d/4\pi)^{1/3} = (73.9 \pm 0.3)\cdot10^{-8}\,\text{cm}$$

The value of the magnetite particle radius will be:

$$R_p = R_d - \delta = (5.39 \pm 0.03)\,\text{nm}$$

where δ is the length of the $C_{18}H_{34}O_2$ molecule adsorbed on the particle surface.
The rather high error in γ is caused by the particle size polydispersity. It should be noted, on the other hand, that in the present case, due to the very small value of ρ_s it is not possible to distinguish, in the total volume of the surfactant layer, the volume fraction ϕ of the surfactant and, correspondingly, the volume fraction $(1-\phi-\gamma)$ of the empty space. However, theoretical methods of statistical mechanics (see, e.g., reference [1]) give $\phi \approx 0.3$.
From the expression of the magnetic component of the differential scattering cross-section for $Q = 0$ and the value of $d\Sigma(0)_m / d\Omega = 0.635\,\text{cm}^{-1}$, the magnetic volume V_m and magnetic radius R_m may be inferred:

$$R_m = (3V_m / 4\pi)^{1/3} = (4.32 \pm 0.05)\,\text{nm}$$

These estimations indicate the existence of a nonmagnetic layer of thickness $t_{nm} = 1\,\text{nm}$ at the surface of the magnetite particle. Similar results have been reported in references [41,59].

4. Conclusions

Finally, from the results presented above, some conclusions may be drawn.

Within the Guinier scattering range, where the magnetic form factor, *f*, of the magnetic nanoparticle is approximately 1 with an error less than 1%, using the contrast variation method when nuclear contrast changes the sign inside the concentration variation range, the contribution of nuclear and magnetic components of the experimental intensity can be established separately even in the case of unpolarised neutrons. This result, i.e. that these components can be distinguished with a contrast variation small angle neutron scattering, method, was first time reported in 90's in the following papers [14-18] on SANS data from experiments accomplished at MURN instrument functioning at IBR-2 high pulsed reactor.

The values of physical and magnetic dimensions of the magnetic nanoparticles (from the same type of ferrofluids) obtained with the method presented above where confirmed lately by EM, magnetic measurements, SANSPOL and muon spectroscopy analysis and reported in several papers [44,61-65].

The surfactant layer has a rather complex structure. So, it is only partially occupied and may be penetrated by the carrier liquid. However, the negative coherent scattering length of the hydrogen leads to a relative small overall scattering length of the surfactant (which is mostly hydrogenated material). In some cases, as in the present one, this value is extremely small hindering the determination of the surfactant volume fraction. A solution of this problem in further SANS contrast variation investigations could be the deuteration of the surfactant.

The unavoidable size polydispersity of particles causes rather high errors in the determination of surfactant, carrier and empty space volume fractions in the outer shell of the magnetite core.

The results are compatible to the models where is supposed that the particle surface has a nonmagnetic layer. This might be caused accordingly to reference [59], by magnetite-surfactant interactions. In latest publications, however, such behaviour is treated as an intrinsic property of nonstructured magnetic materials (see, e.g., reference [60]).

Acknowledgements
This review is dedicated to the memory of Dr. Boris Grabcev, Dr. Doina Bica and Prof. Dorel Bally.

References

[1] Rosensweig R E 1985 *Ferrhydrodynamics* (Cambridge University Press)
[2] Bashtovoj V G, Berkovskij B M and Vislovich A N 1985 *Magnitnie Zhidkosti* (Minsk)

[3] Fertman V E 1988 *Magnitnie Zhidkosti* (Minsk:Visheisheia Shkola)
[4] Blums M Ya, Maiorov M M and Tsebers A O 1989 *Magnitnije Zhidkosti* (Riga: Zinatne)
[5] Taketomi S, Tikadzumi S 1993 *Magnitnie Zhidkosti* (Moskva: Mir)
[6] Luca E, Calugaru, Badescu R, Cotae C and Badescu V 1978 *Ferofluidele si aplicatiile lor in industrie* (Bucuresti:Ed.Tehnica, Bucuresti)
[7] S. Odenbach 2002 Ferrofluids. Magnetically Controllable Fluids and Their Applications *Lecture Notes in Physics* 594 (Berlin:Springer) p 233
[8] Vekas L, Avdeev M V and Bica D 2008 Magnetic Nanofluids: Synthesis and Structure *Nanoscience and Its Applications in Biomedicine* ed Donglu Shi (Berlin: Springer Verlag and Tsinghua University Press) chapter **25** p 645
[9] Bica D, Vekas L, Avdeev M V, Marinica O, Socoliuc V, Balasoiu M, Garamus V M 2007 *J. Magn. Magn. Mater.* **311** 17
[10] Cebula D J, Charles S W and Popplewell J 1981 *Colloid. Polym. Sci.* **259** 395
[11] Cebula D J, Charles S W and Popplewell J 1983 *J. Magn. Magn. Mater.* **39** 67
[12] Itoh S, Endoh Y, Taketomi S and Chikazumi S 1992 *J. Magn. Magn. Mater.* **103** 126
[13] Boue F, Cabuil V, Bacri J-C and Perzynski R 1993 *J. Magn. Magn. Mater.* **122** 78
[14] Grabcev B, Balasoiu M, Bica D and Kuklin A I 1994 *Magnitnaia Gidrodinamica* **10**(2) 156
[15] Balasoiu M, Grabcev B and Bica D 1995 *Rom. Rep. Phys.* **47**(3-5) 319
[16] Balasoiu M 1998 On the microstructure of magnetic fluids *PhD Thesis* (Institute of Atomic Physics, Bucharest)
[17] Grabcev B, Balasoiu M, Tarziu A, Kuklin A I and Bica D 1999 *J. Magn. Magn. Mater.* **201** 140
[18] Balasoiu M, Avdeev M V, Aksenov V L, Ghenescu V, Ghenescu M, Torok Gy, Rosta L, Bica D, Vekas L and Hasegan D 2004 *Rom. Rep. Phys.* **56**(4) 601
[19] Balasoiu M, Avdeev M V, Aksenov V L, Hasegan D, Garamus V M, Schreyer A, Bica D and Vekas L 2006 *J. Magn. Magn. Mater.* **300**(1) e225
[20] Aksel'rod L A, Gordeev G P, Drabkin G M, Lazebnik I M and Lebedev V T 1986 *Zh. Eksp. Teor. Fiz.* **91** 531 [1986 *Sov. Phys. JETP* **64** 312]
[21] Avdeev M V, Balasoiu M, Bica D, Rosta L, Torok Gy and Vekas L 2001 *Mat. Sci. Forum* **373-376** 457
[22] Avdeev M V, Balasoiu M, Torok Gy, Bica D, Rosta L, Aksenov V L and Vekas L 2002 *J. Magn. Magn. Mater.* **252** 86
[23] Aksenov V L, Avdeev M V, Balasoiu M, Rosta L, Torok Gy, Vekas L, Bica D, Garamus V and Kohlbrecher J 2002 *Appl. Phys. [Suppl.]* **A 74** 943
[24] Balasoiu M, et al 2004 *Magnetohydrodynamics* **40**(4) 359
[25] Avdeev M V, Balasoiu M, Aksenov V L, Garamus V M, Kohlbrecher, Bica D and Vekas L 2004 *J. Magn. Magn. Mater.* **270** 371
[26] Balasoiu M, Avdeev M V, Kuklin A I, Aksenov V L, Hasegan D, Garamus V, Schreyer A, Bica D, Vekas L and Almasan V 2006 *Rom. Rep. Phys.* **58**(3) 30
[27] Török Gy, Len A, Rosta L, Balasoiu M, Avdeev M V, Aksenov V L, Ghenescu V, Hasegan D, Bica D and Vekas L 2006 *Rom. Rep. Phys.* **58**(3) 293
[28] Avdeev M V, Aksenov V L, Balasoiu M, Garamus V M, Schreyer A, Torok Gy, rosta L, Bica D and Vekas L 2006 *J. Colloid Interface Sci.* **295**(1) 100
[29] Balasoiu M, Avdeev M V and Aksenov V L 2007 *Crystallogr. Rep.* **52**(3) 505
[30] Avdeev M V 2007 *J. Appl. Cryst.* **40** 56
[31] Vekas L, Bica D and Avdeev M V 2007 China Particuology **5** 43; see also: Vekas L 2004 *Rom. J. Phys.* **49**(9-10) 707
[32] Avdeev M V, Dubois E, Meriguet G, Wandersman E, Garamus V M, Feoktystov A V and Perzynski R, 2009 *J. Appl. Cryst.* **42** 1009
[33] Avdeev M V, Bica D, Vekas L, Aksenov V L, Feoktystov A V, Marinica O, Rosta L, Garamus V M and Willumeit R 2009 *J. Colloid Interface Sci.* **334** 37; see also: Avdeev M V, Bica D, Vekas L, Aksenov V L, Feoktystov AV, Rosta L, Garamus V M and Willumeit R 2009 *Solid State*

Phenomena **152-153** 182; see also Feoktystov A V, Bulavin L A, Avdeev M V, Vekas L, Garamus V M and Willumeit R 2009 *Ukr. J. Phys.* **54** 266; see also Feoktystov A V, Avdeev M V, Aksenov V L, Bulavin L A, Bica D, Vekas L, Garamus V M and Willumeit R 2009 *Poverkhnost. Rentgen. Sinkhrotron. Neitron. Issled.* **1** 3 [2009 *J. Surf. Invest. X-Ray, Synchr., Neutron Tech.* **3** 1]; see also Feoktystov A V, Avdeev M V, Aksenov V L, Petrenko V I, Bulavin L.A, Bica D, Vekas L, Garamus V.M and Willumeit R 2009 *Solid State Phenomena* **152-153** 186
[34] Avdeev. M V and Aksenov V L 2010 Uspehi Fizicheskih Nauk **180**(10) 1009 [2010 *Physics Uspekhi* **53** (10) 971]
[35] Avdeev M V et al 2010 *J. Appl. Cryst.* **43** 959
[36] Cebula C D, Charles S W and Popplewell J 1983 *J. Phys.(Paris)* **44** 207.
[37] Pynn R, Hayter J B and Charles S W 1983 *Phys.Rev.Lett.* **51** 710
[38] Hayter J B, Pynn R, Charles S W, Skjeltorp A T, Trewhella J, Stubbs G and Timmins P 1989 *Phys.Rev.Lett.* **62** 1667
[39] Rosman R, Jansen J J M and Rekveldt M Th 1990 *J.Magn. Magn. Mater.* **85** 97
[40] Aksenov V L, Avdeev M V, Balasoiu M, Bica D, Rosta L, Torok Gy and Vekas L 2003 *J. Magn. Magn. Mater.* **258-259** 452
[41] Lebedev V T, Gordeev G P, Panasiuk E A, Kiss L, Cser L, Rosta L, Torok Gy and Farago B 1993 *J. Magn. Magn. Mater.* **122** 83
[42] Balasoiu M, Vekas L, Avdeev M V, Aksenov V L, Khokhryakov A A, Bica D, Hasegan D, Torok Gy, Rosta L 2005 *Rom.Rep. Phys.* **57**(2) 261
[43] Odenbach S., Schwahn D, Stierstadt K 1995 *Z Phys* B **96** 567
[44] Avdeev M.V, Bica D, Vekas L, Marinica O, Balasoiu M, Aksenov V L, Rosta L, Garamus V M and Schreyer A 2007 *J. Magn. Magn. Mater.* **311** 6
[45] Aksenov V L, Avdeev M V, Shulenina, Zubavichus V V, Velizhanin A A, Rosta L, Garamus V M and Vekas L 2011 *Crystallogr. Rep.* **56**(5) 792
[46] Avdeev M V, Mucha B, Lamszus K, Vekas L, Garamus V M, Feoktystov A V, Marinica O, Turcu R and Willumeit R 2010 *Langmuir* **26**(11) 8503
[47] Stuhrmann H B, and Kirste R G 1965 *Z. Phys. Chem.* **46** 247
[48] Stuhrmann H B 1976 J.Appl.Cryst. **7** 173 [see also Stuhrmann H. B. 1995 *Modern Aspects of Small-Angle Scattering* edited by Brumberger H (Dordrecht: Kluwer Academic Publishers) p 221
[49] Jacrot B 1976 *Rep.Progr.Phys.* **39** 911
[50] Neutron Experimental Facilities at JINR 1991 *User Guide* (Dubna USSR)
[51] Bezzabotnov V Yu and Ostanevich Yu M 1989 *Physica B* **156-157** 595
[52] Kuklin, A I, Islamov A Kh and Gordely V I 2005 *Neutron News* **16**(3) 16
[53] Kuklin A I, Islamov A Kh, Kovalev Yu S, Utrobin P K and Gordely V I 2006 *J. Surf. Invest. X-Ray, Synchr., Neutron Tech..* **6** 74
[54] Kuklin A I et al 2011 *J. Phys: Conference Series* **291**(1) 012013
[55] Guinier A and Fournet G 1955 *Small-Angle Scattering of X-rays* (New York: John Wiley and Sons)
[56] Bacon G 1975 *Neutron Diffraction. (*Publishing House Claredon Press Oxford)
[57] Bica D *Rom. Rep. Phys.* **47** 265; see also Bica D 1985 *RO Patent 90078*
[58] Gorter E M 1954 *Phylips Res. Rep.* **9** 403
[59] Nunes A C, Majkrzak C F and Berkowitz A E 1983 *J. Magn. Magn. Mater.* **39** 59
[60] Wagner W, Wiedenmann A, Petry W, Geibel A and Gleiter H 1991 *J. Mater. Res.* **6** 2305
[61] Rasa M 2000 *Eur. Phys. J.* **E 2** 265
[62] Vekas L, Rasa M and Bica D 2000 *J.Colloid.Int.Sci.* **231** 247
[63] Caizer C 2003 *J. Phys.:Condens. Matter.* **15** 765
[64] Balasoiu M et al 2008 *JETF Lett* **88**(3) 243
[65] Stan C, Cristescu C P, Balasoiu M, Perov N, Duginov V, Mamedov T and Fetisov L 2011 *U.P.B. Bull. A* **73**(3) 117

Journal of Physics: Conference Series **351** (2012) 012013 doi:10.1088/1742-6596/351/1/012013

A study of formation of iron nanoparticles in aluminium matrix with helium pores.

S E Kichanov[1], D P Kozlenko[1], A V Belushkin[1], V F Reutov[2], S O Samoilenko[1,3], Z Jirák[4], B N Savenko[1], L A Bulavin[3] and Y V Zubavichus[5]

[1]Frank Laboratory of Neutron Physics, Joint Institute for Nuclear Research, 141980, Dubna, Moscow Reg., Russia
[2]Flerov Laboratory of Nuclear Reactions, Joint Institute for Nuclear Research, 141980, Dubna, Moscow Reg., Russia
[3]National Taras Shevchenko University of Kyiv, Physics Department, 01601, Kiev, Ukraine
[4]Institute of Physics, Cukrovarnická 10, 162 53 Prague 6, Czech Republic
[5]RRC "Kurchatov Institute", 123182 ,Moscow, Russia

ekich@nf.jinr.ru

Abstract. The structural and magnetic properties of iron nanoparticles formed in an aluminium matrix by combined helium and iron ions irradiation have been studied by small-angle X-ray scattering and susceptibility measurements. The formation of nanoparticles with average size of about 6.8 nm under iron ions irradiation has been found. The additional posterior irradiation with helium ions leads to the iron nanoparticles ordering with correlation length of 7.2(1) nm. The characteristic radius of iron nanoparticles and fractal dimension of the Al matrix were determined and magnetic properties of the Fe-Al nanomaterials were analysed.

1. Introduction

In recent years nanostructured materials have received increased attention due to their enhanced magnetic, mechanical and optical properties when the particle size is reduced to a few nanometers [1]. Improved magnetic properties like large coercitivity, giant magnetoresistance, superparamagnetism and high Curie temperatures are observed in nanostructured magnetic materials compared to their bulk counterparts [2]. These materials have significant potential for a wide range of applications such as catalysis, magnetic recording media, optoelectronics, magnetic fluids, fuel cells, sensors, etc. [3].

It is important to have tools for searching efficient and inexpensive techniques for production of nanostructured magnetic materials with control over composition, nanoparticles sizes and distribution. One of promising methods is the irradiation of materials by high energy charged particles [4]. This allows obtaining uniform volume concentration of the implanted ions in the matrix and provides possibility of doping any matrix by practically any element including chemically non-soluble in this matrix. Because the formation of nanoparticles is governed by fluctuation mechanism, one of the drawbacks of this method is the high-dispersed of nanoparticles formed [4]. Recently this was solved [5] by the additional pre-irradiation of the host matrix by inert gases ions forming the uniform in size and spatially ordered pores in the matrix which can effectively trap the metal ions thus forming the monodispersed ordered nanoparticles [5]. The size and the density of nanopores can be controlled by varying the irradiation conditions such as a dose, intensity of the ions or temperature of the matrix.

The aluminum is a preferable material for the matrix due to its excellent corrosion and wear resistance, chemical stability. In the present work we focused on the study of structure aspects of iron nanoparticles formation in aluminum foils by helium and iron ions irradiation processes.

2. Experimental

The irradiations of aluminium matrix by helium and iron ions were carried out at ECR-1 cyclotron (Joint Institute for Nuclear Research, Flerov Laboratory of Nuclear Reactions, Dubna). The sample irradiation characteristics are presented in Table 1.

Table 1. The irradiation characteristics of different aluminium samples.

Sample	The irradiation process characteristics	The corresponding ions energy and doses	
1	Helium ions only	$E(He^+) = 20$ keV, Dose $=5x10^{17}cm^{-2}$	
2	Helium ions and subsequent iron ions	$E(He^+) = 20$ keV, Dose $=5x10^{17}cm^{-2}$	$E(Fe^+) = 150$ keV, Dose $=1x10^{15}cm^{-2}$
3	Iron ions only	$E(Fe^+) = 150$ keV, Dose $=1x10^{15}cm^{-2}$	
4	Iron ions and subsequentt helium ions	$E(He^+) = 20$ keV, Dose $=5x10^{17}cm^{-2}$	$E(Fe^+) = 150$ keV, Dose $=1x10^{15}cm^{-2}$

The small-angle X-rays scattering (SAXS) experiments were carried out using the Structural Materials Science end-station [6] at the Siberia-2 synchrotron radiation source in RRC "Kurchatov Institute" (Moscow, Russia). The vacuum small-angle scattering chamber has a few sample loading ports at different distances to the detector for covering different scattering vector range. The minimum value of the accessible scattering vector is about 0.05 nm^{-1}. The SAXS data were recorded using a one-dimensioned gas-filled detector. In all spectra obtained, the background and detector anisotropy correction were carried out.

The magnetic DC susceptibility measurements were performed using a SQUID magnetometer. The temperature dependences of the susceptibility were measured in the magnetic field of 1 T. The magnetic field dependences of the susceptibility up to 5 T were measured at $T = 5$ K.

3. Results and discussion

The small-angle X-rays scattering curves of measured samples Nos. 1-4 (fig. 1) exhibit clear distinctions. For the fitting of the SAXS data we used structural model [7, 8] involving contribution from both Al matrix and nanoparticles formed inside the matrix:

$$d\Sigma(q)/d\Omega = Aq^{-\alpha} + B\exp(-\frac{1}{3}q^2R_g^2) + D \quad (1),$$

where the first term is due to fractal structure of the matrix with fractal dimension α. The second term is the Guinier term and depends on iron nanoparticles size R_g. The B is Guinier prefactor which depends on number density of nanoparticles and A, the scaling prefactor, signifies the strength of power-law scattering arising from the fractal structure [8, 9]. The calculated values of α and R_g for different aluminium foils are listed in Table II.

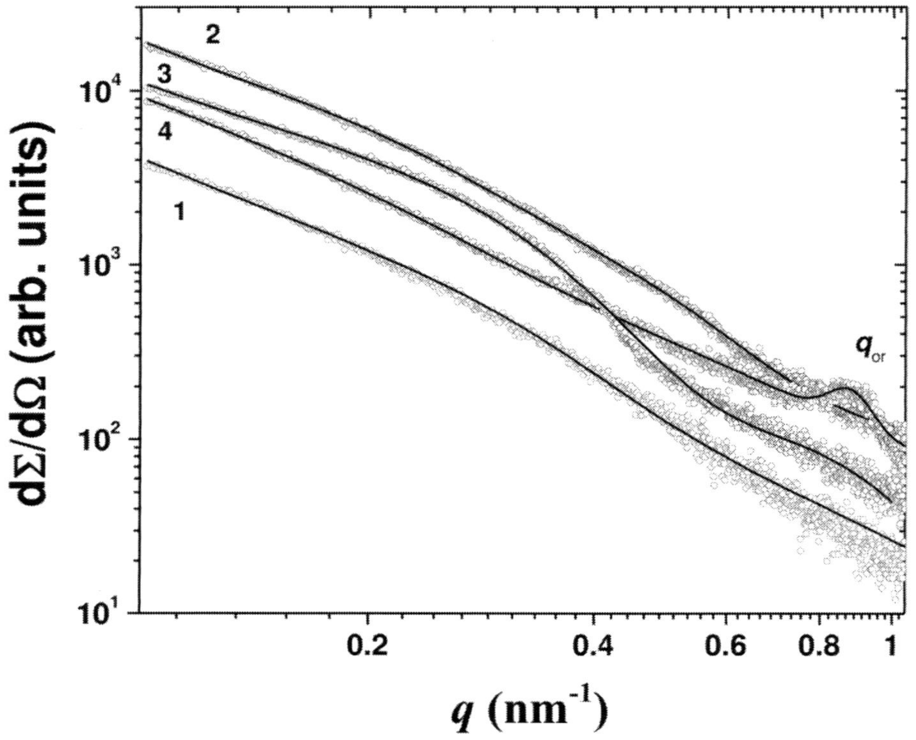

Figure 1. The small angle X-ray scattering curves for irradiated aluminium foils. The solid lines are experimental data fit by equation (1) (curves 1-3) and (2) (curve 4). The "q_{or}" indicate the ordering peak for sample 4.

The scattering curve related to aluminium matrix irradiated by He^+ ions only (sample 1) corresponds to scattering from pure helium pores.

Table 2. The structural and magnetic properties of Fe nanoparticles in aluminium foils

Sample	Gyration radius R_g, nm	Fractal dimension α	Bloch degree *b*
Sample 1	6.95(1)	2.43(2)	1.27(4)
Sample 2	10.4(1)	1.94(1)	1.62(3)
Sample 3	7.10(2)	2.10(3)	1.58(3)
Sample 4	6.80(2)	2.12(4)	1.58(2)

The application of Fe ions irradiation increases the total scattering intensity due a scattering length change [10] in the nanosystems formed. The SAXS curve corresponding to the sample 2 with aluminium matrix irradiated with helium ions followed by irradiation with iron ions is characterized by maximal scattering intensity. This corresponds to the maximal iron nanoparticles characteristic size (Table 1).

The SAXS curve of the sample 4 with aluminium matrix irradiated by iron ions with subsequent helium ions irradiation demonstrates the additional peak at q=0.87 nm^{-1} (fig. 1). This feature can be explained by iron nanoparticles ordering in aluminium matrix. For the fitting of this curve the modified equation (1) have been used:

$$d\Sigma(q)/d\Omega = Aq^{-a} + B\exp(-\frac{1}{3}q^2R_g^2) + C\exp\left(\frac{-\frac{1}{2}\left(\frac{q-q_0}{\sigma}\right)^2}{\sigma\sqrt{2\pi}}\right) \quad (2),$$

where the third term describes the peak related to the ordering of Fe nanoparticles in Al matrix. The calculated values of peak width and position are σ =0.05(2) nm^{-1} and q_0=0.87(2) nm^{-1}. The corresponding correlation length for the ordered nanoparticles is D=$2\pi/q_0$≈7.2(1) nm.

The magnetic susceptibility curves as functions of magnetic field and temperature are presented in fig. 2. The maximum χ value at T=300 K corresponds to the sample 2, as one would expect due to larger characteristic size of iron nanoparticles (table 1).

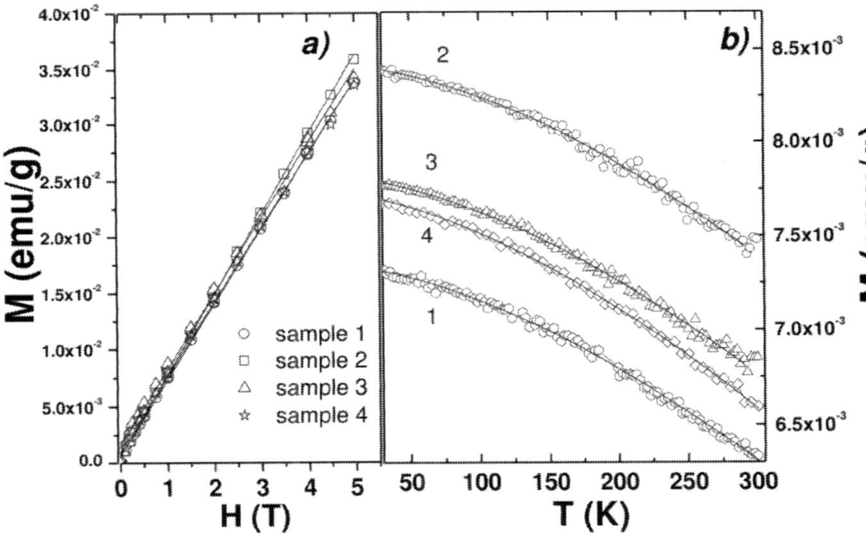

Figure 2. The magnetic susceptibility curves as functions of magnetic field (a) fitted by liner function, and temperature (b), fitted by function (3) for the studied samples 1-4.

For the temperature range well below the Curie temperature, the temperature dependence of magnetization can fitted in frame of a spin-wave fluctuations theory by Bloch equation [11]:

$$M = M_0(1 - BT^b) \quad (3),$$

where B and b are the Bloch constants [11]. The obtained b values are listed in Table 2. For the samples 2, 3 and 4 they are close to ideal "Bloch degree" value b ~ 3/2 and indicate the presence of small-sized magnetic nanoparticles [11]. The existence of small magnetization in sample 1 irradiated by helium ions only can be explained by the presence of some impurities. This sample is characterized by the lowest Bloch degree b=1.27(4).

Conclusions

The formation of iron nanoparticles in aluminium matrix by irradiation of iron and helium ions has been studied by means SAXS and magnetic measurements. The irradiation of Al matrix by He ions with a subsequent Fe ions irradiation leads to a formation of the disordered Fe nanoparticles array with the largest characteristic nanoparticle size of 10.4 nm. The opposite treatment, including irradiation of Al matrix by Fe ions with a subsequent He ions irradiation leads to a formation of the ordered Fe nanoparticles array with a correlation length of 7.2 nm and characteristic nanoparticle size of 6.8 nm.

The work has been supported by contracts 02.740.11.0542 and 16.518.11.7029 within the Russian Federal Programs “The priority activities in researches and developments of scientific-technological complex of Russia in 2007-2012” and "Research and scientific- educational personnel of innovation Russia in 2009-2013”.

References

[1] Schmid G 2010 Nanoparticles: From Theory to Application, Wiley-VCH, Technology & Engineering, 533 p.
[2] Gubin S P 2009 Magnetic nanoparticles Wiley-VCH, Technology & Engineering, 466 p.
[3] Bhushan B, Sundararajan S 1998 Acta Materialia **46** 3793
[4] Raineri V, Fallica P G, Percolla G, et al. 1995 J. Appl. Phys. **78** 3727
[5] Dmitriev S N, Reutov V F 2002 The ions doping of solids Patent of Russian Federation N2193080
[6] Chernyshov A A, Veligzhanin A A, Zubavichus Y V 2009 Nuclear Instruments and Methods in Physics Research A **603** 95–98
[7] Kammler H K, Beaucage G, Kohls D J, Agashe N and Ilavsky J 2005 J. Appl. Phys. **97** 054309
[8] Beaucage G 2004 Phys. Rev. E **70** 031401
[9] Bera M K, Sanyal M K, Yang L, Biswas K, Gibaud A, Rao C N R 2010 Phys. Rev. B **81** 115415
[10] Brumberger H 1995 Modern Aspects of Small Angle Scattering, Dordrecht, Kluwer Acad. Publ. 221 p.
[11] Zhang D, Klabunde K J, Sorensen C M, Hadjipanayis G C 1998 Phys. Rev. B **58** 14167

Journal of Physics: Conference Series **351** (2012) 012014 doi:10.1088/1742-6596/351/1/012014

SANS investigation of a ferrofluid based silicone elastomer microstructure

Maria Balasoiu[1,2], Vasily T Lebedev[3], Diana N Orlova[3], Ioan Bica[4] and Yuriy L Raikher[5]

[1]Joint Institute of Nuclear Research, Dubna, Russia
[2] Horia Hulubei National Institute of Physics and Nuclear Engineering, Bucharest, Romania
[3]Petersburg Nuclear Physics Institute, Gatchina, Russia
[4]West University of Timisoara, Department of Electricity and Magnetism, Timisoara, Romania
[5]Institute of Continuous Media Mechanics, Ural Branch of RAS, Perm, Russia

E-mail: balasoiumaria@yahoo.com; balas@jinr.ru

Abstract. Structural peculiarities of ferroelastomers composed of polydimetylsiloxane with magnetite particles embedded during polymerization are studied by small-angle neutron scattering. The effects of ferroparticle concentration in the range up to 6 mass.%. and external magnetic fields of induction up to 1 kG, applied while curing, on the structure of the polymer and the particle spatial distribution in matrix are analyzed.

1.Introduction

The study of the properties of filled elastomers is a challenging and exciting topic for both fundamental science and industrial application. It is know that adding of hard particulates to a soft elastomer matrix results in the properties that do not follow a straightforward rule of mixtures. The progress of the structure vs. properties relationships research of such systems evolves into several directions: filler type and structure, magneto(electro)-mechanical reinforcement and interactions between the fillers and elastomers. Development of novel technologies of magnetic nanomaterials shows the trends of creation of the composites with anisotropic molecular and magnetic structures such as elastomers filled with fine ferroparticles. The combination of polymers with magnetic materials displays new and often enhanced properties. Due to the presence of the magnetic particles, the shape of the elastomer composite responds the external magnetic field. In general, the cross-influence of the magnetic and elastic properties of such compounds leads to a number of striking phenomena occurring in response to imposed magnetic and stress fields [1-42]. On the other hand, the ability to control and characterize the multiple forming structures inside the elastomer is an example of morphological engineering for reinforced and modified polymers valuable for a variety of applications.

The synthesis and study of structure and physical properties of these advanced materials, which combine the functional properties of elastic polymers and ferromagnetic substances should be considered as a perspective way to provide the understanding of construction principles of a wide class of materials for electronics, electrical engineering, medicine, aero- and cosmic industries. From the fundamental point of view, a comprehensive analysis of the relationship between macromolecular and disperse phase structures and their ordering under the action of magnetic field and by the variation of magnetic component content is needed. Neutron and X-ray scattering techniques are highly useful for determining the morphology of the emerging filler structures [43-52].

The aim of present work is the small-angle neutron scattering examination of subtle structural features of the polymeric matrix and the ensemble of embedded ferroparticles as resulting from the conditions of preparation of ferroelastomers by the variation of ferroparticle concentration and the strength of external magnetic field applied during polymerization.

2. Experimental

Ferroelastomers were prepared [34] using the radical polymerization of dimetylsiloxane with addition of ferrofluid based on magnetite. In the samples, the concentration of the magnetic component (ferrofluid) [53-59] was equal 1.27; 3.9; 5.88 mass.%, respectively. The magnetic field ($B = 0$; 280; 560; 1120 Gauss) was applied perpendicular to the plane of the polymer film of thickness ~ 0.5 mm, see Table 1. A reference sample, i.e., a polymer film without ferroparticles was also was prepared. The small-angle neutron scattering experiments (SANS) were carried out at ambient temperature (20°C) on the "Membrane" diffractometer (PNPI) in the range of momentum transfer $q = (4\pi/\lambda)\sin(\theta/2)$ = 0.03-0.8 nm^{-1}, where θ is the scattering angle and $\lambda = 0.3$ nm is the neutron wavelength ($\Delta\lambda/\lambda$) = 0.25 [52]. The incident beam was orthogonal to the plane of the elastomer films. In all the experiments, the samples, which are isotropic in scattering plane, studied. Thus, the scattering intensities were dependent just on the modulus q of the scattering vector. Since the neutron beam transmission for the samples was relatively high (Tr ~ 0.8), the multiple scattering processes were neglected.

Table 1. Magnetite content, C, in the samples, and the magnetic induction, B, applied during polymerization

№ of the sample	Content of Fe_3O_4, C mass.%	B, Gauss
P_1, matrix	0	0
P_{12}	5.88	0
P_{13}	5.88	280
P_{14}	5.88	560
P_{15}	5.88	1120
P_{22}	3.9	0
P_{23}	3.9	280
P_{24}	3.9	560
P_{25}	3.9	1120
P_{32}	1.27	0
P_{33}	1.27	280
P_{35}	1.27	1120

3. Results and discussion

The scattering patterns for the elastomers containing 5.88; 3.9 and 1.27 mass.% of magnetite are presented in Figures 1-3, where the data for the matrix are shown also. It is evident (Figure 1) that the

Journal of Physics: Conference Series **351** (2012) 012014 doi:10.1088/1742-6596/351/1/012014

scattering from original polymeric matrix is relatively strong at $q \leq 0.2$ nm^{-1}. The addition of ferroparticles into matrix does not cause any substantial contribution to the total scattering intensity at low momenta even at the highest concentration of magnetite (5.88 mass.%). The scattering from ferroparticles dominates only at larger momenta, $q \geq 0.4$ nm^{-1}, due to their small size as compared to the inhomogeneities in the polymer matrix. It should be noted that application of a magnetic field ($B = 1120$ G, sample P_{15}) during polymerization, did not initiate any dramatic structural changes. Indeed, the observed increase of scattering is moderate: the ratio of intensities for samples P_{15} (strong field) and P_{12} (no field) does not exceed factor 2 in the interval $q = 0.4 - 0.8$ nm^{-1}. This means that the polymer network prevents aggregation of ferroparticles and ensures their more homogeneous spatial distribution to be achieved for required material quality. Large-scale inhomogeneities in the polymer matrix are visible despite of low scattering ability of the polymer having the density of coherent length $K_S = 6.3 \times 10^8$ cm^{-2}. It is by two orders in magnitude lower than that for magnetite having $K_M = 6.97 \times 10^{10}$ cm^{-2}. Thus, the contrast factor for magnetite regarding to matrix is also high, $\Delta K_M = K_M - K_S = 6.91 \times 10^{10}$ см$^{-2}$, but it is compensated by small characteristic volume of a particle as compared to the large inhomogeneities in matrix, as it is seen in Figures 2 and 3.

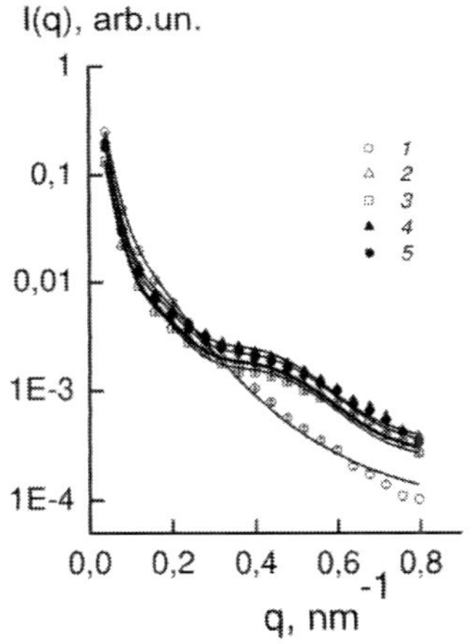

Figure 1. Scattering intensities $I(q)$ vs. momentum transfer q for matrix P_1 (1) and ferroelastomers $P_{12} - P_{15}$ (2–5) with high content of magnetite (synthesis under the magnetic field perpendicular to the plane of the polymer film); the beam is orthogonal to the plane of sample; lines are approximation functions (1) and (2) for the matrix and ferroelastomers.

The decrease of the ferroparticle concentration makes the difference in the scattering intensities between the ferroelastomers and matrix progressively smaller. A detailed analysis of q -dependencies of scattering intensity for matrix enables us to propose a two-level model structure and the related scattering function

$$I(q) = I_{01} \exp[-(qR_g)^2 / 3] + I_{02}[1 + (qR_c)^2]^{-2} . \qquad (1)$$

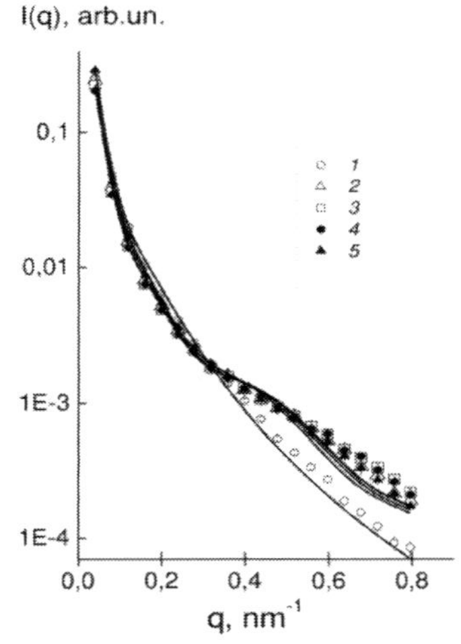

Figure 2. Scattering intensities $I(q)$ vs. momentum transfer q for matrix P_1 (1) and ferroelastomers P_{22} – P_{25} (2–5) with moderate content of magnetite; lines are the approximation functions (1) and (2) for the matrix and ferroelastomers.

n

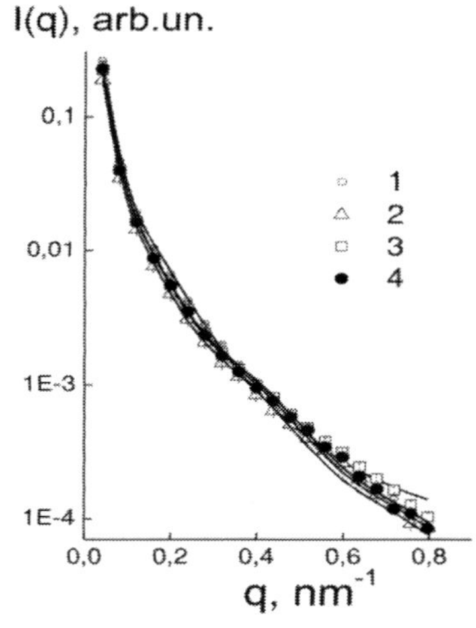

Figure 3. Scattering intensities $I(q)$ vs. momentum transfer q for matrix P_1 (1) and ferroelastomers P_{32}, P_{33}, P_{35} (2–4) with a low content of magnetite; lines are the approximation functions (1) and (2) for the matrix and ferroelastomers.

Here the first Guinier-term describes the large scale inhomogeneities (domains of the gyration radius R_G), and the second Debye-term is related to the smaller globular objects (domains) with correlation radius R_C. The coefficients I_{01} and I_{02} are proportional to the squared contrast factors, volume contents and characteristic volumes of the corresponding objects. This model function approximates the data for matrix satisfactory (Figures 1–3) with parameters given in Table 2.

Table 2. Parameters of functions (1), (2) for matrix and ferroelastomers.

Sample N	B, Gauss	I_{01}, arb.un.	R_G, nm	$I_{02}\times 10^5$, arb.un.	R_C, r_C, nm	n	L, nm
P_1	0	0.487±0.002	39.3±0.1	4590 ±60	6.37±0.04	-	-
P_{12}	0	0.249±0.001	38.1±0.1	617±3	2.33±0.01	1.21±0.01	15.6±0.1
P_{13}	280	0.273±0.001	38.8±0.1	563±5	2.35±0.01	1.30±0.02	15.4± 0.1
P_{14}	560	0.436±0.001	39.2±0.1	895±5	2.43±0.01	1.26±0.01	15.9± 0.1
P_{15}	1120	0.363±0.002	38.8±0.1	762±8	2.35±0.02	1.27±0.02	15.5± 0.1
P_{22}	0	0.490±0.002	39.0±0.1	1160±19	3.42±0.03	1.30±0.03	15.7± 0.1
P_{23}	280	0.448±0.001	38.9±0.1	1007±7	3.21±0.01	1.32±0.02	15.6± 0.1
P_{24}	560	0.399±0.001	38.6±0.1	985±10	3.19±0.02	1.33±0.02	15.6± 0.1
P_{25}	1120	0.603±0.001	39.9±0.1	1406±11	3.71±0.02	1.22±0.01	16.2± 0.1
P_{32}	0	0.353±0.002	39.6±0.1	1993±32	4.95±0.04	0.76±0.02	18.4± 0.2
P_{33}	280	0.422±0.001	39.4±0.1	2146±24	4.77±0.02	0.78±0.01	17.6± 0.1
P_{35}	1120	0.440±0.001	40.0±0.1	2242±26	4.94±0.03	0.74±0.01	17.8± 0.1

Therefore, the radius of a large polymer domain is about $R_G \sim 40$ nm that is approximately seven times greater than that for a small domain: $R_C \sim 6$ nm. As we assume, these large structures in the matrix are composed of small domains, whose centers are situated at the minimum distance $\sim 2R_C \sim 13$ nm corresponding to characteristic domain diameter. It can be expected that the ferroparticles penetrating between domains should be spaced at the same distance. For the ferroelastomers we have used a model that takes into account the scattering from large-scale polymer domains (gyration radius R_G) and ferroparticles (correlation radius r_C) separated in space by the distance L comparable to the small polymer domain diameter $\sim 2R_C$:

$$I(q) = I_{01}\exp[-(qR_g)^2/3] + I_{02}[1+(qR_c)^2]^{-2}[1+n\sin(qL)/(qL)]. \quad (2)$$

Here the parameters I_{01} and I_{02} represent the contributions of large polymer domains and ferroparticles in the scattering intensity at $q \to 0$, while n is the average number of the particles correlated with a given particle at the characteristic distance L Function (2) describes in a satisfactory way the behaviors of the scattering intensities for ferroelastomers at all the concentrations and the magnitudes of induction B; the fitting parameters are listed in Table 2.

In Figure 4 the obtained dependencies of the correlation radius $r_C(B)$ of the magnetic inclusions (single particles, clusters) and the forward scattering intensities $I_{02}(B)$ vs. magnetic induction B for sample series P_{22}–P_{24} are represented.

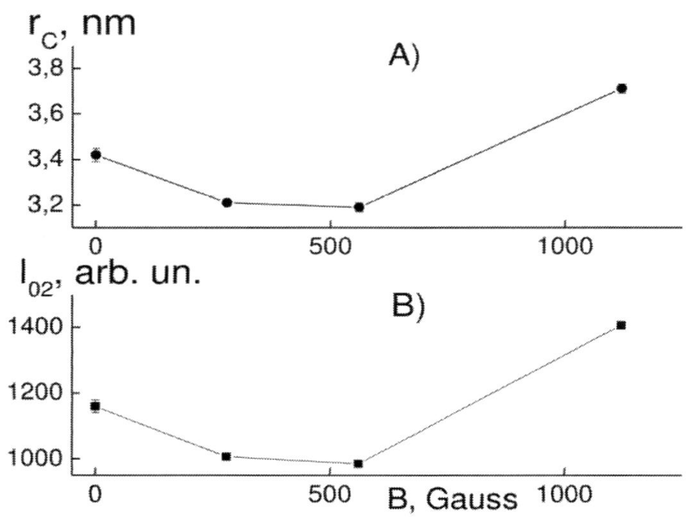

Figure 4. Effective correlation radius of magnetic inclusions $r_C(B)$ and forward intensities $I_{02}(B)$ vs. the induction B of magnetic field applied during the samples synthesis.

Similar behavior of r_C and I_{02} curves vs. the magnetic field induction B (Figure 4) occurs because these functions are correlated through the relation:

$$I_{02} \sim (\Delta K_M)^2 \phi V_1, \tag{3}$$

where ΔK_M is the contrast factor between magnetite and the polymer, φ is the volume fraction of the magnetic phase, V_1 is the volume of a single object. In Eq. (3) we assume that the scattering object is a characteristic cluster with the reference size $r_C(B)$ and volume $V_1(B)$. The latter is defined by the relationship

$$V_1 \sim r_C{}^D, \tag{4}$$

where parameter D refers to the geometry (fractal dimension) of the cluster [60-62]. For example, a linear chain has the fractal dimension $D=1$, for a flexible (polymer-like) chain $D=2$. Branched chain-like entities are characterized by a larger parameter, $2<D<3$; the exponent for globular objects is $D=3$.

In the case of ferroelastomers we have found the dimension $D\approx 2.3$ using the data approximation with function (4) in log-log coordinates, see Figure 5. Note that on the base of the presented data we can only estimate the exponent D. The variation of magnetic induction should vary this exponent, but to evaluate this fine effect, more detailed studies are needed.

In the presented series of SANS-experiments, the field-induced structural changes in the ensembles of ferroparticles during the polymerization process were observed, which increased the strength of the resulting magnetic elastomer samples; the field range for these observations is from zero to $B \sim 1$ kG. Application of weak and moderate fields ($0<B<500$ G) leads to an unusual clusterization effect easing, therefore, the cluster characteristic size and volume decrease for 6% and 20%, respectively.

Application of stronger magnetic fields $B \sim 1$ kG enhances the cluster by 15% in size and 40% in volume in comparison with the minimal obtained level (Figure 4).

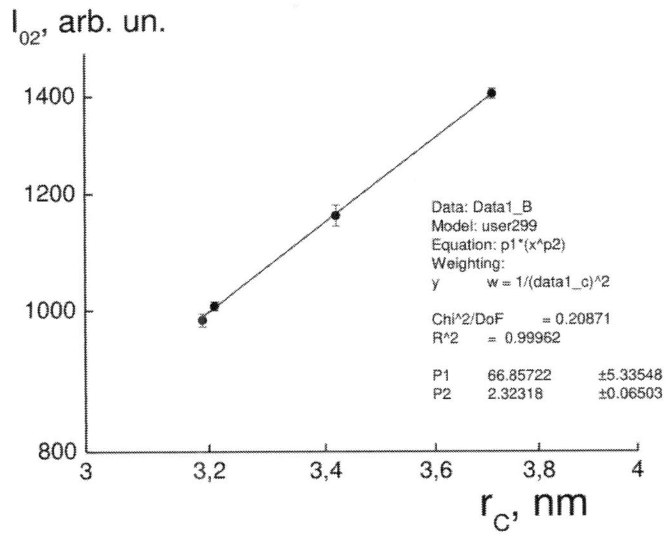

Figure 5. Forward scattering intensity I_{02} vs. correlation radius r_C (in log-log coordinates); line displays the approximation function (4).

These nonmonotonic cluster size and volume changes with the increase of the applied magnetic field at relatively weak field values $B \leq 500$ G, indicate the field-induced breaking of the original (at zero field) distribution of the ferroparticles inside the elastomer. The more intense fields, make the particles to chain along the field direction.

Without external field, the particles strive to organize in the associates with minimal free energy that requires magnetic closure inside the cluster (e.g. ring-shaped structures). When imposed, the external field favors the alignment of the particle magnetic moments, thus "opening" the above-mentioned structures. This competition leads to the weakening of the original local magnetic order. However, when the magnitude of external field $B \sim 1$ kG is exceeded, the particles with aligned moments organize in chain-like structures. The aggregates of such a type are usually observed in ferrofluids under magnetic field.

4. Conclusions

The parameters in Table 2, if summarized, enable us to compare the structural features of the ferroelastomers prepared at different concentrations of magnetite and magnitudes B of the magnetic field. The data evidences that embedding of ferroparticles into a polymer matrix does not disturb its large structures significantly. The gyration radius still remains at the level $R_G \sim 38-40$ nm, while the forward scattering intensity I_{01} from these entities becomes lower. Thus, the ferroparticles loosen initial polymer structures. However, the application of an external field induces the opposite effect, i.e. at a stronger field the structures are more pronounced (parameter I_{01} increases).

At a high content of magnetite (5.88 mass.%), under the change of the applied field we observe practically the same size of magnetic inclusions, $r_C \sim 2.3-2.4$ nm. At the same time, the forward scattering intensity I_{02} demonstrates a growth by 40% that should be attributed to the ordering in the particle ensemble (strengthening correlations along the field direction). A similar effect is found at moderate concentration of magnetite (samples P_{22} – P_{25}): at a practically constant size of scattering objects the forward intensity is enhanced by 40% (Table2). Note, the characteristic spacing of ferroparticles remains at the same level $L \sim 15 - 16$ nm. So, it is practically independent on the concentration and the field strength. At a low content of the particles, their observation is complicated due to the scattering from small-size polymeric matrix inhomogeneities which mask the ferroparticles.

Acknowledgements
This study was supported from the Romanian Governmental Plenipotentiary at JINR Grant No.328/29.05.2009 item 6 and WUT-JINR Cooperation scientific projects No.58/08.02.2010 item 18 and No.57/08.02.2010 item 18; No.146/15.03.2011 item 23 and No.151/15.03.2011 item 23. YLR acknowledges RFBR grant 11-02-96000.

References

[1] Rosensweig R E, Jerauld G R and Zahn M 1981 *Continuum Models Discrete Syst.* **4** 137
[2] Jin S, Tiefel T H, Wolfe R, Sherwood R C and Mottine J J 1992 *Science* **225** 446
[3] Schiga M, Hirose M and Okada K 1992 *Japanese Patent* 4266970
[4] Schiga M, Okada A and Karauchi T 1995 *J. Appl. Polym. Sci.* **58** 787
[5] Zrinyi M, Barsi L and Buki A 1996 *J. Chem. Phys.* **104**(20) 8750
[6] Jolly M R, Carlson J D, Munoz B C, Bullions T J 1996 *J. Intell. Mater. Syst. Struct.* **7** 613
[7] Zrinyi M, Barsi L, Buki A 1997 *Polym. Gels Networks* **5** 415
[8] Szabo D, Barsi L, Buki A and Zrinyi M 1997 *Models Chem.* **134**(2-3) 155
[9] Zrinyi M 1997 *Trends Polym. Sci.* **5**(9) 280
[10] Zrinyi M, Barsi L, Szabo D and Kilian H G 1997 *J. Chem. Phys.* **108**(13) 5685
[11] Bernadek S 1997 *J. Magn. Magn. Mater* **166** 91
[12] Martin J E and Anderson R A 1999 *J. Chem. Phys.* **111**(9) 4273
[13] Zrinyi M, Szabo D and Barsi L 1999 *Magnetic field sensitive polymeric actuators, polymer sensors and actuators* (Berlin Springer Verlag) 385
[14] Bernadek S 1999 *Appl. Phys. A* **68** 63
[15] Ginder J M, Nichols M E, Elie L D and Tardiff J L 1999 *SPIE Conf. Smart Mater. Technol.* **131** 3675
[16] Carlson J D and Jolly M R 2000 *Mechatronics* **10** 555
[17] Raikher Yu L, Stolbov O V 2000 *Tech.Phys.Lett.* **26**(2) 156
[18] Borcea L and Bruno O 2001 *J. Mech. Phys. Solids* **48** 2877
[19] Ginder J M, Clark S M, Schlotter W F and Nichols M E 2002 *Int. J Modern Phys. B* **16**(17-18) 2412
[20] Zrinyi M, Szabo D, Filipcsei G and Feher J 2002 *Electric and magnetic field smart polymer gels. Polymer gels and networks* **11** ed. Y Osada and A R Khokhlov (New York Marcel Dekker Inc) 309
[21] Mitsumata T, Furukawa K, Juliac E, Iwakura K and Koyama K 2002 *Int. J.Modern Phys. B* **16**(17-18) 2429
[22] Zhou G Y 2003 *Smart Mater Struct* **12** 139
[23] Zhou G Y and Jiang Z Y 2004 *Smart. Mater. Struct.* **13** 309
[24] Lanotte L, Ausanio G, Hison C, Ianotti V, Luponio L, Luponio C Jr. 2004 *J. Optoelectron. Adv. Mat.* **6**(2) 523
[25] Varga Zs, Filipcsei G and Zrinyi M 2005 *Polymer* **46** 7779
[26] Varga Zs, Filipcsei G, Szilagyi A and Zrinyi M, 2005 *Macromol. Symp.* **227** 123
[27] Abramchuk S, Kramarenko E, Stepanov G, Nikitin L V, Filipcsei G, Khokhlov A R, Zrínyi M 2007 *Polym. for Adv. Tech.* **18**(11) 883
[28] Raikher Yu L, Stolbov O V and G.V.Stepanov, 2008 *Tech. Phys.* **53**(9) 1169

[29] Raikher Yu L, Stolbov O V 2008 *J. Phys.: Condens. Matter* **20**(20) 204126
[30] Stepanov G V, Borin D Yu, Raikher Yu L, Melenev P V, Perov N S, 2008 *J.Phys.: Condens.Matter* **20**(20) 204121
[31] Zhang X, Peng S, Wen W and Li W 2008 *Smart Mater. Struct.* **17** 045001
[32] Feinblum E, Kosso R, Kohls D J and Schafer D W 2008 *Current Topics in Elastomer Research* **Chapter 17** Bhowmick A K ed. (London New York CRC Press Taylor and Francis Group LLC) 503
[33] Bica I 2009 *J. Ind. Eng. Chem.* **15** 605; Bica I 2009 *J. Ind. Eng. Chem.* **15** 769; Bica I 2009 *J. Ind. Eng. Chem.* **15** 773
[34] Bica I 2009 *Matter. Lett.* **63** 2230
[35] Melenev P V, Raikher Yu L and Rusakov V V 2010 *Polym.Sci. A* **52**(4) 430
[36] Stolbov O V, Raikher Yu L, Stepanov G V, Chertovich A V, Kramarenko E Yu and Khokhlov A R 2010 *Polym Sci. A* **52**(12) 1344
[37] Bica I 2010 *J. Ind. Eng. Chem.* **16** 359; Bica I 2011 *J. Ind. Eng. Chem.* **17** 83
[38] Melenev P, Raikher Yu, Stepanov G, Rusakov V, Polygalova L 2011 *J. Intel. Mater. Syst. Struct.* **22** 531
[39] Papaphilippou P C, Pourgouris A, Marinica O, Taculescu A, Athanasopoulos G I, Vekas L and Krasia-Christoforou T 2011 *J. Magn. Magn. Mater.* **323**(5) 557
[40] Ward MA and Georgion T 2011 *Polymers* **3** 1215
[41] Xu Y, Gong X, Xuan S, Zhang W and Fan Y 2011 *Soft Matter* **7** 5246
[42] Stolbov O V, Raikher Yu L, Balasoiu M 2011 *Soft Matter* **7** 8484
[43] Torok Gy, Lebedev V T, Cser L and Zrinyi M 2000 *Physica* **B 276** 396
[44] Lebedev V T, Torok Gy, Cser L, Buyanov A L, Revelskaya L G, Orlova D N and Sibilev A I 2000 *Mater. Sci. Forum* **321-324** 773
[45] Torok Gy, Lebedev V T, Cser L, Kali Gy and Zrinyi M 2001 *Physica* **B 297** 40
[46] Balasoiu M, Craus M L, Kuklin A I, Plestil J, Haramus V, Islamov A H, Erhan R, Anitas E M, Lozovan M, Tripadus V, Petrescu C, Savu D, Savu S and Bica I 2008 *J. Optoelectron. Adv. Mater.* **10**(11) 2932
[47] Balasoiu M, Anitas E M, Bica I, Osipov V A, Orelovich O L, Savu D, Savu S, Erhan R and Kuklin A I 2008 *Optoelectron. Adv.Mater.-RC* **2**(11) 730
[48] Wang B, Rozynek Z, Zhou M and Fossum J O 2009 *J. Phys.:Conf. Series* **149** 012032
[49] Anitas E M, Balasoiu M, Bica I, Osipov V A and Kuklin A I 2009 *Optoelectron. Adv. Mater. – RC* **3**(6) 621
[50] Balasoiu M, Craus M L, Anitas E M, Bica I, Plestil J and Kuklin A I 2010 *Phys.Solid State* **52(5)** 917
[51] Balasoiu M, Bica I, Raikher Yu L, Dokukin E B, Almasy L, Vatzulik B, Kuklin A I 2011 *Optoelectron. Adv. Mater. – RC* **5**(5) 523
[52] Balasoiu M, Lebedev V T, Orlova D N, Bica I 2011 *Crystallogr. Rep.* **56(7)** 1177 (DOI 10.1134/S1063774511070054)
[53] Bica D 1985 *RO Patent* 90078
[54] Bica D 1993 *PhD Thesis Polytechnic University of Bucharest*
[55] Grabcev B, Balasoiu M, Bica D and Kuklin A I 1994 *Magnetohydrodynam.* **10**(2) 156
[56] Grabcev B, Balasoiu M , Tarziu A, Kuklin A I and Bica D 1999 J. Magn. Magn. Mater. **201** 140
[57] Aksenov V, Avdeev M, Balasoiu M, Rosta L, Torok Gy, Vekas L, Bica D, Garamus V, Kohlbrecher J 2002 *Appl. Phys. A* **74** 943
[58] Avdeev M, Balasoiu M, Torok Gy, Bica D, Rosta L, Aksenov V L and Vekas L 2002 *J. Magn. Magn. Mater.* **252** 86
[59] Avdeev M V, Balasoiu M, Aksenov V L, Garamus V M, Kohlbrecher J, Bica D and Vekas L 2004 *J. Magn. Magn. Mater.* **270** 371
[60] Beaucage G 1996 *J. Appl. Cryst.* **29** 134
[61] Cherny A Yu, Anitas E M, Kuklin A I, Balasoiu M, Osipov V A, 2010 *J. Appl. Cryst.* **43** 790
[62] Cherny A Yu, Anitas E M, Osipov V A and Kuklin A I. 2011 *Phys. Rev. E* **84** 036203

Journal of Physics: Conference Series **351** (2012) 012015 doi:10.1088/1742-6596/351/1/012015

Non-ionic block copolymers leading to a highly ordered organosilica material

V Rebbin, A Rothkirch, U Vainio and S S Funari

HASYLAB/DESY, Notkestr. 85, 22607 Hamburg, Germany

Vivian.Rebbin@desy.de

Abstract. Phenylene-bridged periodic mesoporous organosilica (PMO) materials with hexaethylene glycol mono hexadecylether ($C_{16}(EO)_6$) as structure directing agent (SDA) in acidic solution were synthesised and the reaction was studied in different conditions, including *in situ* small angle X-ray scattering at room temperature and at 60 °C. The *in situ* SAXS investigations at 60 °C show the formation of a 2D hexagonal mesostructure after 3 hours reaction time. In contrast to these results the same reaction performed in the laboratory produced a powder that, after drying at room temperature, shows a large number of diffraction peaks, allowing identifying a lamellar structure, possibly in coexistence of a micellar cubic structure (space group *Pm*3*n*). Aposteriori removal of the surfactant from the powder synthesised in the laboratory led to the collapse of the lamellar structure.

1. Introduction

Periodic mesoporous organosilica (PMO) materials are a class of materials showing high ordering of pores in the mesoscopic, high specific surface areas up to 1000 m^2/g and narrow pore size distributions combined with a uniform distribution of organic functionalities in the pore wall matrix [1-3]. Hence, the PMO materials can be used in a wide field of applications like catalysis, separation and adsorption [4-7]. The synthesis, based on hydrolysis and condensation of an organically modified silica precursor of the formula $(OR)_3Si\text{-}R'\text{-}Si(OR)_3$ (R: methyl, ethyl, R': any organic unit) in presence of an amphiphile structure directing agent (cationic, anionic or non-ionic surfactants or block copolymers) offers the opportunity to tune the resulting material to the designated properties, by choosing the appropriate organic unit R'.

By varying the synthesis parameters like temperature, concentration, reaction time and pH of the solution, different mesostructures can be obtained although the formation process is not yet fully understood. Most common structures of these materials are the lamellar (space group: *p*2), the two dimensional (2D) hexagonal (space group: *p*6*mm*) and several bicontinuous cubic (e.g. *Im*3*m*, *Fm*3*m*, *Pn*3*m*, *Ia*3*d*) and micellar cubic structures (e.g. *Pm*3*n*, $P4_332$). A comparatively very low number of publications about the synthesis of PMOs report the observation of a three dimensional (3D) hexagonal structure ($P6_3/mmc$) [1] and a cubic (*Fd*3*m*) [8, 9].

In 2002, an improvement of the structures of PMO materials was introduced by Inagaki et al. [10] with syntheses using cationic surfactants in basic solution. Additionally to the well-ordered 2D hexagonal pore system, an ordering of the organic units in the pore walls became possible. This was shown by powder X-ray diffraction (P-XRD) and transmission electron microscopy (TEM). These measurements showed a repeat distance of 0.768 nm between the aromatic units in the pore wall framework and were in good agreement with modelling results. Similar findings were then reported for materials containing highly conjugated units but showing a decrease in the mesopore ordering with increasing size of the organic unit of the organosilica precursor [11-13]. A PMO material showing a similar ordering achieved under acidic conditions was introduced recently [14].

2. Experimental section

2.1. Materials

1,4-bis(triethoxysilyl)benzene (BTEB) and hydrochloric acid (32 %) were purchased from Sigma-Aldrich and used without further purification, hexaethylene glycol mono hexadecylether ($C_{16}(EO)_6$) was purchased from Nikko Chemicals, Japan.

2.2. Synthesis

For the *in situ* measurements at a synchrotron a solution of $C_{16}(EO)_6$ in diluted hydrochloric acid (pH = 1) was prepared in a four neck vessel with a reflux condenser. The vessel was connected to a flow through capillary placed in the X-ray beam by flexible tubing and the solution was pumped through the setup with a speed of 5 mL/min. After measuring the pure surfactant solution, the organosilica precursor BTEB was added and data acquisition was started immediately afterwards. Measurements were carried out at 25 °C (room temperature) and 60 °C.

For the synthesis in the laboratory the surfactant $C_{16}(EO)_6$ was dissolved in diluted hydrochloric acid and BTEB was added. The solution was stirred for 24 h at room temperature and then transferred into a Teflon lined steel autoclave and kept for another 24 h at 100 °C. After cooling the resulting solid was filtered off, washed with water and dried in air. Removal of the surfactant was accomplished by solvent extraction using a mixture of ethanol and hydrochloric acid (32 %) (100:3; v:v) in a Soxhlet apparatus for 12 h at 120 °C with subsequent drying in air.

2.3. Instrumentation

In situ small angle X-ray scattering (SAXS) measurements were performed at the Soft Condensed Matter Beamline A2 at DORIS III synchrotron at DESY. Data acquisition was started after the addition of BTEB to the reaction solution with 60 s exposure time followed by a 60 s break between the measurements. The measurement was interrupted when the capillary was blocked by the occurring solid in the solution.

Powder-X-ray diffraction measurements were carried out at Beamline G3, small angle X-ray scattering measurements of the powders at Beamline B1 (both DORIS III at DESY).

3. Results

3.1. In situ experiments

The evolution of the *in situ* SAXS patterns of a reaction containing $C_{16}(EO)_6$ as structure directing agent (SDA) and BTEB as organosilica precursor measured at 60 °C is shown in **Figure 1a**. After about 55 min reaction time a transition structure showing only one reflection starts evolving and remains present for about 30 minutes. Upon decreasing of this single reflection a new diffraction peak evolves and another phase transition takes place, leading to a 2D hexagonal structure after 100 min (**Figure 1b**). Due to the appearance of only one reflection of the metastable transition state, an indexing to a certain structure is not possible. We assume this structure to be lamellar due to the concept of charge density matching introduced by Monnier et al. [15], which says that the surface curvature of the liquid crystal changes with the changing number of charges on the silica surface due to hydrolysis and condensation reactions. A low number of charges on the silica surface (e.g. at the

beginning of the hydrolysis) forces the liquid crystal into a highly curved surface (e.g. micellar cubic) while a high number of charges on the silica surface leads to a low curved surface (e.g. lamellar). For the described results this let us assume that a lamellar structure is formed at the time many silica species are hydrolysed and with on-going condensation the number of charges decreases and the curvature of the liquid crystal increases until the formation of the final 2D hexagonal structure.
The observation of a selected frame after 100 min reaction time illustrates the quality of the scattering pattern. It can be seen, due to the peaks broadness, that the structures were "loosely" formed, however still enabling a structure assignment.

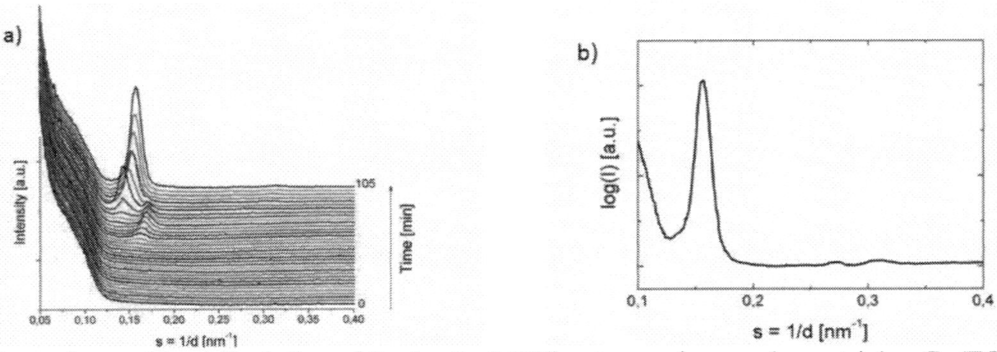

Figure 1 a and b. a) Evolution of the *in situ* SAXS patterns of a sample containing $C_{16}(EO)_6$ as SDA and BTEB as organosilica precursor at 60 °C and pH 1; b) SAXS pattern collected after 105 min showing the (100), (110) and (200) reflections of the 2D hexagonal structure.

In order to test the temperature relevance for the formation of the lamellar intermediate structure and its stability, we repeated the reaction at room temperature. Selected frames of the *in situ* measurement are shown in Figure 2a. Due to the slower kinetics, the on-going chemical reaction could be monitored for more than 180 min without blockage of the capillary. However, as can be seen from **Figure 2a** no clear structure could be identified. Phase transitions are on-going until the last recorded pattern showing one broad peak and one more intense sharp peak. This last pattern remains unchanged over a long time period. A single pattern recorded after 24 hours reaction time did not show significant changes compared to the last frame of the sequence taken at 180 min (**Figure 2b**). This clearly illustrates a significant influence of the reaction temperature on the resulting structure.

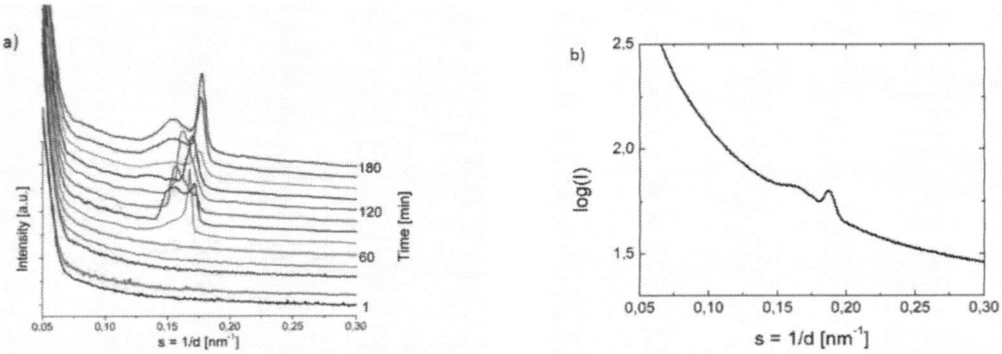

Figure 2a and b. a) Selected frames of the sample measured in situ at 25 °C and pH 1; b) SAXS pattern of the solution collected after 24 h at 25 °C.

3.2. Powder samples

The investigation of powder samples as synthesised (as), or upon extraction of the surfactant (ex) show different results and leads to the assumption that the structure found at the end of the *in situ* measurements not necessarily represents the final and stable structure.

Prior to the surfactant extraction, the dry powder shows some rather broad and weak features, **Figure 3a**, that can tentatively be attributed to a micellar cubic phase with the space group *Pm*3*n*. However, the pattern is too poor for an unquestionable structure determination. The powder diffraction at wide angles reveals a well-defined set of sharp peaks, despite low intensity (**Figure 3b**) that index a lamellar structure. It is noticeable that we can identify up to 14 orders of diffraction from this structure. As far as we are aware, such large number of diffraction orders has not been reported before. Moreover, the coexistence of two different structures with so different lattice parameters; a large cubic phase seen at small angles and a lamellar at wide angles is remarkable. Similar samples synthesised with $C_{12}(EO)_6$ or $C_{18}(EO)_6$ did not show diffraction patterns with so many peaks suggesting that such degree of ordering to be surfactant specific. The removal of the surfactant from such samples leads to a collapse of the lamellar structure which was expected (**Figure 3b**), considering the absence of the surfactant layers between the silica layers. However, the small angle diffraction pattern remains almost the same, showing the poor ordering in the small angle region.

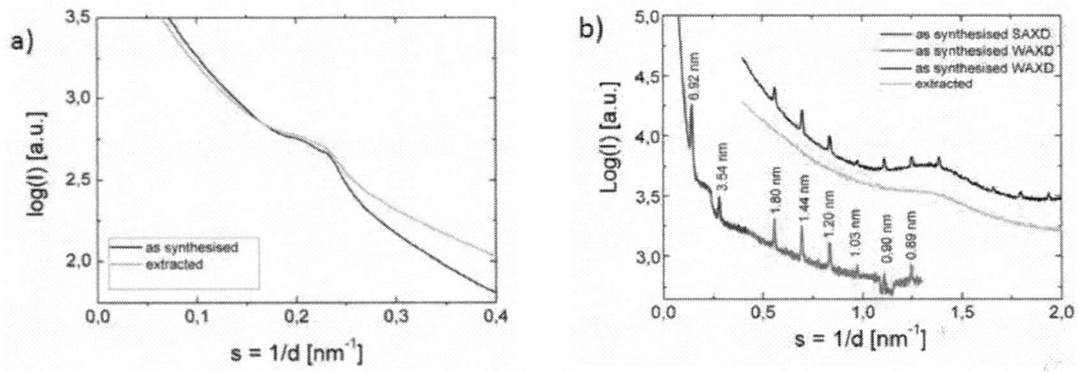

Figure 3 a and b. a) Small angle powder diffraction measurement of the sample synthesised in the laboratory; b) Small and wide angle diffraction measurements of the dry powder sample (as: as synthesised, ex: extracted).

In **Table 1** we list the lamellar peaks identified, noting that some of them are common to both lamellar and cubic phases, suggesting an epitaxial relationship between the structures.

Table 1. s-values of the experimental diffraction peaks, the respective repeat distances and associated Miller indices.

Lamellar structure (space group: p2)		
s [nm^{-1}]	d [nm]	hkl
0.144	6.94	001
0.284	3.52	002
-	-	003
0.560	1.79	004
0.697	1.43	005
0.835	1.19	006
0.972	1.03	007
1.110	0.90	008
1.248	0.80	009
1.385	0.72	0010
1.523	0.66	0011
1.660	0.60	0012
1.798	0.56	0013
1.936	0.52	0014

The sequence of the diffraction peaks belonging to the lamellar structure listed in **Table 1** is plotted with the s-value versus diffraction order (Figure 4). The fit of this data leads to a perfect linear correlation confirming the structure assignment.

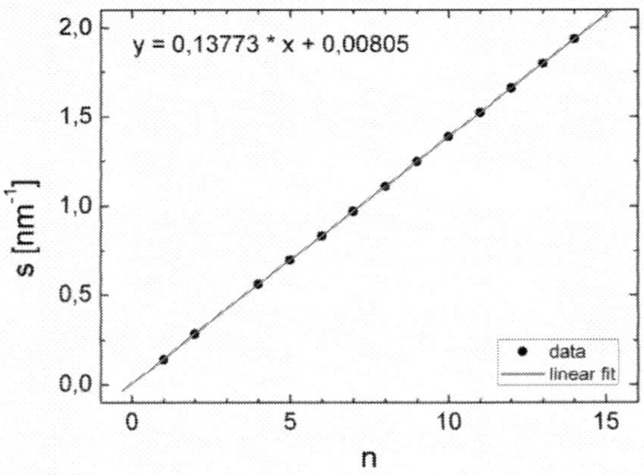

Figure 4. Linear correlation between the s-values and lamellar diffraction order of the detected peaks.

4. Discussion

The results obtained from the dry powder samples first lead to the assumption that the diffraction peaks in the wide angle region originate from an additional ordering in the pore walls as already described by Inagaki et al. [10]. However, two aspects disagree with this assumption. First, the calculated distance of the first wide angle reflection should have been around 0.768 nm, which corresponds to the length of one condensed precursor unit. One would expect a diffraction peak with this repeat distance as well as higher orders of it. This is, as visible from **Table 1**, not the case. The second aspect is, as can be concluded from **Figure 4**, the correlation of the diffraction peaks in the small and the wide angle regions excluding the peaks specific for the micellar cubic structure mentioned above. These diffraction peaks all belong to the same lamellar system. That means that all appearing reflections are related to the existing mesostructure and not to an additional ordering of organic units in the pore walls.

The existence of two different structures in the final powder material, a possible micellar cubic and the lamellar, would suggest that we have an epitaxial growth [16] during the chemical reactions leading to the final structures observed.

A diffraction pattern of the reaction solution that was recorded after a stirring time of about 24 hours is shown in **Figure 2b**. This was the last pattern collected before heating the material in an autoclave to obtain the final powder. As one can see there is coexistence of two different, not very well ordered structures. Comparing this pattern with the patterns of the final powder after heating the reaction mixture to 100 °C for 24 hours, the ordering process is significantly changed by increasing the reaction temperature. Following the reaction at room temperature on time (**Figure 2a**) shows that after 3 hours no significant changes take place. Thus, the evolution of the structure is clearly temperature dependent. The high temperature during the *in situ* experiment (**Figure 1a + b**) and for the synthesis of the powder samples the heating step is necessary if high ordering of the structure is desired. The temperature difference between the *in situ* measurement and the heating temperature of the powder sample of 40 °C leads to different final structures, the 2D hexagonal for the *in situ* measurement (60 °C) and the cubic plus the lamellar for the powder sample (100 °C).

5. Conclusion

From our results we conclude that under the given synthesis conditions such as acidic medium with pH 1 and the non-ionic surfactant hexaethylene glycol mono hexadecylether different synthesis temperatures lead to different structures of the final material. The formation of the structure during the powder synthesis in the laboratory passes a really highly ordered liquid crystal leading to a lamellar structure. Such high ordering covering a wide angular range is totally untypical for mesoporous materials and, to the best of our knowledge, has never been observed before. It should also be noticed that the final structure observed in the *in situ* reaction is different from the ones observed in the respective dry powders, illustrating that the dehydration is also a driving force for the structure formation, in addition to the temperature, as mentioned before. Moreover, the extraction of the surfactant leads to a collapse of the lamellar ordering which is at the same time the evidence for the correct indexing of the structure. The less dominating micellar cubic structure remains at the end.

6. Acknowledgment

We thank Stephan Botta and Wolfgang Ternes for technical support at beamlines A2 and G3 and HASYLAB for financial support.

References

[1] Inagaki S, Guan S, Fukushima Y, Ohsuna T, Terasaki O 1999 *J. Am. Chem. Soc.* **121** 9611

[2] Asefa T, MacLachlan M J, Coombs N, Ozin G A 1999 *Nature* **402** 867
[3] Melde B J, Holland B T, Blanford C F, Stein A 1999 *Chem. Mater.* **11** 3302
[4] Weitkamp J, Hunger M, Rymsa U 2001 *Microporous Mesoporous Mater.* **48** 255
[5] Mercier L, Pinnavaia T J 1997 *Adv. Mater.* **9** 500
[6] Rebbin V, Schmidt R, Fröba M 2006 *Angew. Chem. Int. Ed.* **45** 5210
[7] Hoffmann F, Cornelius M, Morell J, Fröba M 2006 *Angew. Chem. Int. Ed.* **45** 3216
[8] Shen S, Li Y, Zhang Z, Fan J, Tu B, Zhou W, Zhao D 2002 *Chem. Commun.* 2212
[9] El-Safty S A, Mizukami F, Hanaoka T 2005 *J. Mater. Chem.* **15** 2590
[10] Inagaki S, Guan S, Ohsuna T, Terasaki O 2002 *Nature* **416** 304
[11] Kapoor M P, Yang Q, Inagaki S 2002 *J. Am. Chem. Soc.* **124** 15176
[12] Cornelius M, Hoffmann F, Froeba M 2005 *J. Chem. Mater.* **17** 6674
[13] Cornelius M, Hoffmann F, Ufer B, Behrens P, Fröba M 2008 *J. Mater. Chem.* **18** 2587
[14] Yang Y, Sayari A 2007 *Chem. Mater.* **19** 4117
[15] Monnier A, Schüth F, Huo Q, Kumar D, Margolese D, Maxwell R S, Stucky G D, Krishnamurty M, Petroff P, Firouzi A, Janicke A, Chmelka B 1993 *Science* **261** 1299
[16] Bastos M, Silva T, Teixeira V, Nazmi K, Bolscher J G M, Funari S S, Uhríková D 2011 *Biophys. J.* **101** L20-L22

Small angle neutron scattering study of the gemini nonionic surfactant in heavy water solutions

A Rajewska[1, 2]

[1] Institute of Atomic Energy, 05-400 Świerk-Otwock, Poland

[2] Joint Institute for Nuclear Research, LNP, 141980 Dubna, Russia

Abstract. The nonionic gemini surfactant α α'-[2,4,7,9-tetramethyl-5-decyne-4,7diyl]bis[ω hydroxyl-polyoxyethylene] (S-10) was investigated in heavy water solutions only for concentrations: 2.3%, 2.5%,3%, 3.4%, 4% and 5% at temperature 25°C with small angle neutron scattering (SANS) method. All of surfactants solutions were prepared using D_2O (99.9% deuterated, Prikladnaia Chimia, St. Petersburg, Russia) as a solvent. The nonionic gemini surfactant S-10 was obtained from Air Products & Chemicals, Inc., and used without further purification. All SANS measurements were performed on V-4 SANS spectrometer at BENSC, Berlin (Germany). Neutrons were used in wavelength range of 0.02 – 4 nm^{-1}. For the measurements quartz cells of were used during experiment. Up to 14 such cells were placed in a holder. Results from experiment was calculated and evaluated with PCG 2.0 program from Graz University (Austria). In the investigated solutions two axis ellipsoidal micelles was observed.

1. Introduction

The gemini in our study, α α'-[2,4,7,9-tetramethyl-5-decyne-4,7diyl]bis[ω hydroxyl-polyoxyethylene] [1] with m+n ethylene oxide (EO) segments. The molecules have two hydrophilic poly(ethylene oxide) (PEO) chains connected to a 14-carbon hydrophobic " hub"and the segmental lengths (m and n) of these two PEO chains are not identical. The surfactant used in this experiment will be named as follows: S-m+n,(e.g.,S-10 has a total of 10 EO monomers per molecule, m+n=10). Sato and Kashimoto first investigated aqueous S-10 solutions using vapour pressure depression, viscosity measurements [2], enthalpy of micellation [3], osmotic pressure [4], and NMR spectroscopy [5]. They observed two cmc values (0.9 and 2 wt. %) for water solutions of the S-10 gemini surfactant.

2. Experimental method and sample preparation

The nonionic gemini surfactant α α'-[2,4,7,9-tetramethyl-5-decyne-4,7diyl]bis[ω hydroxyl-polyoxyethylene] (S-10) (with first CMC = 0.9 wt%, second CMC = 2.0%)) was obtained from Air Products & Chemicals, Inc., and used without further purification. S-10 has consistency of a viscous and transparent fluid. This surfactant is solvated using heavy water D_2O instead of H_2O

Published under licence by IOP Publishing Ltd

Journal of Physics: Conference Series **351** (2012) 012016 doi:10.1088/1742-6596/351/1/012016

(99%, Prikladnaia Chimia, St. Petersburg, Russia) to the better contrast the surfactant from the solvent. The surfactant concentration used in experiment was varied from 2.5% to 5%. All measurements were carried out at temperature 25°C. Experiment was performed using SANS instrument located at the Hahn-Meitner-Institute, Berlin (Germany) [6]. The SANS instrument V4 [7] covers q range from 10^{-2} to 8.5 nm^{-1} allowing density composition. Neutrons during this experiment were used in wavelength range of 0.02 – 4 nm^{-1}. The instrument is installed at the curved neutron guide NL3a. In coming neutrons are monochromatized by a mechanical velocity selector with variable wavelength resolution (from 8% to 18%) and collimated on a variable length from 1 to 16 m. The two dimensional position sensitive detectors can be positioned at any distance between 1m and 16 m from the sample in the horizontal direction. Additional at 1m distance it can be moved vertically by 0.3m extending the q range to higher values. The ^3He detector of 64x64 cm^2 is equipped with new read-out electronics for 128x128 cells. For the measurements quartz cells were used during experiment. Up to 14 such cells were placed in a holder. The data treatment was done by the standard software BerSANS provided by the HMI [8].

3. IFT method for determine of PDDF function

The intensity of scattered radiation as a function of the magnitude of the scattering vector q is represented for monodisperse particles by [9]

$$I\,(q) = n\,P\,(q)\,S\,(q) \qquad (1)$$

where $q = 4\pi / \lambda \sin(\theta/2)$, λ is the wavelength of radiation, θ is the scattering angle, and n is the number density of particles, P (q) is the particle form factor and contains the effect of particle size, shape and scattering power on scattered intensity, S (q) is the structure factor and accounts for interparticle interactions, S (q) $\rightarrow 1$ for no interacting particles or in the limit of a very dilute solution. From the particle form factor it is possible to obtain the pair distance distribution function. The micelle shape determination for the experimental results was achieved using the generalized indirect Fourier transformation (GIFT) method [10]. The conventional Fourier transformation of I (q) involves the integral

$$p(r) = \frac{1}{2\pi^2}\int I(q)qr\sin(qr)dq \qquad (2)$$

which yields the pair distance distribution function $p(r)$,where r is the distance in real space. The point, at which the $p\,(r)$ falls to zero,is indicative of the particle maximum dimension.

4. Results and discussion

The nonionic gemini surfactant α α' – [2,4,7,9-tetramethyl-5-decyne-4,7diyl]bis[ω hydroxyl–polyoxyethylene] (S-10) (with first CMC = 0.9 wt% and second CMC = 2.0wt%) was investigated in heavy water solutions only for concentrations: 2.3%, 2.5%,3%, 3.4%, 4% and 5%. The measured results of SANS experiment distribution ($d\Sigma$ / $d\,\Omega$ vs q) after standard corrections and normalizations are shown on figure 1. The curves with big plateau were observed for all investigated solutions. The preliminate results and calculations of experiment are presented in this paper.

The PDDF (pair distance distribution function) for all investigated binary systems S-10 with heavy water shown figure 2. The curves are asymmetrical (symmetrical curves are for spherical micelles [11-12]) for all concentrations and micelles in these investigated solutions are slightly elongated spheres with diameter D=4.7nm. The size of micelles in these solutions is rather the same but with increase of concentration was observed increasing amount of micelles in investigated solutions. The micellization behaviour of dimeric surfactants has been the subject of several publications [13-16]. Depending on the chain and spacer lengths, various types of aggregates, such as spheres, ellipsoidal micelles, cylindrical or threadlike micelles, membranes and vesicles have been observed. Small-angle neutron scattering (SANS) is an ideal technique to study the micellar structures of surfactants. These studies clearly showed that micelle sizes and shapes for S-10 surfactant are approximately the same (slightly elongated spheres) for all investigated water solutions.

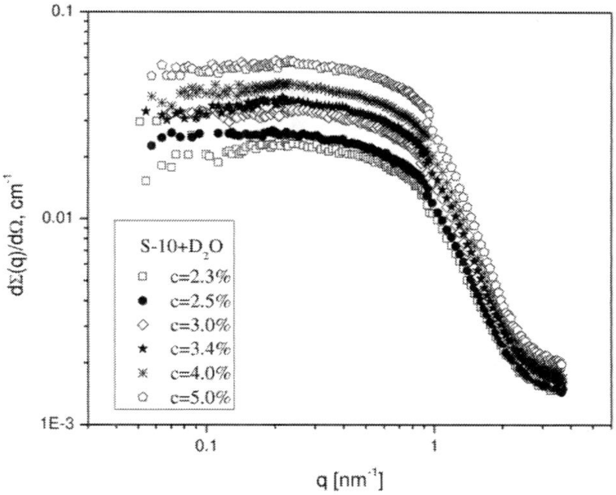

Figure 1. Intensity of neutron scattering vs scattering vector q for nonionic gemini surfactant S-10 in heavy water only for concentrations: 2.3%, 2.5%, 3.0%, 3.4%, 4.0% and 5.0%.

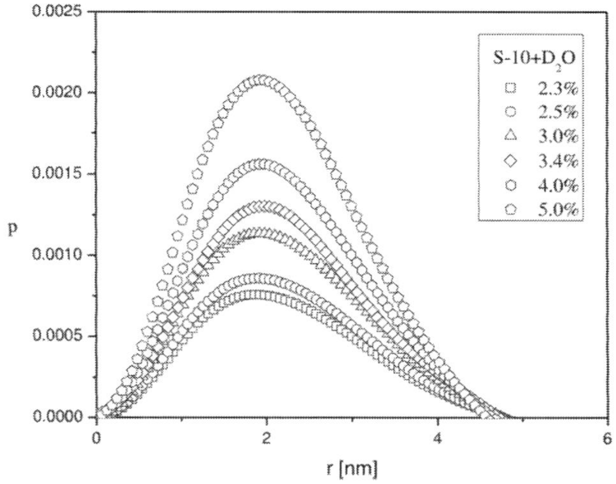

Figure 2. PDDFs of S-10 in D_2O at 25°C as a function concentration.

References

[1] Nieh M P, Sanat K, Kumar S K, Fernando R H, Colby R H and Katsaras J 2004 *Langmuir* **20** 9061
[2] Sato S and Hashimoto H 1985 *Bull. Chem. Soc. Jpn.* **58** 282
[3] Sato S and Hashimoto H 1988 *J. Colloid Interface Sci.* **123** 216
[4] Sato S and Hashimoto H 1988 *J. Colloid Interface Sci.* **126** 108
[5] Sato S 1989 *J. Phys. Chem.* **93** 4829
[6] Fe Lisi R, Gradzielski M, Lazzara G, Milioto S, Muratore N and Prevost S 2006 *J. Phys. Chem. B* **110** 2 5883
[7] Keiderling U and Wiedenmann A 1995 Phys. *B* **213 and 214** 895
[8] Keiderling U 2002 *Appl. Phys. A* **74** S1455
[9] Glatter O, Fritz G, Lindner H, Brunner-Popela J, Mittelbach R, Strey R and Egelhaaf S U 2000 *Langmuir* **16** 8692
[10] Fritz G and Glatter O 2006 *J. Phys.: Condens. Matter* **18** *S2403*
[11] Glatter O 1977 *Acta Phys. Austriaca* **74** 83
[12] Glatter O and Kratky O (ed) 1982 *Small-angle X-ray scattering (*London: Academic Press)
[13] De S, Asval V K, Goyal P S and Bhattacharya S 1996 *J. Phys. Chem.* **100** 11664
[14] Glatter O 1977 *J. Appl. Crystallogr.* **10** 415
[15] Alami E, Abrahamse`n-Alami S, Eastoe J and Heenan R K 2003 *Langmuir* **19** 18
[16] Wang X, Wang J, Yan H, Li P and Thomas R K 2004 *Langmuir* **19** 53

The studies of nanoparticles formed in silicate glasses doped by cerium and titanium oxides by means of small angle neutron scattering

S A Samoylenko[1,2], S E Kichanov[1], D P Kozlenko[1], A V Belushkin[1], V M Haramus[3], E A Trusova[4], G P Shevchenko[5], V S Gurin[5], L A Bulavin[2], S K Rakhmanov[5], B N Savenko[1]

[1]Joint Institute for Nuclear Research, 141980, Dubna, Russia

[2] National Taras Shevchenko University of Kyiv, Physics Department, 01601, Kiev, Ukraine

[3] Helmholtz-Zentrum Geesthacht, D-21502, Geesthacht, Germany

[4]The Belarusian State Technological University, 220006 Minsk, Belarus

[5]Research Institute for Physical Chemical Problems of the BSU, 220080, Minsk, Belarus

ekich@nf.jinr.ru

Abstract. The structure aspect of Ti-Ce-O nanoparticles, which is forming in silicate glasses doped by TiO_2 and CeO_2 oxides have been studied by means of a small angle neutron scattering. It was found, the complex oxide nanoparticles forms in these glasses in sizes range 30-38 nm and its average sizes depends on initial oxides concentration ratio. The correlation between nanoparticles structure features and silica glasses optical properties are discussed.

1. Introduction

At the present time one of the topical problems in optics is a search for nanosystems, which could efficiently absorb light in a given spectrum range [1]. The promising materials are silicate glasses doped by cerium and titanium oxides [1-3]. These glasses have yellow–orange color and its optical absorption edge can be shifted by varying the CeO_2/TiO_2 concentrations ratio [2]. Such glasses are characterized by hard UV-radiation protection and thermal stability, and they are suitable for numerous technological applications, such as laser optics elements, light filters and gem imitation [3].

It was assumed [4], that these unique physical properties of these glasses can be result of complex oxide nanoparticles Ce-Ti-O forming. The nanoparticles structural characteristics define the fine color tuning and UV radiation and temperature stability of such glasses. In those nanoparticles the titanium ions could be tetravalence, but the cerium ions can be both tetravalence and trivalence [3-5].

The physical and functional properties of these glasses depend on both the chemical composition of oxide system and the structure futures of Ti-Ce-O nanosystem constituents. The previous small angle X-ray scattering experiment [6] was detected the forming of nanoparticles in silicate glasses but its structural futures didn't obtain. In the present work most attention is focused on the research of structure aspects of nanoparticles formation in glasses with different initial oxides concentration by means of small angle neutron scattering and finding of correlations between nanoparticles structural futures and optical properties of glasses.

2. Experimental

The present silicate glasses were obtained from the melting of inorganic glass matrix SiO_2-Al_2O_3-MgO-CaO-BaO-SrO-Na_2O-K_2O-Li_2O with the adding of titanium and cerium oxides with necessary concentrations by means of high temperature gas furnace. The collection of glasses with relative molar concentrations of TiO_2/CeO_2 oxides: 0.0/2.0; 1.0/2.0; 5.0/2.0; 6.5/2.0; 10.0/2.0; 0.0/0.3; 5.0/0.3 and 10.0/0.3 have been obtained.

The small-angle neutron scattering (SANS) experiments were carried out with the spectrometer SANS-1 [5] on the research reactor FRG-1 (HZG, Germany) [7]. The neutron wavelength in experiments was 8.1 Å with average resolution $\Delta q/q$=1%. The four sample–detector distances to provide available in experiments q-range 0.005-25 nm^{-1}. The standard background and detector anisotropy corrections have been done.

3. Results and discussion

The SANS curves for pure silicon glass and glasses with different molar concentration ratio of CeO_2/TiO_2 oxides are shown in figure 1. The scattering curves from pure and titanium- free glasses are correspond with scattering from large objects and can be approximated by the simple power law [8]:

$$d\Sigma(q)/d\Omega = Aq^{-\alpha} + B, \qquad (1)$$

where A and B are scaling prefactors, α - the curves slope coefficient, which characterized the fractional dimension of scattering objects [8]. The obtained value of the slope coefficient for pure glass is α = 2.99(5). The microscopic air bubbles and pores forming at glasses manufacturing can be such large scattering objects. Previously [5] it was observed, that these pores does not appreciably influence on optical properties of obtained silicate glasses.

The small angle neutron scattering curves from silicate glasses doped by CeO_2/TiO_2 oxides are significant differ when for pure glasses (figure 1b-c). For SANS data treatment the complex model with consideration of the scattering from large objects and additional scattering from formed small Ti-Ce-O nanoparticles [6] have been used. In order to obtain the values of the gyration radius R_g corresponded with average size of the formed oxides nanoparticles, the experimental data were fitted by a function [8]:

$$d\Sigma(q)/d\Omega = Aq^{-\alpha} + C\exp(-\frac{1}{3}q^2R_g^2) + D, \qquad (2)$$

where A, C and D – scaled prefactors coefficients and R_g- gyration radius of forming nanoparticles. The obtained values of R_g and curves slope coefficients α are listed in Table 1.

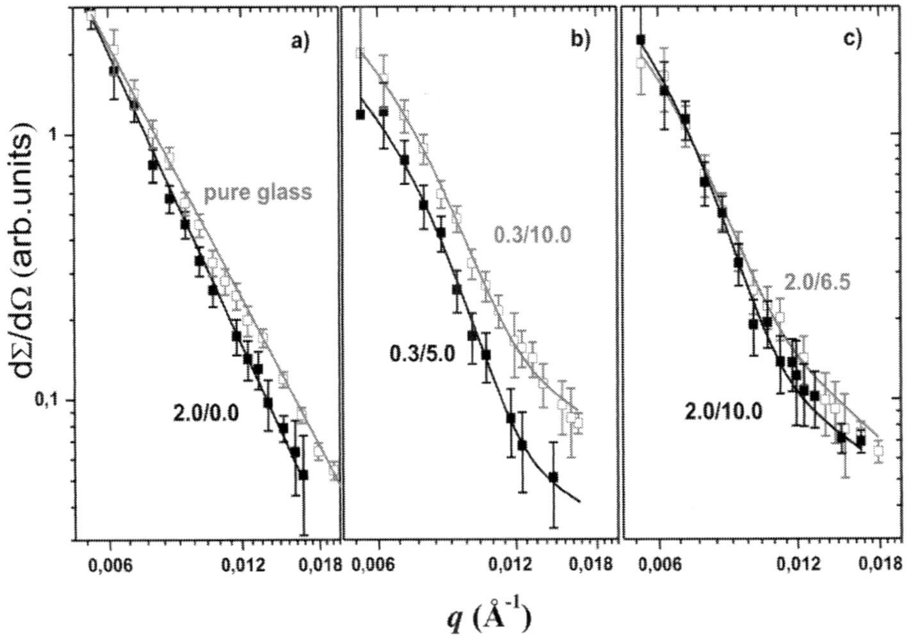

Figure 1. The small angle neutron scattering curves for pure and doped silicate glasses with relative molar concentration CeO_2/TiO_2 2.0/0 (a), 0.3/5.0 and 0.3/10 (b), 2.0/6.5 и 2.0/10.0 (c). The solid lines are experimental data fit by equation (1) (for curves at a part) and (2) (for curves at b and c parts).

Table 1. The gyration radius and curve slope coefficients for silicate glasses.

The molar ratio of initial oxides TiO_2/CeO_2	**Gyration radius R_g, Å**	**The curves slope coefficient α**
0.3/2.0	344±2	1.79±0.11
1.0/2.0	347±2	1.91±0.03
5.0/2.0	360±4	1.63±0.02
6.5/2.0	364±4	1.6±0.1
10.0/2.0	371±4	1.28±0.02
5.0/0.3	331±5	1.06±0.16
10.0/0.3	329±4	1.16±0.09
0/2.0	-	3.05±0.08
0/0.3	-	3.09±0.11

The nanoparticles formation in the glasses doped by cerium oxides only did not observed. The previous optical measurements [4, 5] did not find some significant changes in optical properties of glasses doped without TiO_2 inclusions.

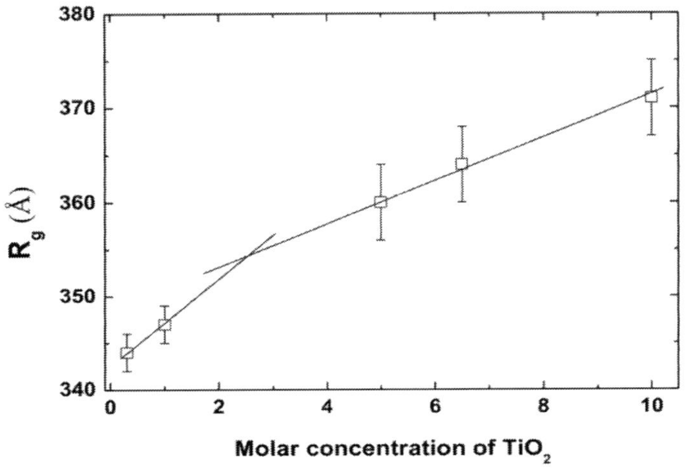

Figure 2. The gyration radius R_g as function of relative molar concentration of TiO_2 at fixed $x(CeO_2)$=2.0

The oxides nanoparticles gyration radius us function of the relative TiO_2 molar concentration at fixed cerium oxide concentration $x(CeO_2)$ = 2.0 are shown in Figure 2. The obtained data can be described by a combination of two linear functions: for low and for high TiO_2 concentrations regions. At low molar concentration of titanium oxide range from 0.3 to 2.0 gyration radius R_g of formed nanoparticles increases with coefficient $dR_g/dx \approx 4.3(1)$, while in high concentration range of TiO_2 this coefficient decreased to 2.2(2).

The previous obtained [3, 4] optical transmission coefficient K and corresponding wavelength λ_{max} of silicate glasses as a function of Ti-Ce-O nanoparticles gyration radius are shown at figure 3. For smaller nanoparticles with R_g<350 Å the significant decreasing of transmission coefficient K was found but the border λ_{max} ~ 390 Å characterized of glasses color is unchanged. For larger nanoparticles with R_g> 350 Å the shaper increasing of λ_{max} to 470 Å was found (figure 3).

We can propose the model of complex oxide nanoparticles forming based on the present SANS data and color centers nature studies [5, 6]. At small molar concentrations ratio of initial CeO_2/TiO_2 oxides the nanoparticles are formed of predominately from oxygen, trivalent cerium ions Ce^{3+} and tetravalent titanium Ti^{4+} [5]. At relative concentration increasing a valence state of cerium ions changes to tetravalence Ce^{4+} [5, 6] and this fact are reason to formation of another nanoparticles type, which characterized by larger average size and differ mechanism of optical properties forming.

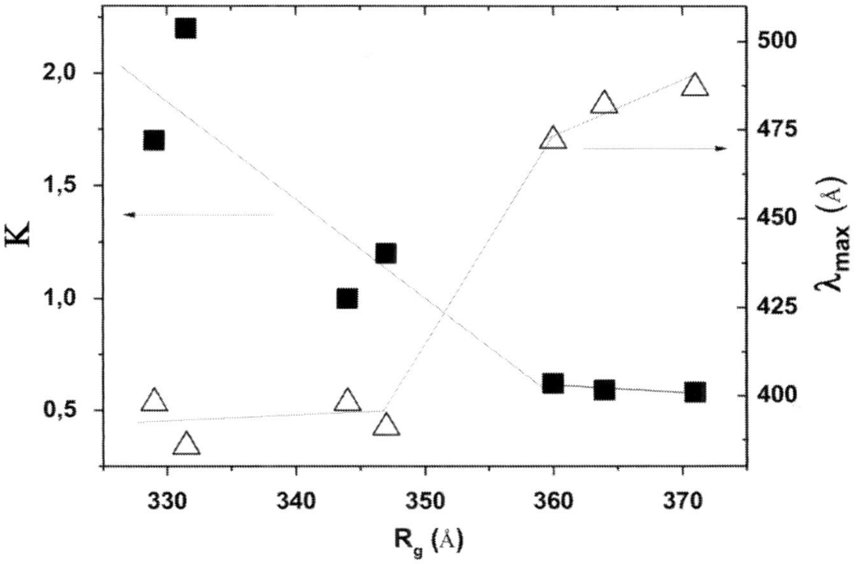

Figure 3. The optical transmission coefficient K and border wavelength λ_{max} (data from [3]) as function of gyration radius R_g

Conclusions

The results of our studies shown that in silicate glasses doped by cerium and titanium oxides the complex oxide Ti-Ce-O nanoparticles formed. The variation of TiO_2 relative concentration in low molar concentration range is reason the average size of formed nanoparticles increases as $dR_g/dx \approx 4.3(1)$, but at high TiO_2 doping concentration this coefficient approximately twofold decreased. This fact can be explain by model with two different type nanoparticles forming in glasses at different initial oxides ratio and corresponding with different valence state of cerium, and as result the noticeable influencing on glasses optical properties are observed.

The work has been supported by BRFBR-JINR grant №X10Д-007, contracts 02.740.11.0542 and 16.518.11.7029 within the Russian Federal Programs "The priority activities in researches and developments of scientific-technological complex of Russia in 2007-2012" and "Research and scientific- educational personnel of innovation Russia in 2009-2013".

References

[1] Kutolin S A and Neich A I 1988 *Physical Chemistry of Colored Glass*, Stroiizdat, Moscow, 296 p.
[2] Bogdanova G S, Antonova S L and Dzhurinskii B F 1970 Izv. Akad. Nauk SSSR, Neorg. Mater. **6** 943–948
[3] Trusova E E, Bobkova N M, Gurin V S 2009 Jurnal of Applied Spectroscopy **76** 2 211−217
[4] Trusova E E, Bobkova N M, Gurin V S, Tyavlovskaya E A 2009 *Glass and Ceramics* **66** 7–8
[5] Zenkovets G A, Shutilov A A, Gavrilov V Yu, Tsybulya S V, Kryukova G N 2007 Kinet.

Katal. **48** 742-748

[6] Trusova E E, Bobkova N M, Gurin V S, Golubkov V V 2009 in book Physics, Chemistry and Application of Nanostructures: Reviews and Short Notes to Nanomeeting-2009, Eds. V.E.Borisenko, World Scientific, Singapore 345-348 p.

[7] Stuhrmann H B, Burkhardt N, Dietrich G, Jünemann R, Meerwinck W, Schmitt M, Wadzack J, Willumeit R, Zhao J, Nierhaus K H. 1995 Nucl. Instr. & Meth. A **356** 133

[8] Brumberger H 1995 Modern Aspects of Small Angle Scattering, Dordrecht, Kluwer Acad. Publ. 221 p.

Small-Angle X-ray and neutron scattering from diamond single crystals

A A Shiryaev[1], P Boesecke[2]

[1]Institute of Physical Chemistry and Electrochemistry RAS, Leninsky pr. 31, 1199991, Moscow, Russia

[2] ESRF, 6 rue Jules Horowitz, BP220, 38043 Grenoble Cedex, France

E-mail: shiryaev@phyche.ac.ru

Abstract. Results of Small-Angle Scattering study of diamonds with various types of point and extended defects and different degrees of annealing are presented. It is shown that thermal annealing and/or mechanical deformation cause formation of nanosized planar and three-dimensional defects giving rise to Small-Angle Scattering. The defects are often facetted by crystallographic planes 111, 100, 110, 311, 211 common for diamond. The scattering defects likely consist of clusters of intrinsic and impurity-related defects; boundaries of mechanical twins also contribute to the SAS signal. There is no clear correlation between concentration of nitrogen impurity and intensity of the scattering.

1. Introduction

Small-angle X-ray (SAXS) and neutron (SANS) scattering are well-known methods of investigation of nanosized heterogeneities in solids and is widely employed for studies of aging of carious compounds and alloys. However, systematic studies of defects in diamonds are lacking. The first significant SAS study of synthetic diamonds using SAXS is described in [1]. It was shown that some synthetic diamonds possess anisotropic scattering ascribed to nanosized metal lamellae, entrapped during crystal growth. However, the progress in diamond growth under High-Pressures – High-Temperatures (HPHT) conditions is remarkable and modern crystals are virtually free from macro- and microinclusions (e.g., [2]). In several papers [3-8] we have demonstrated that prior to thermal annealing diamonds belonging to types IIa (nitrogen-free) and Ib (with single substitutional N atoms) show negligible intensity of small-angle scattering, but annealing and/or deformation of diamonds at HPHT-conditions lead to appearance of the scattering without relationship to presence of nitrogen defects in the lattice. Intensity and symmetry of the scattering depend on conditions of the sample treatment. The scattering is caused by defects with sizes in the range from 8-9 till ~60 nm; some of the defects are oriented in diamond lattice and are thus facetted or planar. Some of these defects, likely, influence thermal conductivity and mechanical properties of diamonds [9]. There is no unique correlation between nitrogen concentration in diamond lattice and the SAS properties, but a general trend is clear: the size of the scatterers decreases and their contrast with the diamond matrix increases with advancement of the annealing. The scattering from diamonds with low or moderate degree of nitrogen aggregation is generally weak and is due to large defects with broad size distribution. The scattering from heavily annealed diamonds of IaB type is strong and is probably related to voidites-like defects.

In this paper we present results of investigation of natural and synthetic diamonds using Small-Angle X-ray Scattering. This paper extends results obtained using neutron scattering for the same set of diamonds [6].

2. Experimental

Natural and synthetic diamonds with various concentrations of the principal chemical impurity - nitrogen – were studied. Concentration and type of N-defects were estimated using standard approaches by employing Infra-red spectroscopy [9-11]. Some of the specimens preserved crystallographic faceting; some were represented by flat plates with (100) orientation laser cut through the crystal center. Conditions of HPHT deformation and annealing as well as information about microstructure of the samples are described in detail in [7].

Small-angle scattering was studied at beamline ID01 at ESRF, Grenoble. The sample-detector distance was 4192 cm, the scattering path was under vacuum. Raw experimental data were treated using software described in [12]; calibration of the scattering intensity was performed using a Lupolen standard. Size distribution of the scatterers was calculated using GNOM software [13].

Already in early papers on SAXS studies of irradiated diamonds [14] it was shown that in case of certain orientation of the crystal towards the incoming beam double Bragg scattering (DBS, see details in [15]) may appear. An example of the SAXS pattern with significant contribution of the double scattering is shown on figure 1. It is obvious that interpretation of such patterns is very problematic. To avoid the double scattering the measurements were performed at several different energies around 8 keV. Additional control of the potential presence of the DBS was performed by comparison of anisotropies of X-ray and neutron scattering patterns. In the SANS experiments the DBS effect was fully absent since the employed wavelength (6 Å [6]), is more than two times larger than the largest interplanar distance in diamond (d_{111} = 2.06 Å).

It is important to note that qualitatively the 2D SAXS and SANS patterns are fairly similar, but the statistics of the X-ray data is much higher. In addition, in the X-ray experiments averaging of the scattering curves obtained at several energies was performed. Therefore, the main part of quantitative results was obtained from the SAXS measurements.

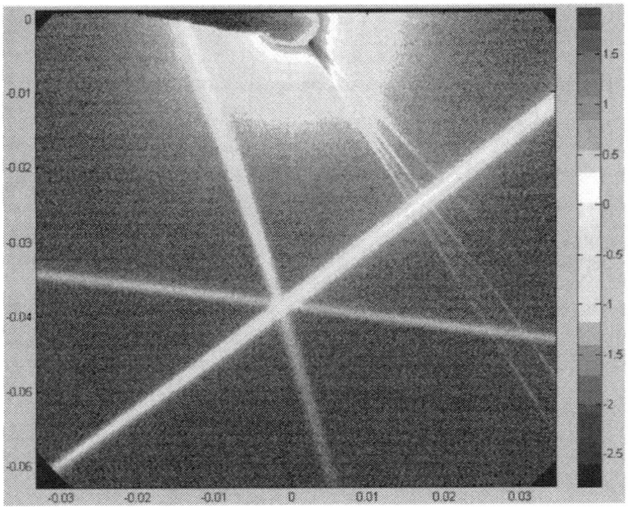

Figure 1. An example of 2D SAXS pattern with pronounced contribution of Double Bragg Scattering

3. Results

The small-angle scattering patterns from the majority of the studied diamonds with moderate degrees of nitrogen aggregation (i.e., with moderate post-growth annealing) are usually isotropic. This indicates that the shape of the scatterers is rather isometric. In principle, the isometric character of the scattering could result from random mutual orientation of the incident X-ray beam and anisotropic particles as it happens in solutions. However, our dedicated experiments demonstrate that in diamonds the anisotropy of the SAS patterns, if present at all, is pronounced even when the misalignment is significant.

Typical SAXS curves for moderately annealed diamonds are shown at figure 2a. The size distribution of the spherical scatterers is shown on figure 2b. In such diamonds the observed defects are heavily polydisperse with relatively abundant defects with radii between 50 and 100 nm. These results are in accordance with our previous studies of diamonds with low and moderate degrees of nitrogen aggregation [3-5, 8].

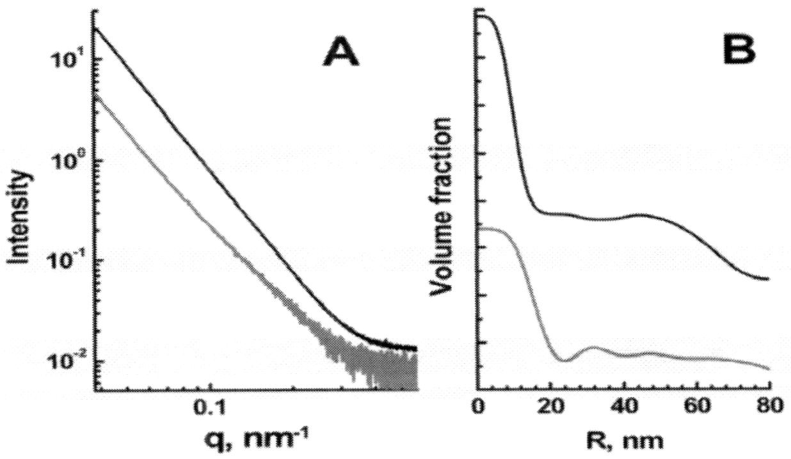

Figure 2. SAXS curves and (A) and size distribution of the scatterers in diamonds with moderate degree of nitrogen aggregation

Small-angle scattering from some specimens is markedly anisotropic. In most cases such diamonds are either deeply annealed in nature (diamonds of pure IaB type) or heavily deformed mechanically. For the IaB diamond (580 at. ppm of nitrogen) the X-ray and neutron scattering is strong and markedly anisotropic (figure 3a). The "rays" of the scattering patterns are directed along directions [110], [111], [311]. The scattering curves along these directions are shown on figure 3b. Broad Bragg peaks along the [111] indicate ordering of the scattering defects with the period approx. 13.6 nm. Moreover, weak peaks with periodicities approx. 25 and 36 nm are also present. Taking into account their considerable width and uncertainty in determination of the peak positions, it seems likely that these peaks correspond to doubled and tripled period of 13.6 nm. The existence of the periodicity in spatial distribution of the scatterers is rather unusual. The gyration radii are equal to 36 nm in [111] and 25 nm in [110]. The slope of the scattering curve in [110] on double logarithmic plot suggests that the shape of the scatterers in this direction is intermediate between disks and cylinders, but due to obvious polydispersity of the system such interpretation should be treated with caution.

One of the specimens was mechanically deformed at HPHT conditions (6.5 GPa, 1600 °C); a mixture of SiC powder and hard silicates was employed as a pressure transmitting medium in direct contact with the diamond sample. Transmission electron microscopy study of this sample showed that the deformation has produced high dislocation density with certain degree of polygonisation as well as mechanical twins [7]. The SAS patterns of this diamond are markedly anisotropic. The gyration radii

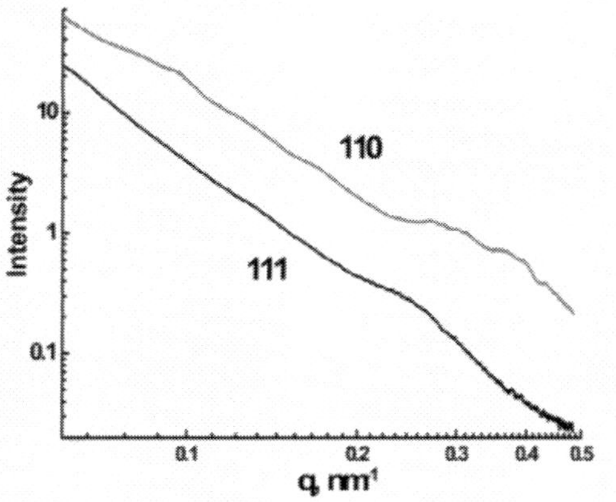

Figure 4. Deformed diamond: SAXS curves along different crystallographic directions.

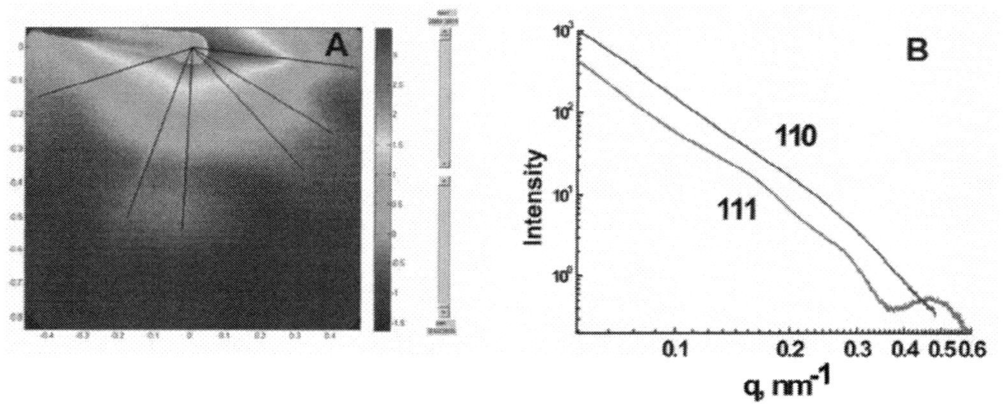

Figure 3. 2D SAXS pattern (A) the scattering curves (b) along two crystallographic directions for the type IaB diamond (the highest N aggregation state).

of the scatterers are similar for different directions and are between 32 and 36 nm. The most remarkable peculiarity of this sample is the existence of relatively weak Bragg peaks in several crystallographic directions (figure 4). In [111] the peak (probably an unresolved doublet) corresponds to the period 23-26 nm; possibly doubled period (46-47 nm) also exists. The curve in [110] contains numerous peaks with periodicities between 14.5 and 23 and 65-77 nm.

One of the studied samples - BR2 – belongs to so-called fibrous diamonds. Natural diamonds of this variety contain numerous submicroscopic inclusions of the growth medium and are of considerable interest for geosciences [9, 16]. The scattering pattern from this sample is obviously anisotropic and is characterised by eight-fold symmetry (figure 5). The scatterers are bound by (100) and (110) planes. Most likely, the scattering is mostly caused by facetted microinclusions. Existence of several subpopulations of the microinclusions - faceted oriented and irregular - in fibrous diamonds was discovered in TEM study [17]. These subpopulations differ in composition and origin. Large slope of the scattering curve on a log-log plot (~-5) indicates that the sizes of the scatterers exceed the resolution of the employed setup.

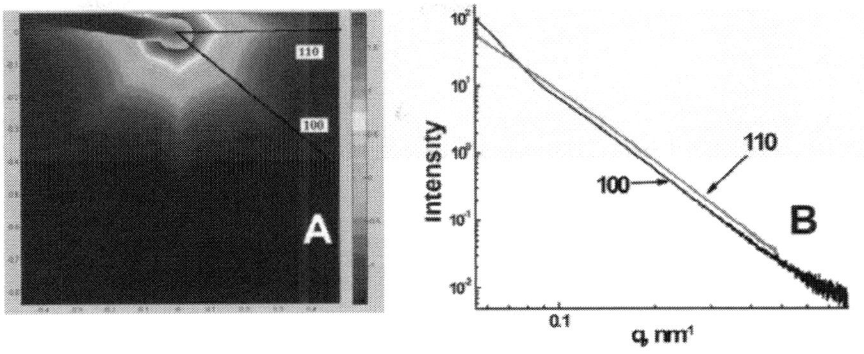

Figure 5. 2D SAXS pattern (A) and scattering curves (B) for a fibrous diamond.

4. Discussion

Results of the current study support conclusion of our previous works that many diamond crystals contain defects with sizes up to several tens nanometers which are formed as a result of diffusion of intrinsic and extrinsic defects during high temperature annealing. The SAXS and SANS [6] results also show that mechanical deformation also leads to formation of strongly scattering defects. Application of two-dimensional detectors reveals that some of the defects responsible for the SAS are highly anisotropic. The scattering anisotropy indicates that the scatterers produced during long-term annealing or mechanical deformation are facetted and/or highly anisotropic. The preferred directions – [100], [110], [111], and [113] – correspond to the most important crystallographic axes in diamond lattice. The faceting of macroscopic inclusions of foreign phases in diamond is a well-known phenomenon. Direct microscopic information about nanosized inclusions in diamonds is limited, but data on voidites shows that the majority of these defects are bounded by (111) planes. Most likely, this indicates achievement of an equilibrium shape. The SAS data shows that nanosized defects in diamonds could be facetted not only by common octahedral and cubic planes, but also by (110) and (113). This variability can be explained by changes in chemical composition of the trapped matter and corresponding changes in wettability of the diamond planes.

Most likely, voidites are the principal scatterers in heavily annealed diamonds. These defects are characterized by significant contrast of electron density with the diamond matrix. The SAS signal from type IaB diamonds, which usually contain voidites, is strong; therefore, SAS techniques represent a method of express selection of voidites-rich stones for subsequent detailed studies of voidites. Chemical composition of the scatterers in diamonds with moderate degrees on annealing remains poorly defined. Their sizes are considerable (tens of nanometers) and the contrast with the matrix should be small, since the SAS intensity is low-to-moderate. SANS and SAXS data [4, 6] show that even annealing of nitrogen-free diamonds leads to appearance of the scattering. Therefore, presence of chemical impurities such as nitrogen is not necessary for formation of the scatterers. Presumably, clusters of vacancies and interstitials may also give rise to the SAS signal. Existence of such clusters is indirectly supported by diffuse X-ray scattering [19], electron microscopy [20] and positron annihilation spectroscopy [21].

We also observe ordering in the spatial distribution of the extended defects. Similar ordering is often observed in aged alloys and semiconductors (e.g., [22]). However, in aged metal alloys the ordering results from high concentrations (several at.%) of chemical impurities, segregating into the new phases. Concentration of impurities in diamonds is always much smaller. It is thus not very probable that the ordered scattering defects fill the whole volume of a diamond crystal. More plausible model implies existence of localized lattice volumes with high concentration of the defects. As examples of such situation one may recall that voidites are usually concentrated around degraded

platelets [18] or on dislocations [23]. Such localization can be explained by directed diffusion of intrinsic and extrinsic point defects in strain fields surrounding the extended defects and by decay of supersaturated solid solution.
What is the nature of the scatterers produced by mechanical deformation? We have earlier demonstrated that microstructure of deformed diamonds depends on the treatment conditions and the deformation may lead to mechanical twinning or to appearance of dislocations with variable degrees of polygonisation. Mechanical twinning of diamond lattice may create lamellae of other (hexagonal) diamond polytypes. The differences in electron density of the diamond polytypes is small, but certain configurations of the interpolytype boundary may give rise to strong anisotropic Small-Angle Scattering.

5. Acknowledgments

The work was partially supported by Alexander von Humboldt foundation (AAS). We thank Drs. V.V. Volkov and E. Stoyanov for help in the experiments and for discussions.

References

[1] Woods G S 1970 *Philos. Mag.* **22** 1081
[2] Pal'yanov Yu N., Khokhryakov A F, Borzdov Yu M, Sokol A G, Gusev V A, Rylov G M and Sobolev N V 1997 *Russ. Geol. Geophys.* **38** 882
[3] Shiryaev A A, Klyuev Yu A, Naletov A M, Dembo A T, Feigelson B N 2000 *Diam. Relat. Mater.* **9** 1494
[4] Shiryaev A A , Hutchison M T, Dembo K A, Dembo A T, Iakoubovskii K, Klyuev Yu A, Naletov A M 2001 *Physica* **B308–310** 598
[5] Shiryaev A A, Dembo K A, Klyuev Yu A, Naletov A M, Feigelson B N 2003 *J. Appl. Cryst.* **36** 420
[6] Shiryaev A A 2007 *J. Appl. Cryst.* **40** 116
[7] Shiryaev, Frost D J, Langenhorst F 2007 *Diam. Relat. Mater.* **16** 503
[8] Shiryaev A A, Wiedenbeck M and Hainschwang T 2010 *J. Phys.: Condens. Matter* **22** 045801
[9] Bokii G B, Bezrukov G N, Klyuev Yu A, Naletov A M and Nepsha V I 1986. *Natural and Synthetic Diamonds*. Moscow: Nauka.
[10] Boyd S R, Kiflawi I, Woods G S 1994 *Philos. Mag.* **B69** 1149
[11] Boyd S R, Kiflawi I, Woods G S 1995 *Philos. Mag.* **B72** 351
[12] Boesecke P 2007 *J. Appl. Cryst.* **40** 423 (2007); http://www.esrf.eu/UsersAndScience/Experiments/TBS/SciSoft
[13] Svergun D I 1992 *J. Appl. Cryst.* **25** 495
[14] Perret R G and Keating D T 1967 *Small-angle X-ray scattering*, ed. Brumberger H, Gordon & Breach, New York 373-380.
[15] Terminasov Yu S, Tuzov L V 1964 *Physics-Uspekhi* **7** 434
[16] Shiryaev A A, Izraeli E S, Hauri E H, Zakharchenko O D, Navon O 2005 *Russ. Geol. Geophys.* **46** 1207
[17] Klein-BenDavid O, Wirth R, Navon O 2007 *Earth Planet. Sci. Lett.* **264** 89
[18] Bursill L A and Glaisher R W 1985 *Am. Mineral.* **70** 608
[19] Ramanan R R, Goswami S N N, Lal K 1998 *Acta Cryst.* **A54** 163
[20] Bangert U, Barnes R, Gass M H et al., 2009 *J. Phys.: Condens. Matter* **21** 364208
[21] Shiryaev A A, Iakoubovskii K, Schut H, van Veen A, Escobar Galindo R, Zakharchenko O D, Klyuev Yu A, Kaminsky F V, Feigelson B N 2001 *Mater. Sci. Forum* **363-365** 40
[22] Kostorz G 1991 *J. Appl. Cryst.* **24** 444
[23] Chen J H, Bernaerts D, Seo D W, van Tendeloo G, Kagi H 1998 *Phil.Mag.Lett.* **77** 135

Journal of Physics: Conference Series **351** (2012) 012019 doi:10.1088/1742-6596/351/1/012019

Offer search deviations from Newton's law of gravity using low-energy neutrons

Yu Alexandrov[1]

FLNP, JINR, Dubna, Russia

E-mail: alex@nf.jinr.ru

Abstract. Information about the planned experiment at the reactor IBR-2M (installation YuMO) dedicated to finding differs from the Newtonian law of gravitation at small (atomic) distances reported. The experiment is to study the energy dependence of the scattering length on the matter with a very small length of pure nuclear scattering (e.g. by a mixture of isotopes of tungsten), allowing thereby increasing the relative contribution of gravitational scattering. At the concentration of tungsten-186 of about 90% in the mixture, nuclear scattering length may to turn to zero at low energies. Relying on the accuracy of existing experiments it can argued that not Newtonian gravitational scattering can be detected if the scattering length is more than 10-16 or 10-17cm. Estimates show that in the theoretical model associated with the length of the electroweak interaction the amplitude of the gravitational scattering of neutrons in the Born approximation may be in the region from 10-11 to 10-21 cm. The magnitude of usual Newtonian gravitational scattering is 10-32 cm, i.e. significantly less.

The possible existence of multidimensional gravity is an idea that creates an increasing interest among physicists. The topicality of this idea is confirmed, in particular, e.g. by the Russian gravitation conferences (Kazan, September 2007 and Moscow, June 2008) and whose program included a number of questions related to multidimensional gravity. Some papers have appeared [1–3] on possible experimental tests of the idea of extra dimensions with the aid of neutron scattering.

The conventional Newtonian law of gravity reads

$$F = -Gm_1m_2r^{-2} \tag{1}$$

where F is the force of gravity, G is the Newtonian gravitational constant, m_1 and m_2 are the interacting masses and r is the distance between them. The minus sign means that gravity is an attracting force.

The law (1) has been well confirmed by now up to distances of the order of a few millimeters. Meanwhile, at smaller distances one can expect deviations from it, and, even from a purely experimental viewpoint, testing the Newton law at small distances is highly desirable.

Following [1], one can recall some theoretical reasoning connected with the weak interaction length that can lead to a modification of Newton's law at small distances. Suppose that the masses as well as all forces (except the gravitational ones) obey the laws established in our usual three-dimensional space while the gravitational forces should be described in terms of a multidimensional space. Then, according to [1], we can assume that the gravitational force between two particles placed at a distance equal to the electroweak length $R_e=[G_F(\hbar c)^{-1}]^{1/2}$, where $G_F = (1.166\ 10^{-5}\ GeV^{-2}\ (c\hbar)^3 = G_f\ (c\hbar)^3$ is the Fermi constant, from each other, and the usual Newtonian force between the same particles at a distance equal to the Planck length $R_p=[\hbar G(c^3)^{-1}]^{1/2}$, are equal. Then the gravitational force may be written as follows:

$$F_n= - Gm_1m_2\, R_c^n\, r^{-(n+2)} \tag{2}$$

where R_c^n is some characteristic distance, and n = 0, 1, 2, . . . characterizes the multidimensional nature of gravity, so that for n = 0 equation (2) passes over to the conventional Newtonian expression (1). Equation (2) should be valid for $r \le R_c^n$. At $r > R_c^n$ the Newton law of gravity should hold. Following [1], one can also obtain that

$$R_c^n = {}^{2-n/2}G_F^{1+n/2}G^{-1}(\hbar^{2+n/2})^{-1} \tag{3}$$

Calculations lead to the following results:

Table 1. Dependence R_c from n

n	1	2	3	4
R_c, cm	1.2×10^{17}	2.8	8×10^{-6}	1.4×10^{-8}

It should be noted that the above reasoning and equations (2) and (3) do not form a rigorous theory but can only serve for estimation purposes.

The value n = 1 in the above table should be immediately rejected since Rc in this case exceeds the size of the Solar system and would lead to instability of the planetary orbits. The values of Rc smaller than 1cm are not easily checked by macroscopic experiments but tests with the aid of neutrons are basically possible. At any rate, it is an attractive idea to seek deviations from the inverse square law using neutrons even if the above theoretical reasoning proves to be wrong.

For distances $r \le Rc$, the gravitational potential may be written in the form:

$$V(r) = -m_1m_2\, GR_c^n[(n+1)r^{n+1}]^{-1} \tag{4}$$

The gravitational scattering length may be found in Born's approximation. For n = 2 it can be written [4]:

$$a_{coh,grav} = -2/3m_n^2m_TG_f^2c^7\,\hbar \int_0^{R_c} \frac{\sin qr}{qr^2}\,dr \tag{5}$$

where $q=2k\sin(\theta/2)$, k is the wave number (unrelativistic $k=2.197\times10^9\sqrt{E}$ cm if E is in eV), m_n and m_T are the masses of neutron and target, respectively, R is the nuclear radius, $G_f = 1.166\times10^{-5}\ GeV^{-2}$

The quantity (5) grows as the neutron energy decreases, and for small qR, expanding the first four terms in (5) in a series, we obtain

$$a_{coh,grav} = -2/3m_n^2m_TG_f^2c^7\,\hbar\ [0.756 - \ln(qR)] \tag{6}$$

Thus the best conditions for measurements (a comparatively large length $a_{coh,grav}$) take place at small energies of the scattered neutrons. One can obtain

Journal of Physics: Conference Series **351** (2012) 012019 doi:10.1088/1742-6596/351/1/012019

$$a_{coh,grav} = -5.44 \times 10^{-21} \text{cm at } \lambda = 250\text{nm} \quad (7)$$
$$a_{coh,grav} = -4.56 \times 10^{-21} \text{cm at } \lambda = 11\text{nm} \quad (8)$$

For n = 0 holds Newtonian gravitational potential and one can obtain $a_{coh,grav} = 10^{-32}$cm i.e. quite small. It is almost impossible to measure.

For n=1 $V(r) \sim 1/r^2$ and for example in the electroweak theory form [1] one can be obtained that

$$a_{coh,grav} = m_n^2 m_T G_f^{3/2} c^6 \int_0^{R_c} \frac{\sin qr}{qr} dr \quad (9)$$

and

$$a_{coh,grav} = -6.77 \times 10^{-10} \text{cm at } \lambda = 250\text{nm} \quad (10)$$
$$a_{coh,grav} = -3.10 \times 10^{-11} \text{cm at } \lambda = 11\text{nm} \quad (11)$$

For comparison, it should be recalled that the lengths of neutron scattering is 5×10^{-13}cm, i.e. significantly less. The truth according to the work [1] n = 1 option is unlikely to be realized, the solar system probably would not be stable. We should not forget that the electroweak theory of option [1] should be regarded only as estimates - the value of Rc can be changed for example in 137 times (the 1/137 is the fine structure constant) even in the work [1].

At n>2, the quantity $a_{coh,grav}$ decreases strongly, for n=3 $a_{coh,grav} \approx 10^{-28}$cm.

In this paper we look for deviations from the formula (1) using neutron scattering at low energies for nuclei with small nuclear scattering length, for example, a mixture of isotopes of tungsten-186 (its content in the mixture should be ≈ 0.9) The scattering length of one of these mixtures was measured at λ=150nm and $a_{coh} = -0.0466(6) \times 10^{-12}$cm [5].

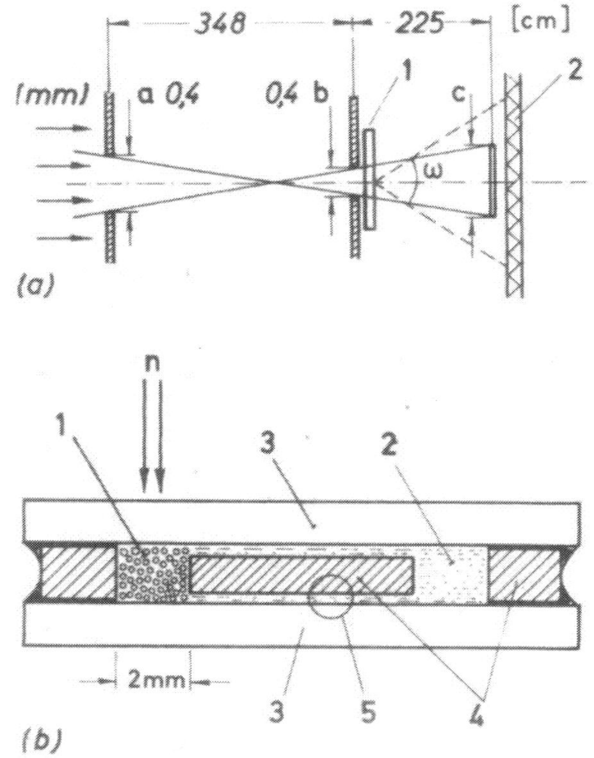

Figures a and b. Experimental set-up

for Christiansen filter measurements.

(a)- beam collimation: a,b-slits; c-stopper, 1-filter, 2-neutron counter. (b) cross cut of the Christiansen filter: 1-powder, 2-liquid, 3-glass, 4-distance holder, 5-liquid split

One of the accepted methods of measuring the scattering length is the method of Christiansen filters (in optics was first used as early as 1884). The method is as follows (see e.g. [5,6]). In a well-collimated neutron beam (divergence in the horizontal plane did not exceed 2-3 minutes of arc) set so-called Christiansen filter (see figure a, b), which is a vertical cavity of the test substance.

The thickness of the cavity in the direction of the beam is a few millimeters and the width of it fully covers the beam. The space between the grains of powder is filled alternately by several liquids with known refractive indices of neutrons.

At a distance of several meters behind the filter on the beam axis a neutron detector establish so wide that it can detect neutrons scattered by the filter at angles of less than 0.5^0 in the horizontal plane. Measurement is to monitor the neutron scattering intensity at small angles by the filter for several values of the refractive index of the liquid filling it. As the liquid can be used such as various mixtures of heavy ($Na_{coh} > 0$) and light ($Na_{coh} <0$) water, where N is the number of nuclei per unit volume.

In principle, this scattering should be completely absent if the refractive index of the liquid is equal to the refractive index of the substance of the powder. We can show that the scattering intensity [5,6]:

$$I \sim (n_{acoh})_L - (n_{acoh})_P \tag{12}$$

where L-refers to the liquid, and P - to the powder.

The absence of small-angle scattering when $(n_{acoh})_L = (n_{acoh})_P$, to determine $a_{coh,P}$. The error of the experiment may not exceed 0.1%.

Coherent scattering length of neutrons by tungsten in the case of low energy neutrons can be written as (see e.g. [5]):

$$a_{coh} = R - \frac{\alpha\Gamma_n}{2k_o E_o}\left(1+\frac{E}{E_o}\right) + Zfa_{ne} + a_{pol} + a_{coh\,grav} \tag{13}$$

where R - radius of the nucleus, α - the concentration of tungsten-186 in the mixture, ko = $2.197\text{x}10^9\sqrt{E_0}$ - wave number (energy E in eV), Γ_n -neutron width of the first resonance of the tungsten-186 (Eo = 18.83 eV), f - atomic form factor (at small scattering angles can be shown that f = 1), a_{ne} -length neutron scattering on an electron (the value of $a_{ne} = -1.60\ (3)\ \text{x}10^{-16}$ cm, it is this line [7] the meson theory of Yukawa). At small scattering angles for tungsten value $Zfa_{ne} = -\ 0.118\text{x}10^{-13}$ cm. However, the value a_{ne} insignificant for consideration following the experiment - the only important thing to Zfa_{ne} - does not depend on the neutron energy. a_{pol} - neutron scattering length induced by the electric polarizability of the neutron. At low energies (λ ≈ 150nm) the

$$a_{pol} = \frac{m\alpha}{2R}\left(\frac{Ze}{\hbar}\right)^2\left(1-\frac{\pi}{4}qR+...\right) \approx 10^{-14}(1-2\text{x}10^{-6}+...) \tag{14}$$

and for small scattering angles is almost independent of neutron energy [8]. If you choose a value of α such that in formula (10) members compensate one another, then the total scattering length is close to zero. At the wavelength of the neutron λ= 150nm value α should be with about 0.9. In the wavelength range from 10 to 150nm size of the scattering length will vary with energy, but not very strongly, since E <<Eo (at λ= 150nm E = 0.00036 eV, and $E \sim 1/\lambda^2$).

A few words about the possibilities of the proposed method of searching for gravitational scattering length. First of all, it is much more straightforward for experimentalists, since it does not need many of systematic corrections, while not well measured at present, but necessary for a number of other options for proposals (see e.g. [3]).

Building on the work [9], we can assume that used a mixture of isotopes of tungsten deviation of the scattering length of the linear energy dependence due to nuclear interaction is unlikely.

Further, taking the smallest error in the famous experiments by the method of Christiansen filters of 10^{-16} cm (see e.g. book [10]), we can assume that the allowable measured length of the gravitational scattering could now make a value of approximately the same order. However, measurement error can be decreased and the method proposed above measurements probably all still close to the most relevant.

The proposed experiment can be performed on YuMO installation. Currently there is a discussion of its details.

Author is grateful to A.I. Kuklin and his collaborators for good discussions.

References

[1] Frank A, Van Isacker P, Gomez-Camacho J 2004 *Phys.Lett.B* **582** 15
[2] Alexandrov Yu 2006 *Preprint JINR* E3-2006-142
[3] Pokotilovski Yu 2006 *Yad. Fiz.* **69** 924
[4] Alexandrov Yu 2008 *Gravitation and Cosmology* **14** 283
[5] Alexandrov Yu, Koester L, Samosvat G 1970 *Communication JINR* E3-5371
[6] Koester L 1977 Neutron Scattering Length. In *Springer Tracts in Modern Physics*, **80** Springer-Verlag,
[7] Alexandrov Yu 2010 *Preprint JINR* P3-2010-63
[8] Alexandrov Yu 1992 Fundamental Properties of the Neutron. (Oxford Science Publications)
[9] Alexandrov Yu, Samosvat G 1990 *VI Intern. School on Neutr. Phys.*, Alushta, **1** 187
[10] Rauch H, Waschkowski W 2000 In Landolt-Bornstein *«Low Energy Neutron Physics»*, p.6-1, Springer.

Journal of Physics: Conference Series **351** (2012) 012020 doi:10.1088/1742-6596/351/1/012020

Small-angle scattering from generalized self-similar Vicsek fractals

Alexander Yu Cherny[1], Eugen M Anitas[1,2], Vladimir A Osipov[1] and Alexander I Kuklin[1]

[1] Joint Institute for Nuclear Research, Dubna 141980, Moscow region, Russian Federation
[2] Horia Hulubei National Institute of Physics and Nuclear Engineering, RO-077125 Bucharest-Magurele, Romania

E-mail: cherny@theor.jinr.ru

Abstract.
An analytical approach for calculating the small-angle X-ray or neutron scattering (SAXS/SANS) from generalized self-similar Vicsek fractals (GSSVF) is presented; each fractal consists of spherical subunits. The system considered is a mass-fractal, generated iteratively from a regular 3D Vicsek fractal structure. Its fractal dimension is controllable and increases with increasing the value of the scaling factor. Small-angle scattering (SAS) intensity is determined from a set of non-interacting, randomly oriented and uniformly distributed GSSVF fractals. It is shown that in the fractal region, the curve $I(q)q^D$ is approximately log-periodic with the period equal to the scaling factor of fractal; here D and $I(q)$ are the fractal dimension and the SAS intensity, respectively. In particular, the positions of deepest minima and highest maxima are log-periodic, and their number coincides with the number of fractal iterations. The log-periodicity of the scattering curves is a consequence of the self-similarity of GSSVF.

1. Introduction

Small-angle scattering (SAS) [1, 2] from fractal systems [3] provides valuable information about the structure at nano-micro scales [4, 5, 6]. The SAS intensity decays in the momentum space according to the power law [7]:

$$I(q) \propto q^{-D}. \tag{1}$$

Here q and D are the absolute value of the scattering vector and the fractal dimension of the fractal, respectively. This type of scattering is often referred to as the power-law decay. Equation (1) is valid within the reciprocal fractal region, given by

$$1/l \lesssim q \lesssim 1/l_{\mathrm{min}}. \tag{2}$$

Here l is the size of the fractal and l_{min} is the minimal distance between the scattering centers.

In the fractal region, the scattering intensity is proportional to the Fourier transform of the pair distribution function $g(r)$, describing the spatial correlations between particles inside a fractal. Equation (1) leads to

$$g(r) \propto r^{D-3}, \tag{3}$$

by Erdélyi's theorem for asymptotic expansion of Fourier integrals [8]. This equation is satisfied within the fractal region in real space

$$l_{\mathrm{min}} \lesssim r \lesssim l. \tag{4}$$

Therefore, three parameters can be extracted from experimental fractal scattering intensities: the exponent and the edges of the fractal region in the q-space, which appear as "knees" in the scattering line on a logarithmic scale. For random fractals, it is rather difficult to extract more information from the scattering, because a fine structure of particle correlations is usually smeared due to the randomness. One can expect that deterministic fractals [9], being more ordered, allow us to obtain additional information from the scattering data.

Analytical calculations of SAS intensity from the 3D triadic Cantor and Vicsek [10] sets were reported in Ref. [11]. In the previous paper [12], the generalized 3D Cantor set was suggested, whose dimension can vary from zero to three by means of varying the scaling factor. If the SAS is considered from monodispersive sets, which are randomly oriented and placed, then the scattering intensities represent minima and maxima superimposed on a power-law decay, with the exponent equal to the fractal dimension of mass fractal. This dependence of the intensity on momentum is called the generalized power-law decay.

Recently, it was shown [13] that in the reciprocal fractal region, the curve $I(q)q^D$ is approximately *log-periodic* with the period equal to the scaling factor of the fractal. In this paper we relate this log-periodicity with the self-similarity of generalized self-similar Vicsek fractals (GSSVF) and explicitly show that the pair distribution function $g(r)$, describing spatial correlations between positions of the fractal subunits, also obeys the generalized power law. As was shown [5, 11, 12], the minima and maxima are damped with increasing polydispersity of the fractal sets. Physical reasons of such a behavior are quite clear: the fractal dimension dictates the power law (3) for the particle correlations *only on the average*, and polydispersity smears the spatial correlations between the units composing the fractal.

We present a scheme, first introduced in Ref. [13], for estimating the iteration number, the scaling factor and the total number of structural units in GSSVF from the scattering curves. The present analysis of GSSVF is based on the method of calculating the scattering amplitude, developed in Refs. [14, 15, 16, 17].

1.1. Background on small-angle scattering

Here we restrict ourselves to two-phase systems, which are composed from homogeneous units of "mass" density $\rho_{\rm m}$. The units are immersed into a solid matrix of "pore" density $\rho_{\rm p}$. Then [1, 2] we can consider the system as if the units were "frozen" in a vacuum and had the density $\Delta\rho = \rho_{\rm m} - \rho_{\rm p}$. The density $\Delta\rho$ is called scattering contrast. The scattering intensity (that is, the cross section per unit volume of the sample) is given by

$$I(q) = n|\Delta\rho|^2 V^2 \left\langle |F(\boldsymbol{q})|^2 \right\rangle, \tag{5}$$

where n is the fractal concentration, V the "mass" volume of each fractal, and $F(\boldsymbol{q})$ the normalized form factor

$$F(\boldsymbol{q}) = \frac{1}{V} \int_V e^{-i\boldsymbol{q}\cdot\boldsymbol{r}} \mathrm{d}\boldsymbol{r}, \tag{6}$$

obeying the condition $F(0) = 1$. The brackets $\langle \cdots \rangle$ stand for the ensemble averaging over all orientations of $\boldsymbol{q}$. Therefore, the SAS intensity at zero angle will be $I(0) = n|\Delta\rho|^2 V^2$. Once a deterministic fractal is composed of the same objects, say, N balls of radii R, then (see, e.g., Ref. [13])

$$F(\boldsymbol{q}) = \rho_{\boldsymbol{q}} F_0(qR)/N, \tag{7}$$

with $\rho_{\boldsymbol{q}} = \sum_j e^{-i\boldsymbol{q}\cdot\boldsymbol{r}_j}$ the Fourier component of the density of the balls, and $\boldsymbol{r}_j$ are the center-of-mass positions of balls. Here the ball form factor of unit radius is given by [2]

$$F_0(z) = 3(\sin z - z\cos z)/z^3. \tag{8}$$

By substituting Eq. (7) into Eq. (5), we obtain for the scattering intensity

$$I(q) = I(0)S(q)|F_0(qR)|^2/N, \tag{9}$$

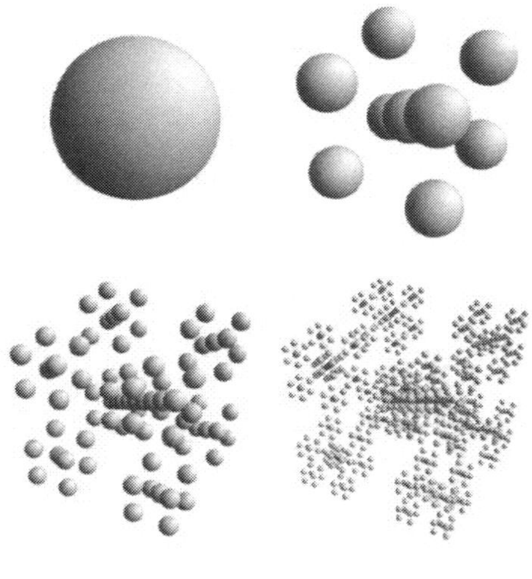

Figure 1. The initiator and first three iterations for the GSSVF at the scaling factor $\beta_s = 1/6$.

where we put by definition

$$S(q) \equiv \langle \rho_{\boldsymbol{q}} \rho_{-\boldsymbol{q}} \rangle / N. \tag{10}$$

The quantity $S(q)$ is called structure factor, and it is intimately connected to the pair distribution function, see Eq. (21) below. In the fractal region (2), $q \lesssim 1/R$ because of $R \leqslant l_{\min}$, and hence $F_0(qR) \simeq 1$. This implies that the scattering intensity is proportional to the structure factor in the fractal region.

1.2. Construction of 3D generalized self-similar Vicsek fractals

We follow the procedure described in Ref. [13] by starting with a cube with edge l_0 and placing in its center a ball of radius $l_0/2$. This is zero fractal iteration, called the initiator. The iteration rule (generator) is to replace the initial ball by nine smaller balls of radius $\beta_s l_0/2$ (Fig. 1). We choose the Cartesian coordinates such that the origin lies in the cube center, and the axes are parallel to the cube edges. The position of one ball is at the origin, while the centers of the eight other balls are shifted from the origin by the vectors

$$\boldsymbol{a}_j = \{\pm\beta_t l_0, \pm\beta_t l_0, \pm\beta_t l_0\}, \tag{11}$$

with all the combinations of the signs, where

$$\beta_t \equiv (1 - \beta_s)/2. \tag{12}$$

Repeating the same operation to each ball m times, we arrive at the mth approximant, which is composed of the $N_m = 9^m$ balls of radius $r_m = \beta_s^m l_0/2$.

The ideal GSSVF is obtained in the limit $m \to \infty$. Its Hausdorff (fractal) dimension can be derived from the condition [3] $N_1 \beta_s^D = 1$. For GSSVF it takes the explicit form

$$D = -\frac{2\log 3}{\log \beta_s}, \tag{13}$$

where the restriction $\beta_s < \sqrt{3}/(\sqrt{3}+2)$ is imposed on the scaling factor in order to avoid overlapping of the balls.

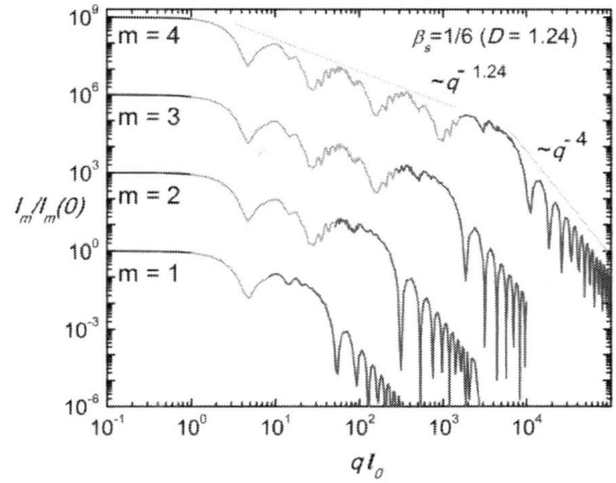

Figure 2. (Color online) Scattering intensities (16) for the first four iterations of monodisperse GSSVF. The scattering curve for the mth iteration is scaled up, by the factor $10^{3(m-1)}$. Black at low q shows the Guinier regions ($q \lesssim 1/l$); light gray (orange) represents fractal regions, given by Eq. (2); dark gray (blue) at high q, Porod regions ($q \gtrsim 1/r_m$).

2. Form factor

One can obtain [11, 12, 13] an analytical formula for the monodisperse form factor

$$F_m(\boldsymbol{q}) = F_0(r_m q) \prod_{i=1}^{m} G_i(\boldsymbol{q}), \tag{14}$$

for $m = 1, 2, \cdots$. Here the generative function reads

$$G_m(\boldsymbol{q}) = \left[1 + 8\cos(u_m q_x)\cos(u_m q_y)\cos(u_m q_z)\right]/9, \tag{15}$$

with $u_m \equiv l_0 \beta_t \beta_s^{m-1}$, $m = 1, 2, \cdots$. Then SAS intensity is obtained with the help of Eq. (5)

$$I_m(q)/I_m(0) = \left\langle |F_m(\boldsymbol{q})|^2 \right\rangle, \tag{16}$$

see Fig. 2.

A real physical system is almost always composed of scatterers of different sizes, and, therefore, a more realistic description should include polydispersity. The distribution function $D_N(l)$ of the scatterer sizes is defined in such a way that $D_N(l)dl$ gives the probability of finding a fractal whose size falls within the interval $(l, l + dl)$. We consider a log-normal distribution function of the fractal size

$$D_N(l) = \frac{1}{\sigma l (2\pi)^{1/2}} \exp\left(- \frac{\left[\ln(l/l_0) + \sigma^2/2\right]^2}{2\sigma^2} \right), \tag{17}$$

where $\sigma = \left[\ln(1 + \sigma_r^2)\right]^{\frac{1}{2}}$. The mean length and its relative variance are

$$l_0 \equiv \langle l \rangle_D, \quad \sigma_r \equiv \left(\langle l^2 \rangle_D - l_0^2\right)^{1/2}/l_0, \tag{18}$$

with $\langle \cdots \rangle_D = \int_0^\infty \cdots D_N(l)\mathrm{d}l$.

The resulted intensity is the average of (16) over (17) [see Figs. 3(a) and 3(b)]

$$I_m(q) = n|\Delta\rho|^2 \int_0^\infty \left\langle |F_m(\boldsymbol{q})|^2 \right\rangle V_m^2(l) D_N(l)\mathrm{d}l, \tag{19}$$

where $V_m = N_m \beta_s^{3m} V_0$ is the total volume of "mass" in the mth iteration.

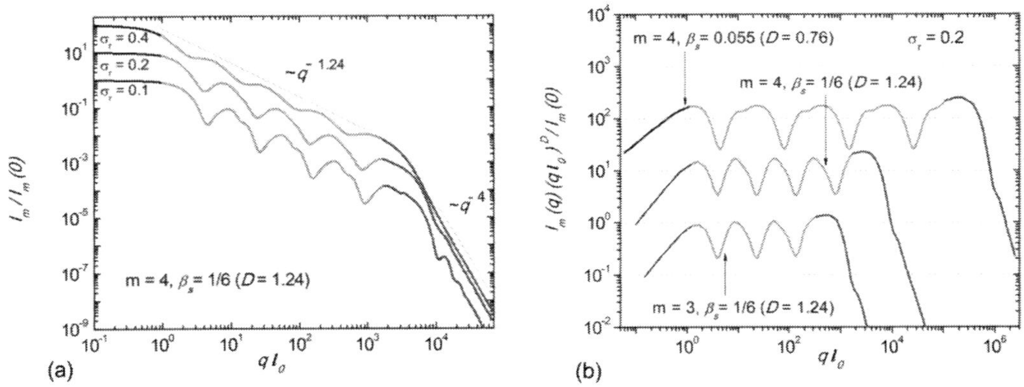

Figure 3. (Color online) Influence of polydispersity on the scattering from GSSVF: (a) intensity (19) for various values of the relative variance (18). Scattering curves for $\sigma_r = 0.2$ and 0.4 are scaled-up by 10 and 100, respectively. (b) The quantity $I(q)q^D$ clearly shows the log-periodicity in the fractal region (2). The period in the log-scale is equal to $\log_{10}(1/\beta_s)$, where β_s is the scaling factor of fractal. The colors are the same as in Fig. 2.

3. Pair distribution function

The pair distribution function gives information about the spatial correlations inside the fractal [2]. It can be written as

$$g(r) \equiv \frac{l_0^3}{4\pi r^2} \frac{2}{N(N-1)} \sum_{k<j} \langle \delta(r - r_{jk}) \rangle_D, \tag{20}$$

where $r_{jk} \equiv |\mathbf{r}_j - \mathbf{r}_k|$ are the relative distances between the ball centers. It gives the probability density to find a particle at the distance r from another particle, provided a position of the latter particle is given. Then, the structure factor $S(q)$ of the fractal system can be related to $g(r)$ through

$$S(q) = 1 + (N-1) \int_0^{+\infty} \mathrm{d}r \, \frac{4\pi r^2}{l_0^3} g(r) \frac{\sin qr}{qr}. \tag{21}$$

The polydispersity effect in Eq. (20) can be taken into account by averaging over (17) [Fig. 4(a) and 4(b)].

4. Discussions of the Results

The properties in real-space of $g(r)$ are shown in Fig. 4(a) and 4(b). The self-similarity of GSSVF manifests itself in the periodicity, on the double logarithmic scale, of the minima and highest maxima. Explicitly, in the fractal region of real space, $g(r)$ has the property (Fig. 4(b)):

$$\frac{g(\beta_s r)}{(\beta_s r)^{D-3}} = \frac{g(r)}{r^{D-3}}. \tag{22}$$

This is a consequence of the log-periodicity of $S(q)q^D$ in the fractal region in reciprocal space, which follows from Eq. (21). Conversely, the log-periodicity of $S(q)q^D$ results from the log-periodicity of $g(r)r^{3-D}$.

5. Conclusions

- We develop a model of deterministic fractal, which allows explicit analytic solutions for the scattering amplitude.

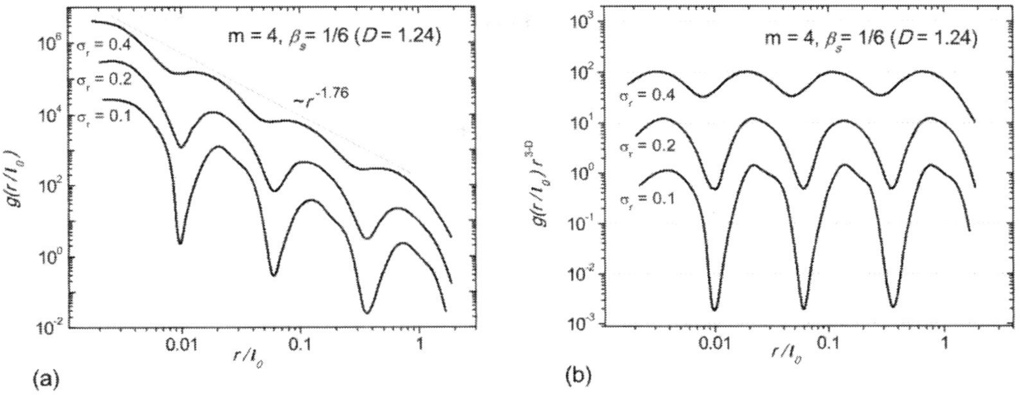

Figure 4. The influence of polydispersity on $g(r)$. Scattering curves for the distribution width $\sigma_{\rm r} = 0.2, 0.4$ are scaled-up for clarity by 10 and 10^2 respectively. (a) expression (20) for various values of the relative variance (18) and (b) the quantity $g(r)r^{3-D}$ clearly shows the log-periodicity. The period in the log scale is equal to $\log_{10}(1/\beta_s)$, where β_s is the scaling factor of the fractal.

- We derived analytical expressions for the form factor and the pair distribution function.
- We suggest a scheme for interpreting the experimental data in the framework of *deterministic mass fractals with a single scale*. Once we *a priori* know that the fractal is of that kind then we can extract a number of parameters from the scattering intensity (Fig. 3):
  (i) The fractal dimension from the generalized power law.
  (ii) The fractal scaling parameter from the period on the logarithmic scale.
  (iii) The number of fractal iteration, which is equal to the number of periods of function $I(q)q^D$.
  (iv) The lower and upper fractal edges from the diagram $I(q)q^D$ as the beginning and end of "periodicity region".
  (v) The total number of structural units, from which the fractal is composed, by the relation $N_m = (1/\beta_{\rm s})^{mD}$.
- The results can be applied for a various number of structures, whose geometries are based on an iteration of a fractal system.

Acknowledgments
The authors acknowledge financial support from Russian state contract No. 02.740.11.0542 and the JINR–IFIN-HH projects.

References

[1] O. Glatter and O. Kratky, *Small-angle X-ray Scattering* (Academic Press, 1982, London).
[2] L. A. Feigin and D. I. Svergun, *Structure Analysis by Small-Angle X-Ray and Neutron Scattering* (Plenum press, 1987, New York).
[3] B. B. Mandelbrot, *The Fractal Geometry of Nature*, (Freeman, 1982, San Francisco).
[4] J. E. Martin and A. J. Hurd, *J. Appl. Cryst.*, **20**, 61 (1987).
[5] P. W. Schmidt, *J. Appl. Cryst.*, **24**, 414 (1991).
[6] M. Balasoiu, M. L. Craus, E. M. Anitas, I. Bica, J. Plestil and A. I. Kuklin, *Physics of the Solid State*, **52**, 917 (2010).
[7] J. Teixeira, *J. Appl. Cryst.*, **21**, 781 (1988).
[8] A. Erdélyi, *Asymptotic Expansions*, (Dover, 1956, New York).
[9] M. F. Barnsley, *Fractals Everywhere: The First Course in Deterministic Fractal Geometry*, (Academic, 1988, London).
[10] T. Vicsek, *J. Phys. A*, **16**, L647 (1983).
[11] A. Yu. Cherny, E. M. Anitas, A. I. Kuklin, M. Balasoiu and V. A. Osipov, *J. Surf. Invest.*, **4**, 903 (2010).

[12] A. Yu. Cherny, E. M. Anitas, A. I. Kuklin, M. Balasoiu and V. A. Osipov, *J. Appl. Cryst.*, **43**, 790 (2010).
[13] A. Yu. Cherny, E. M. Anitas, V. A. Osipov and A. I. Kuklin, *Phys. Rev. E*, **84**, 036203 (2011).
[14] P. W. Schmidt and X. Dacai, *Phys. Rev. A*, **33**, 560 (1986).
[15] C. Allain and M. Cloitre, in *Scaling Phenomena in Disordered Systems*, NATO ASI Series; Series B: Physics, **133**, pp. 193–196, 1985.
[16] R. Pynn and A. Skjeltorp, in *Scaling Phenomena in Disordered Systems*, NATO ASI Series; Series B: Physics, **133**, pp. 141–149, 1985.
[17] C. Allain and M. Cloitre, *Phys. Rev. B*, **33**, 3566 (1986).

Main features of the new software control system for the YuMO instrument

A S Kirilov

FLNP of JINR, Moscow Region, Dubna, Joliot-Curue, 6

E-mail: akirilov@nf.jinr.ru

Abstract. During the last years the new software instrumental complex Sonix+ has been developed at FLNP JINR to replace the former Sonix control system [1]. This complex has been tested at a number of IBR-2 instruments (REMUR, NERA-PR) and on instruments at other centers - KIA, Moscow (MOND), etc. We plan to install the new complex at the YuMO instrument as well. The Sonix+ is implemented on the PC/Windows XP platform, whereas the Sonix is based on the VME/Os-9 obsolete platform. The Sonix+ [1] has been designed considering the experience of long-term operation of the predecessor and recent trends. The paper is devoted to the main features of the new software and the comparison with the former one.

1. Introduction

Since 2006 the YuMO instrument control system has been radically changed. This paper deals with the most important changes from the user's point of view.

First of all the hardware and software platforms have been changed. Please have a look at the table 1.

Table 1. Comparison of the platforms for the YuMO control systems.

	2006	2011
Hardware platform	VME	PC
Processor	Motorola 68010 (486 100MHz)	Intel Core i3 (3.07GHz)
RAM	32 MB	4.096 MB
Disk space	6 GB	1000 GB
Network speed	10 Mb/sec	100 (1000) Mb/sec
Software platform	Os-9 (Unix-like OS)	Windows XP
Instrument software	Sonix	Sonix+
+ PC computer for 2D PSD (Windows XP, Sonix+ subset)		

The Sonix+ was created at FLNP on the basis of the experience obtained with Sonix and in accordance with the recent trends in the field. It inherited some basic solutions from the older system and provides new features to simplify the instrument control.

We will consider:

- changes in the structure of the user interface;
- script comparison;
- data visualization;
- remote access via web;
- fields of possible user activity.

2. Changes in the structure of the user interface

The former interface was actually an interface to the available devices (controllers) — a separate window for each controller. So, an interface for a comprehensive instrument usually consists of too many windows, which is not good.

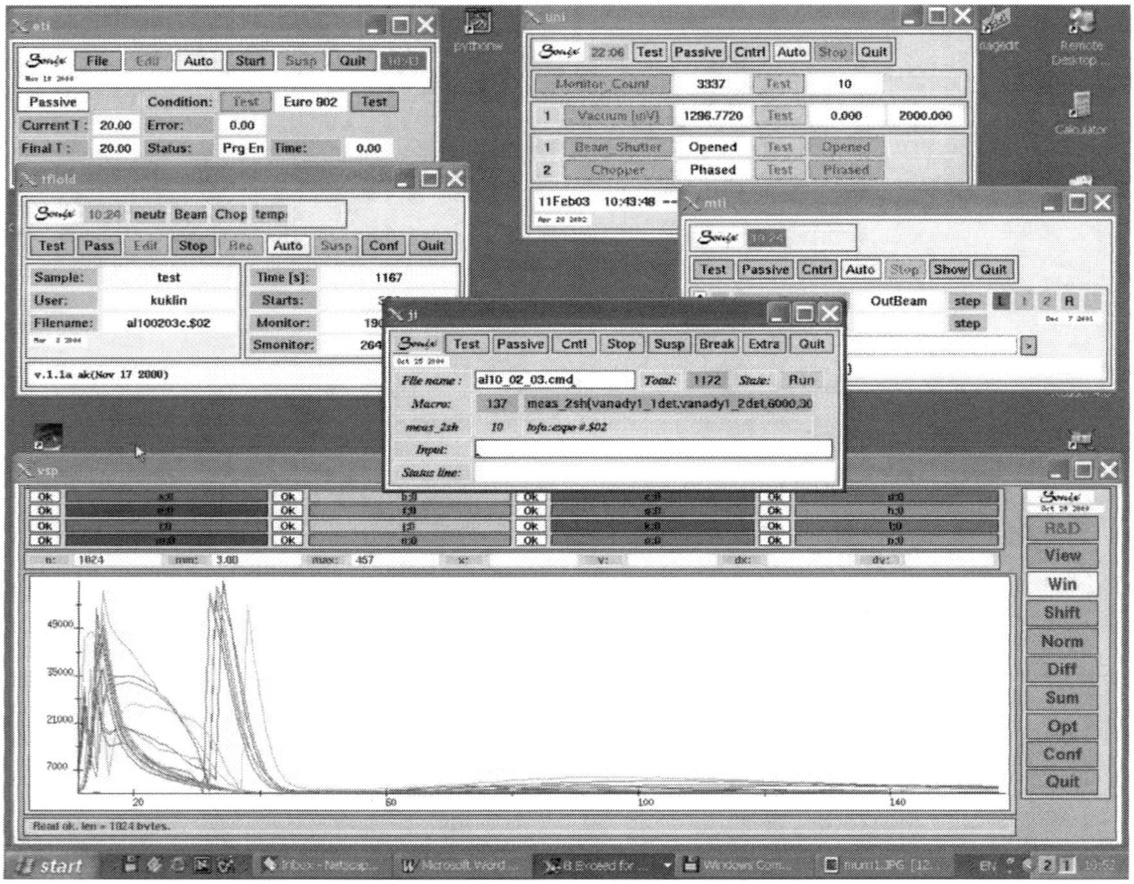

Figure 1. The former interface was actually an interface to the available devices.

The new interface is organized according to another structural principle – each main user need has its own window. There are three main needs to watch: current state, measurement history (log file), picture of spectrum. The fourth need is to control the measurement being performed. Thus, four programs (windows) are sufficient to conduct experiments generally. There are some additional programs as well.

Table 2. Basic user interface conmponents.

Main Programs	Function
Reflector	Watch current values of parameters
LogViewer	Watch history of an experiment
Is_client	Script interpretation control
SpectraViewer	Spectra visualization

Important note. All parameters of the instrument controlled by the software are accessible. In addition, the user can select any subset of them to watch in the special list.

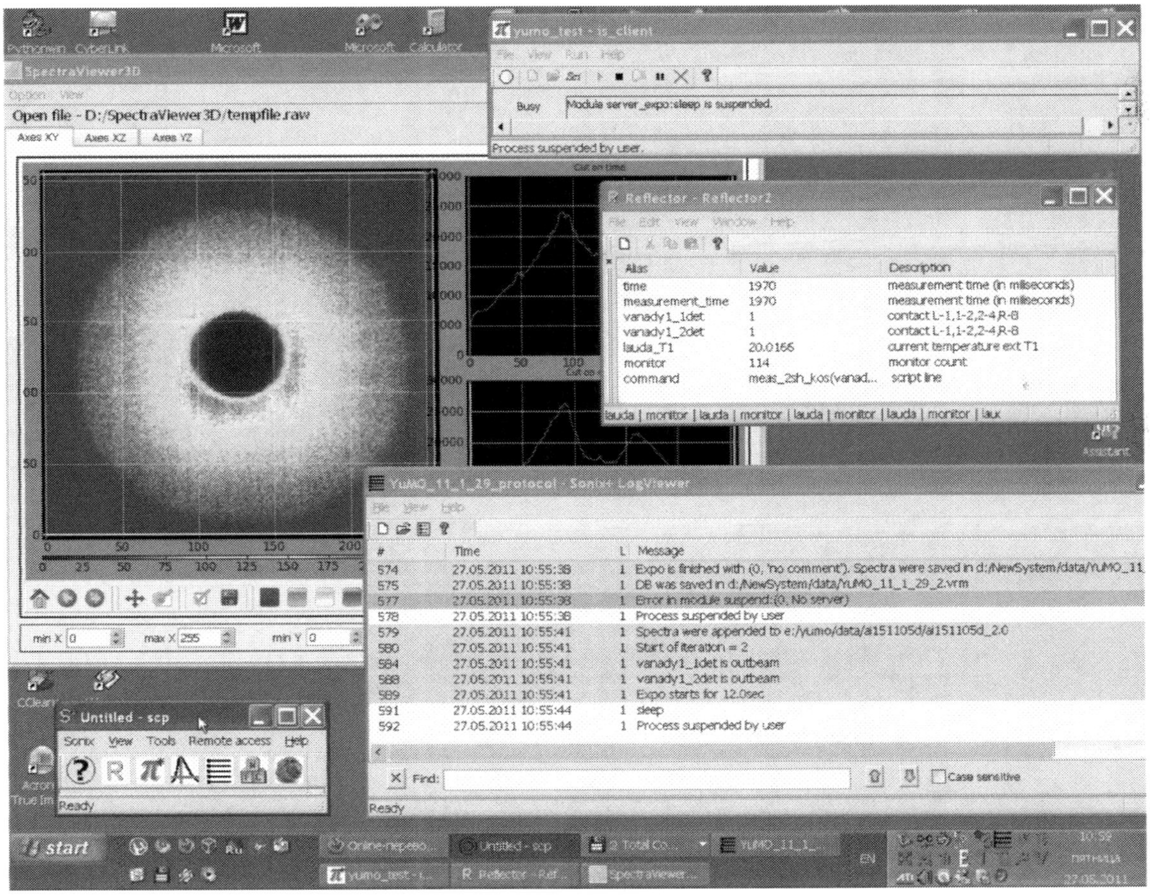

Figure 2. The new interface is organized according to another structural principle – each main user need has its own window.

Journal of Physics: Conference Series **351** (2012) 012021 doi:10.1088/1742-6596/351/1/012021

3. Script comparison

Whereas the script language in the Sonix system was custom-made and has rather poor functionality, the Python language [2] is used as a script language in the Sonix+ and all nice features and packages of Python are available to control experiments and for preliminary data processing.

Python is an open-source widespread powerful interpretative programming language that allows you to work more quickly and integrate your systems more effectively.

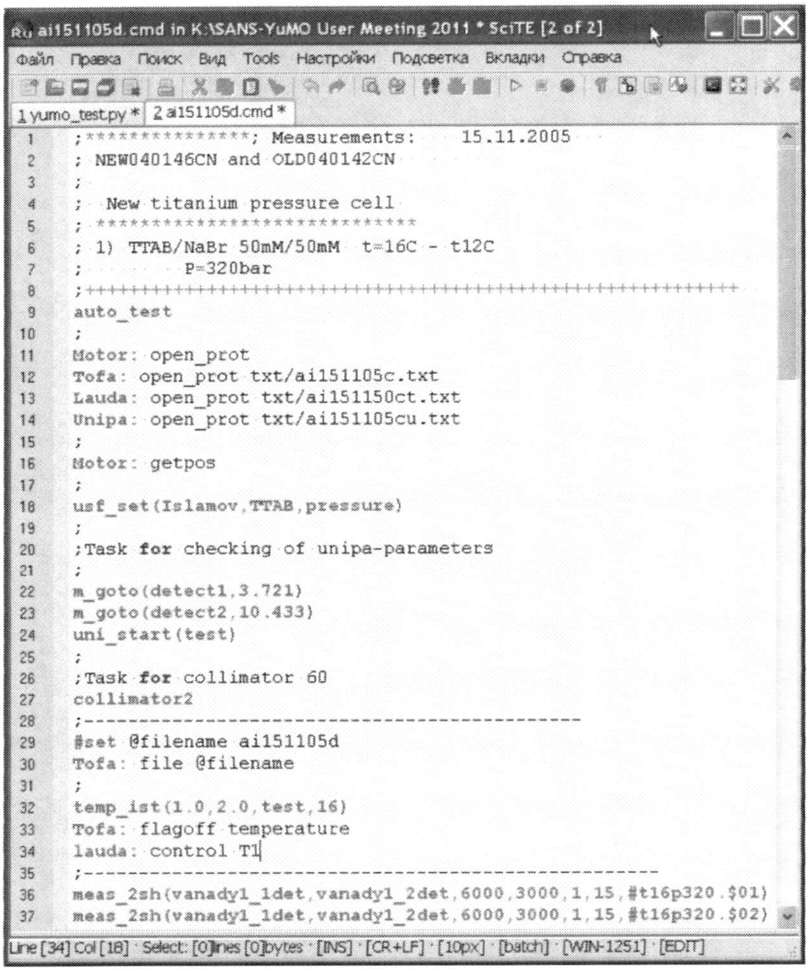

Figure 3. An example of a script for the old system.

To simplify the creation of the experiment program, the operation library dedicated to the YuMO specific features was created. In accordance with the wishes of the people in the YuMO group the old list of functions was preserved in the new library.

Important note. Using the Python as a script language opens a wide door for the user to program experiments and to include preliminary data processing components without the obligatory assistance of the authors of the instrument software.

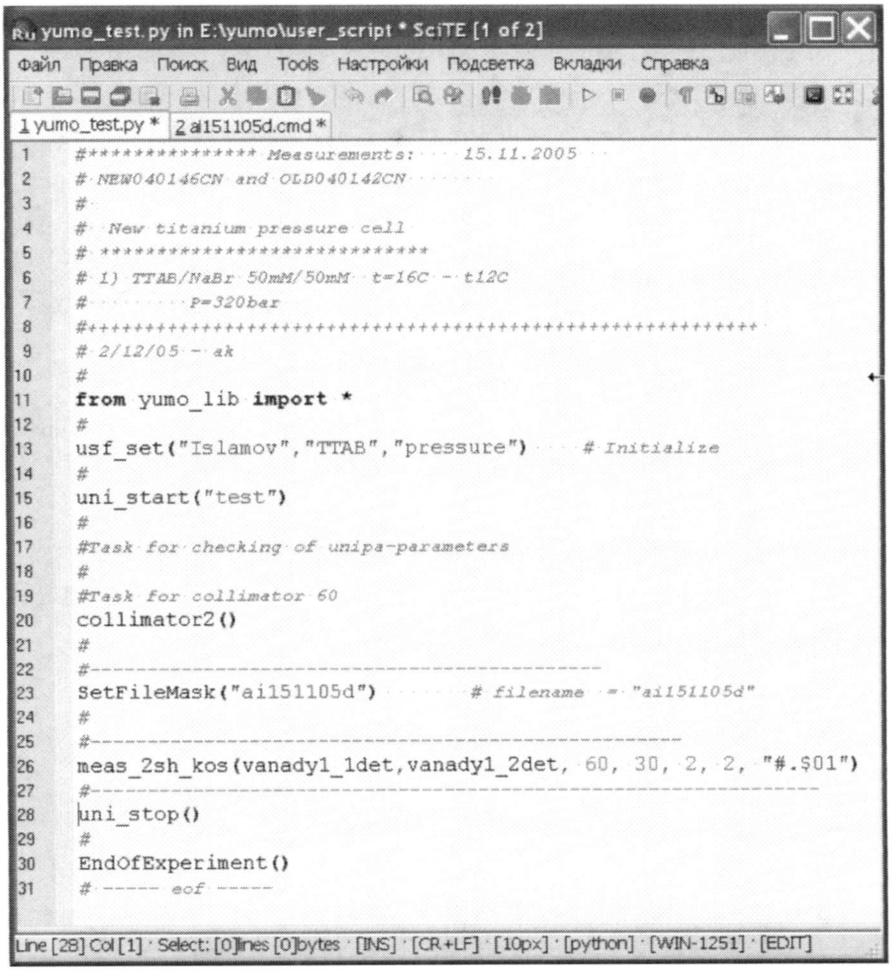

Figure 4. An example of a Python script for the new system.

4. Data visualization

At the moment there is a program (SpectraViewer) written in C++ to visualize data from point detectors and 1D PSD.

The new program tools to visualize data from all kinds of detectors including 2D PSD are developed in the PyQt environment using mathploplib [3] as a graphical library.

Matplotlib is a python 2D plotting library, which produces publication quality figures in a variety of hardcopy formats and interactive environments across platforms. Matplotlib can be used in python scripts. Matplotlib tries to make things easy and hard things possible. You can generate plots, histograms, power spectra, bar charts, errorcharts, scatterplots, etc, with just a few lines of code.

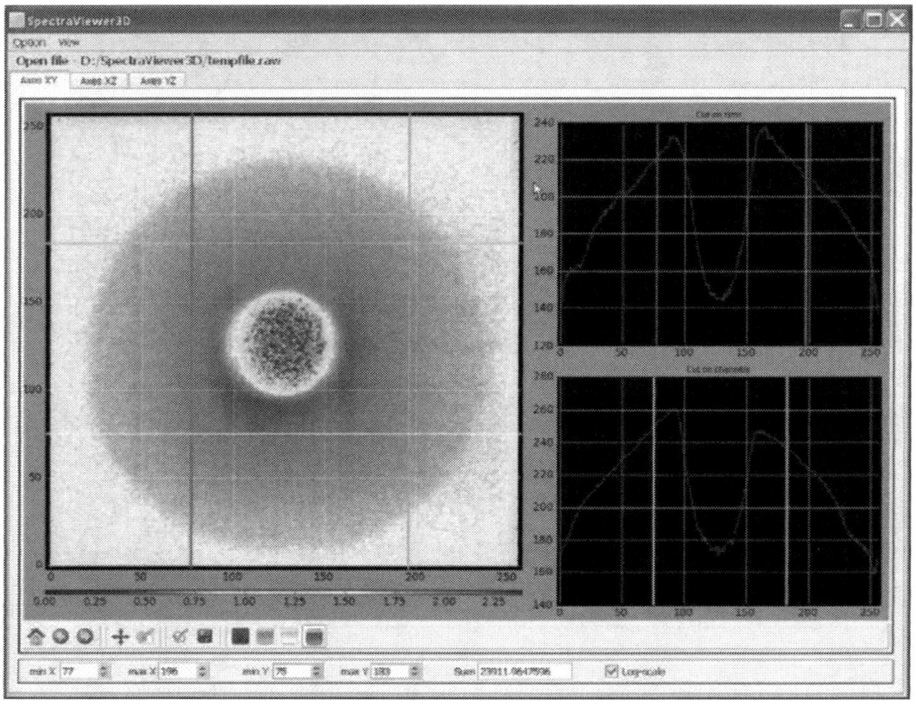

Figure 5. Visualization of data from 2D PSD with SpectraViewer 3D.

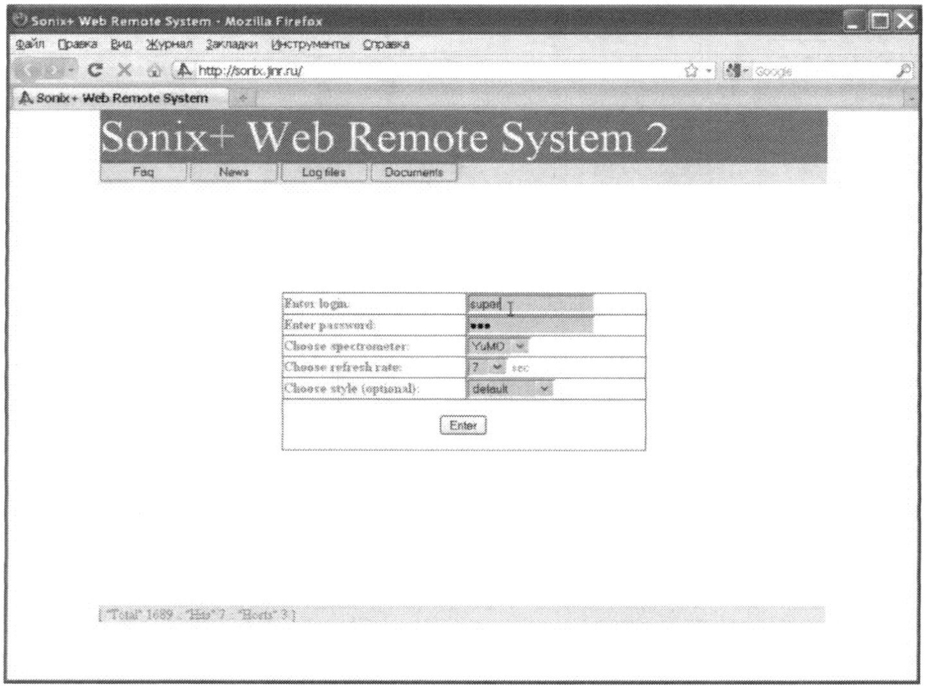

Figure 6. The start page for the Sonix+ Web Remote System.

5. Remote access via web

The WebSonix site is now available for remote supervising and controlling experiments at the IBR-2M instruments, in particular at the YuMO instrument. The site is organized in the same manner as the new user interface. There is also the Sonix+ help page implemented as a wiki.

6. Fields of possible user activity

The new YuMO control system could be a field of possible cooperation of programmers and instrument users to make this system more useful and easy to use. This user's activity could be in the form of proposal inclusion of existing programs or creating new ones. We expect possible activity of the YuMO users in the following directions:

- more convenient operation library (proposal);
- script generator, which simplifies preparation of scripts for experiments specific to YuMO (proposal);
- software components for preliminary data processing, if necessary (proposal and/or participation);
- spectra visualization methods, especially for 2D PSD (proposal and/or participation).

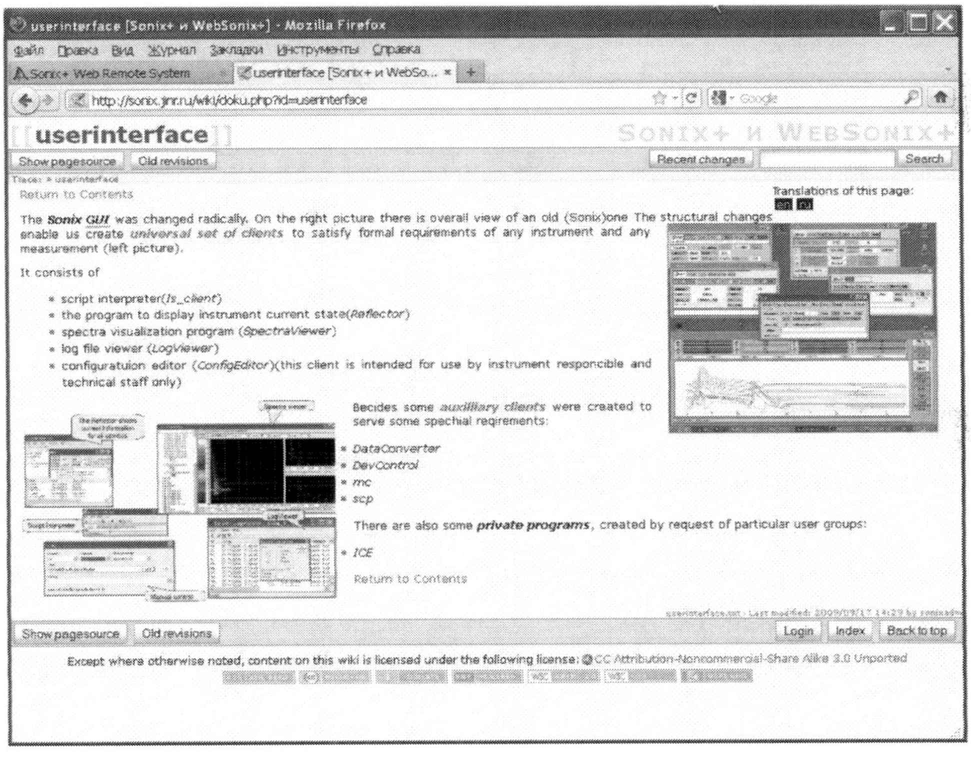

Figure 7. One of the Sonix+ help pages.

References

[1] http://sonix.jinr.ru/
[2] http://www.python.org/
[3] http://matplotlib.sourceforge.net/users/intro.html

Magnetic system for small angle neutron scattering investigations of nanomaterials at YuMO-SANS instrument

A I Kuklin[1,2], M Balasoiu[1,3,*], S A Kutuzov[1], Yu S Kovalev[1], A V Rogachev[1,4], R V Erhan[1,3], A A Smirnov[1], A S Kirilov[1], O I Ivankov[1], D V Soloviov[1,5], W Kappel[6], N Stancu[6], M Cios[7], A Cios[7], V I Gordeliy[1,2,8,9]

[1] Joint Institute for Nuclear Research, Dubna, Russia
[2] Bionanocenter, Moscow Institute of Physics and Technology, Dolgoprudny, Russia
[3] Horia Hulubei National Institute for R&D in Physics and Nuclear Engineering, Bucharest, Romania
[4] Skobeltsyn Institute of Nuclear Physics of Moscow State University, Moscow, Russia
[5] Kyiv National Taras Shevchenko University, Kyiv, Ukraine
[6] National Institute for R&D in Electrical Engineering ICPE-CA, Bucharest, Romania
[7] CIPEC SRL 5 Mizil, Bucharest, Romania
[8] ICS-5, Forschungszentrum Juelich, Germany
[9] Institute of Structural Biology J.P. Ebel, Grenoble, France

[*]E-mail: balasoiumaria@yahoo.com; balas@nf.jinr.ru

Abstract. SANS measurements using unpolarized neutron beams are able to provide quantitative information on the magnetic microstructure and the magnitude and microstructure of the magnetic anisotropy of nanomagnetic materials. A new magnetic system for SANS at the YUMO spectrometer is presented. The system includes 2.5 T electromagnet established on a two-axes goniometric table, a power supply, cooling system, PC based control equipment. Main features of magnetic system are: changeable gap for samples (up to 130 mm size), computer controlled horizontal and vertical rotation and sufficiently big space for the sample holders. First results of SANS experiments of ferrofluids and magnetic elastomers obtained at YuMO spectrometer equipped with the new magnetic system are presented.

1. Introduction

Neutron beams produced by different reactors are used for fundamental research on the structure and dynamics of the matter. Research efforts seek a better understanding of phase transitions, crystal structures, magnetic properties, superconductors, quantum liquids, the fundamental properties of new materials. Neutron scattering facilities have unique state-of-the-art capabilities for investigating structures and excitations of solid state matter for research in solid state physics, polymer science, and biology. The small angle neutron diffractometer is widely used by colloid, polymer, biology scientists, scientists involved in investigation of nano-materials and nano-composites.

The instrument YUMO (Figure 1) is located on the beamline 4 of the high pulsed IBR-2 reactor [1-3]. The useful wavelength range is $0.7 < \lambda/\text{Å} < 8$, the range of momentum transfers is $0.007 < Q/\text{Å}^{-1} < 0.5$. The time averaged neutron flux at the sample is up to 4×10^7 cm^{-2}s^{-1}.

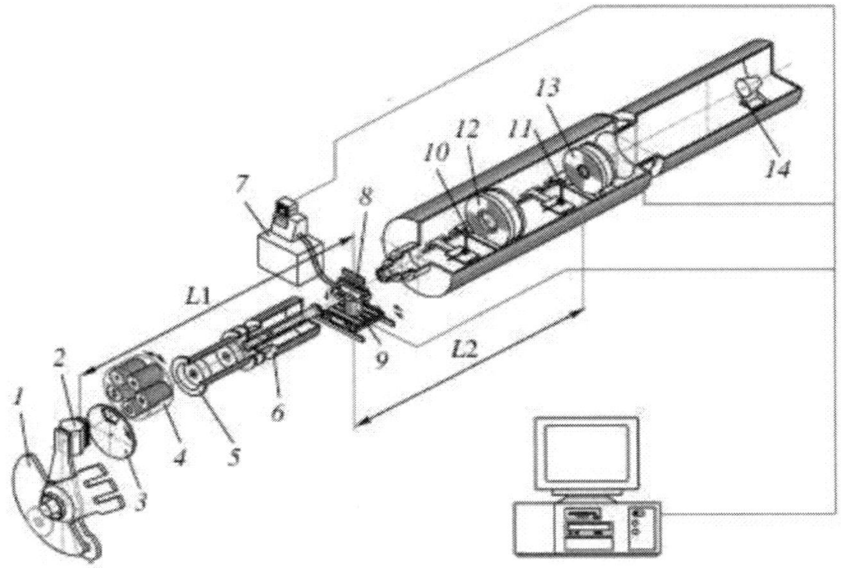

Figure 1. YuMO SANS TOF spectrometer in function at IBR-2 reactor [1-3]: 1) movable reflector; 2) moderator/cold moderator; 3) chopper; 4) first collimator; 5) vacuum tube; 6) second collimator; 7) liquid bath thermostat; 8, 9) place for the electromagnet during SANS experiments on magnetic samples; 10, 11) V, graphite, H_2O standards; 12) circular detector of thermal neutrons; 13) PSD circular detector of thermal neutrons; 14) detector of direct beam.

A variety of auxiliary devices are necessary for materials studies, in special environment conditions including low and high temperature, high pressure and strong magnetic fields.

For investigation of magnetic nanostructures as nanoparticles, ferrofluids, magnetic gels and polymers, nano-carbon composites with magnetic properties there is highly necessary to apply during the experiment magnetic fields of different intensity and orientation to the samples [4-20].

A new magnetic system for small angle neutron scattering at YUMO instrument was put into operation. The system includes about 3T (factory setup) electromagnet fixed on a two-axes goniometric table, power supply, cooling system and PC based control equipment. Main features of magnetic system are: big changeable gap for the samples (up to 130 mm size), computer controlled horizontal and vertical rotation and sufficiently big space for the sample holders [21, 22].

In the present paper a magnetic system special created for the YUMO SANS instrument is presented. The first SANS measurement results on magnetic elastomers and ferrofluids in applied magnetic field performed at YUMO diffractometer are given.

2. Magnetic system at YUMO SANS instrument

The magnetic field, like the temperature or pressure, is an important parameter influencing the state of matter and properties of many materials. The applied magnetic field determine certain effects not only in ferro- and ferrimagnetic materials but also in paramagnetic and diamagnetic ones.

It is very important to control the microstructure of magnetic materials because textured or anisotropic magnetic materials usually show better performance in practical applications. Magnetic field processing is a well-proven technique to impose the desired texture and magnetic properties of magnetic materials. Macroscopic materials comprising a large number of nanostructures in crystalline alignment with one another would constitute highly anisotropic materials with wide potential applications. It has been also found that magnetic field can make inclusions migrate in a melt and align grains in fabrication processes for the magnetic and non-magnetic materials.

The "control of materials" for materials science by means of a magnetic field, during the materials synthesis, chemical reactions or crystal growth in high magnetic fields is useful method for creating and investigating new types of materials [23].

A new dimension in advanced materials (such as magnetic materials, oxide superconductors and organics) processing, and in the microstructure research of nanometer-scaled materials, spintronics, polymer, bio-chemistry and so on, through the possibility of SANS investigation in situ of the phenomena induced by applied magnetic fields are under investigation (see e.g. [24-27]). Other aspect of using magnetic field in small angle neutron scattering experiment refers to the possibility of determination of the nuclear and magnetic elastic scattering contributions to the scattering. When investigating magnetic materials in a SANS with non-polarized neutrons experiment, measuring the scattering intensity without magnetic field and with a saturation magnetic field oriented perpendicular to the scattering momentum transfer, the nuclear and magnetic contributions can be differentiated [28-30]. Previously, this differentiation was possible to be obtained at SANS-YuMO instrument only in the case of magnetic liquids by means of the contrast variation method [31-34]. Magnetic neutron scattering plays a central role in determining and understanding the microscopic properties of a vast variety of magnetic systems, from the fundamental nature of magnetically ordered materials to elucidating the magnetic characteristics essential in applications.

For common goals the magnitude of magnetic field inducing changes in samples, (namely: anisotropy and orientation of magnetic domains in magnetic materials, as well as in the liquid crystal polymers [35]) is about few Tesla.

The new magnetic system constructed for YUMO instrument is presented (figure 2). The magnetic system components are: the electromagnet with the goniometric device, the power supply; computer for the automated command and control of the system.

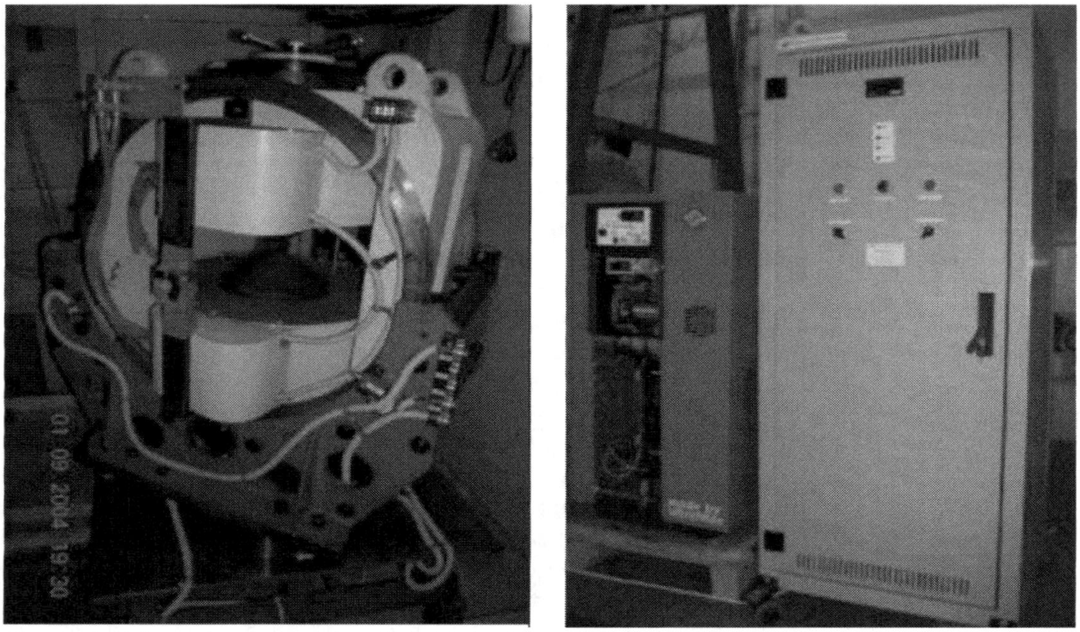

Figure 2. Magnetic system for YUMO SANS instrument (electromagnet fixed on a two-axes goniometric table, power supply, cooling system)

The system was developed and constructed by the INCDIE ICPE CA Bucharest Romania in collaboration with CIPEC SRL Bucharest Romania.

Journal of Physics: Conference Series **351** (2012) 012022 doi:10.1088/1742-6596/351/1/012022

The realization of the new position sensitive [1-3] permits the visualization and investigation of the magnetic field induced anisotropy in the analyzed sample.

2.1. Technical description.

The electromagnet is a rigid system, made of two coils, with distilled water cooling system and two magnetic poles with changeable caps (two types). The pole gap is variable, from 25 to 100 mm.

On the both lateral parts of the electromagnet there is one slit, 100 × 600 mm, for giving the possibility to introduce the neutron flux. The whole electromagnetic system is sitting on one rotating system. There are two rotating planes with a rotation liberty of $\pm 40^0$ in the vertical plane and of $\pm 95^0$ in the horizontal plane. The rotating devices are equipped with electrical limiters, which stop the rotation motors. The motors are step-by-step motors. The positioning accuracy for both rotation planes is 0.1^0.

The assembly electromagnet and rotating system are placed and fixed on a mobile support (for translation) by means of some screws. The highest point of this support is adjustable, with ± 30 mm, for having the facility to the center of electromagnet in the neutron flux. The electromagnet cooling system is a closed system, with distilled water. The cooling of the water is made by a water chiller wit with a capacity of 5000 W. The water flow is about 400 l/h. The electromagnet is supplied by a DC source, with a current adjustment of 10 to 60 A DC. The adjustment accuracy is 0.7% from the end of the scale. Maximum reached intensity of magnetic field is 2.5 T.

The measuring and control systems are connected with a PC. The software is an friendly one, who gives a lot of facilities:

(1) Measuring and control for following parameters: value of the magnetically field; value of current in the coils; cooling water temperature; water flow; electromagnet position in vertical plane; electromagnet position in horizontal plane; turning limits in vertical plane; turning limits in horizontal plane.

(2) Alarms for outrunning of measuring values set points

(3) Curves, diagrams for measuring values variations.

(4) History of measuring values variations, for a period of 30 days.

The technical drawings from Figure 3 present the block diagram of the whole magnetic system [21,22].

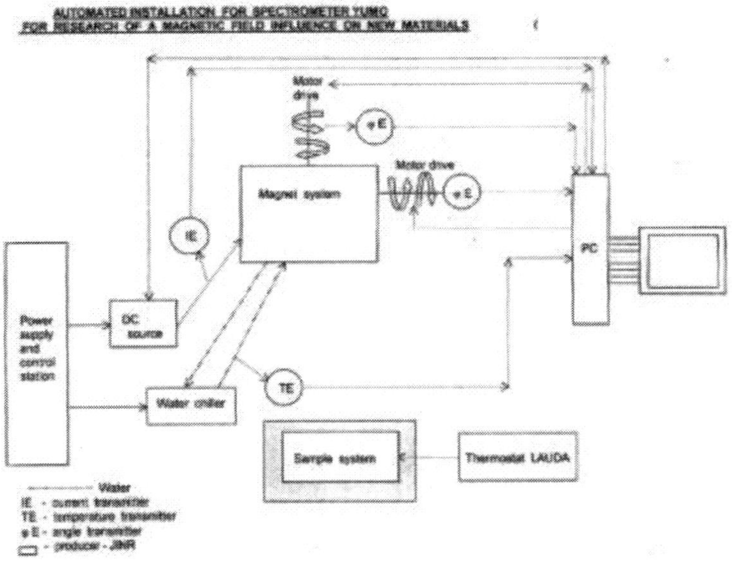

Figure 3. Block diagram of the whole magnetic system.

Standard measurements and data analysis

When investigating magnetic materials in a SANS with non-polarized neutrons experiment, measuring the scattering intensity without magnetic field and with a saturation magnetic field oriented perpendicular to the scattering momentum transfer, the nuclear and magnetic contributions can be differentiated. As example, a ferrofluid sample is examined [36,37] (Figure 4).

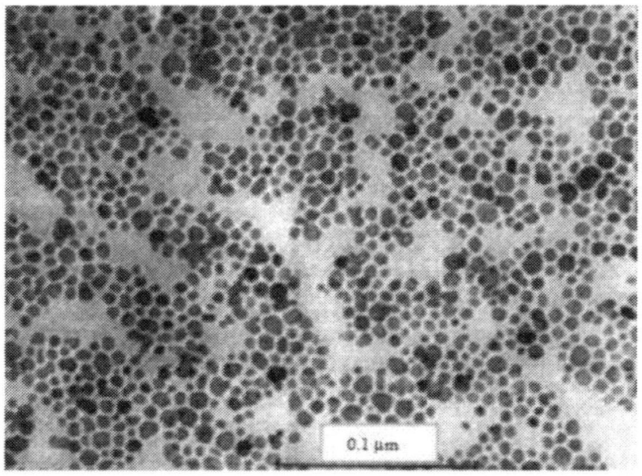

Figure 4. TEM image of a ferrofluid sample.

During standard tests the sample in closed quarts cuvette is put under external magnetic field ($B > 1$ T) so that the saturation magnetization in the sample takes place. Then, the sample is irradiated

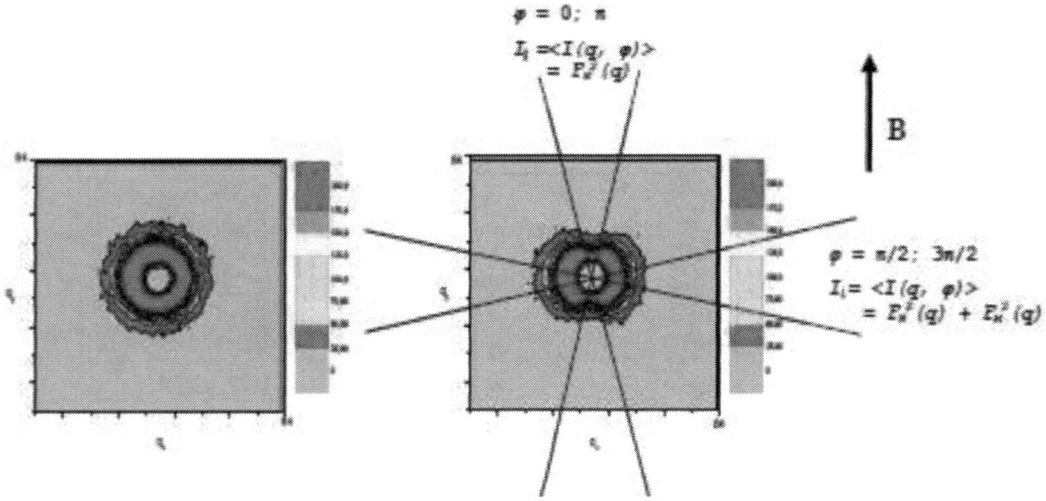

Figure 5. 2-dimensional scattering patterns from magnetic sample in absent (left) and present (right) of external magnetic field. The separation of the nuclear and magnetic scattering contribution for saturated sample is demonstrated.

by the thermal neutron beam followed by registering of neutrons scattered at small angles (< 0.1 rad) by the position sensitive detector behind the sample. The 2-dimensional scattering pattern on the detector has specific features determined by two contributions: nuclear and magnetic scattering.

In the case of non-polarized neutron beam for the saturated sample the scattering intensity is

$$I(Q,\varphi) = F_n^2(Q) + F_m^2(Q)\sin^2\varphi \tag{2}$$

where $Q = (4\pi/\lambda)\sin(\theta/2)$ is the module of the scattering vector with neutron wavelength, λ, and scattering angle, θ, φ is the radial angle on the detector from the zero-direction corresponding to direction of the strength of the applied magnetic field; $F_n^2(Q)$ and $F_m^2(Q)$ are the nuclear and magnetic scattering contributions, respectively. The separation is made by averaging of the scattering pattern for given Q-value over radial angle φ around the vicinities of the directions parallel and perpendicular to the magnetic field (Figure 5). This procedure transforms the 2-dimensional pattern to two 1-dimensional scattering curves:

$$I_{||} = <I(Q,\varphi)>_{\varphi=0;\pi} = F_n^2(Q) \tag{3a}$$

$$I_{\perp} = <I(Q,\varphi)>_{\varphi=\pi/2;3\pi/2} = F_n^2(Q) + F_m^2(Q) \tag{3b}$$

These equations are the base to obtain nuclear and magnetic scattering contributions.

If the applied magnetic field does not produce the saturation magnetization in the sample the scattering intensity on the detector has the general form:

$$I(Q,\varphi) = A(Q) + B(Q)\sin^2\varphi \tag{4}$$

where $A(q)$ and $B(q)$ are functions corresponding to isotropic and anisotropic contributions to the scattering. The $B(q)$ function is determined purely by magnetic scattering of the partially oriented magnetic moments in the system. The $A(q)$ function is composed from isotropic nuclear scattering and isotropic part of magnetic scattering. The separation of these functions is made by averaging of the scattering pattern for given q-value over radial angle φ around the vicinities of the directions parallel and perpendicular to the magnetic field (Figure 5). This procedure transforms the 2-dimensional pattern to two 1-dimensional scattering curves:

$$I_{||} = <I(Q,\varphi)>_{\varphi=0;\pi} = A(Q) \tag{5a}$$

$$I_{\perp} = <I(Q,\varphi)>_{\varphi=\pi/2;3\pi/2} = A(Q) + B(Q) \tag{5b}$$

These equations are the base to obtain $A(q)$ and $B(q)$. Analysis of the magnetic system based on changes in these functions with the strength of the applied magnetic field, is possible. The behavior of the standard sample with the change of the applied magnetic field is known.

Conclusion

First experiments have demonstrated that the electromagnet together with the position sensitive detector is suitable for SANS measurements at YUMO instrument.

Further, a data analysis to transform the 2-dimensional pattern to 1-dimensional scattering curves as presented above in the formulae from (2) to (5) will be performed. For this task a new SAS program with a 2D analysis is in progress.

Acknowledgements
The Governmental Romanian Plenipotentiary to JINR Dubna is acknowledged for the financial support through the JINR-INCDIE ICPE CA economic Contract Nr.08626319/021454-72 (2002-2004), responsibles Dr. A.I.Kuklin, Dr. M.Balasoiu and Dr. N. Stancu.

References

[1] Kuklin, A I, Islamov A Kh and Gordely V I 2005 *Neutron News* **16**(3) 16
[2] Kuklin A I, Islamov A Kh, Kovalev Yu S, Utrobin P K and Gordely V I 2006 *J. Surf. Invest. X-Ray, Synchr., Neutron Tech..* **6** 74
[3] Kuklin A I et al 2011 *J. Phys: Conference Series* **291**(1) 012013
[4] Kohlbrecher J and Wagner W 2000 *J.Appl.Cryst.* **33** 804
[5] Klose F and Ehlers G 2002 *Microscopy and Microanalysis* **8** 374
[6] Weissmueller J, Michels A, Barker J G, Wiedenmann A, Erb U and Shull R D 2001 *Phys.Rev. B* **63** 214414
[7] Wiedenmann A 2001 *Physica B* **297** 226
[8] Yusuf S M, De Teresa J M, Mukadan M D, Kohlbrecher J, Ibarra M R , Arbiol J, Sharma P and Kulshreshtha S K 2006 *Phys.Rev. B* **74** 224428
[9] Grabcev B, Balasoiu M, Tarziu A, Kuklin A I and Bica D 1999 *J. Mag. Mag. Mater* **201** 140
[10] Avdeev M V, Balasoiu M, Aksenov V L, Garamus V M, Kohlbrecher J, Bica D and Vekas L, 2004 *J. Mag. Mag. Mater.* **270** 371
[11] Balasoiu M, Avdeev M V, Kuklin A I, Aksenov V L, Bica D, Vekas L, Hasegan D, Torok Gy, Rosta L, Garamus V M and J.Kohlenbrecher J 2004 *Magnetohydrodynamics* **40**(4) 359
[12] Balasoiu M, Avdeev M V, Kuklin A I, Aksenov V L, Hasegan D, Garamus V M, Schreyer A, Bica D, Vekas L and Almasan V 2006 *Romanian Reports in Physics* **58**(3) 305
[13] Dokukin M E, Perov N S, Dokukin E B, Islamov A Kh, Kuklin A I, Kalinin Yu E, Sitnikov A V 2007 *Bull. Russ. Acad. Sci., Physics* **71**(11) 1602
[14] Balasoiu M, Craus M L, Kuklin A I, Plestil J, Haramus V, Islamov A Kh, Erhan R, Anitas E M, Lozovan M, Tripadus V, Petrescu C, Savu D, Savu S and Bica I 2008 *J.Optoelectron Adv. Mater* **10**(11) 2932
[15] Balasoiu M, Anitas E M, Bica I, Osipov V A, Orelovich O L, Savu D, Savu S, Erhan R and Kuklin A I 2008 *Optoelectron. Adv.Mater.-RC* **2**(11) 730
[16] Anitas E M, Balasoiu M, Bica I, Osipov V A and Kuklin A I 2009 *Optoelectron. Adv. Mater. – RC* **3**(6) 621
[17] Balasoiu M, Craus M L, Anitas E M, Bica I, Plestil J and Kuklin A I 2010 *Phys.Solid State* **52(5)** 917
[18] Balasoiu M, Bica I, Raikher Yu L, Dokukin E B, Almasy L, Vatzulik B and Kuklin A I 2011 *Optoelectron. Adv. Mater. – RC* **5**(5) 523
[19] Balasoiu M, Lebedev V T, Orlova D N and Bica I 2011 *Crystallogr. Rep.* **56(7)** 1177
[20] Stolbov O V, Raikher Yu L and Balasoiu M 2011 *Soft Matter* **7** 8484
[21] Balasoiu M, Kirilov A S, Kutuzov S A, Smirnov A A, Kappel W, Cios M, Cios A, Kuklin A I 2008 Magnetic system for small-angle scattering investigation at YuMO instrument of nanomaterials 2008 *JINR Communication E14-2008-193*
[22] Kappel W et al 2008 *Rev.Roum.Sci.Tech.-Series Electroth.Energ.* **53**(2) Suppl. S 79
[23] Ma Y, Xiao L, Yan L 2006 *Chinese Science Bulletin* **51**(24) 2944
[24] Simkin V G and Craus M L 2007 *13-th Conference Progress in Cryogenics and Isotopes Separation* (7-9 November 2007 Calimanesti-Caciulata,Romania)
[25] Dobrea V, Lozovan M, Craus M L and Simkin V 2010 *J.Optoelectron Adv. Mater* **12**(4) 854
[26] Craus M L, Lozovan, Cornei N and Simkin V 2010 *J.Optoelectron Adv. Mater* **12**(4) 868
[27] Burzo E, Balasoiu M, Isnard O, Kuklin A I, Rogachev A V, Cuello G, Cristiglio V and Deme B 2011 SANS-YuMO User Meeting *Book of Abstracts* (Dubna: Publishing House JINR) p 46

[28] Bacon G 1962 *Neutron Diffraction* (Oxford: University Press).
[29] Kohlbrecher J, Wiedenmann A, Hollenberger H 1997 *Z. Phys. B* **104** 1
[30] Loffler J F, Braun H B, Wagner W, Kostorz G, Wiedenmann A 2005 *Phys. Rev.* B **71** 134410
[31] Grabcev B, Balasoiu M, Bica D and Kuklin A I 1994 *Magnitnaia Gidrodinamica* **10**(2) 156
[32] Balasoiu M, Grabcev B and Bica D 1995 *Rom. Rep. Phys.* **47**(3-5) 319
[33] Balasoiu M 1998 On the microstructure of magnetic fluids *PhD Thesis* (Institute of Atomic Physics, Bucharest)
[34] Grabcev B, Balasoiu M, Tarziu A, Kuklin A I and Bica D 1999 *J. Magn. Magn. Mater.* **201** 140
[35] Kuklin A I, Cherny A Yu, Cherezov V G, Kostromin S, Rogachev A V, Shibaev V P 2008 X-ray investigation of comb-like liquid crystal polymers, oriented by a strong magnetic field for SANS measurements *Annual report FLNP* (Dubna: JINR Publishing Department) p 43
[36] Bica D 1985 *RO Patent* **90078**
[37] Bica D 1995 *Romanian Reports in Physics* **47**(3-5) 265

Journal of Physics: Conference Series **351** (2012) 012023 doi:10.1088/1742-6596/351/1/012023

Status and perspective of development of cold moderators at the IBR-2 reactor

S Kulikov, E Shabalin

Joint Institute for Nuclear Research, 141980, Joliot Curie 6, Dubna, Russia

e-mail: ksa@nf.jinr.ru

Abstract. The modernized IBR-2M reactor will start its operation with three water grooved moderators in 2011. Afterwards, they will be exchanged by a new complex of moderators. The complex consists of three so-called kombi-moderators, each of them containing a pre-moderator, a cold moderator, grooved ambient water moderators and post-moderators. They are mounted onto three moveable trolleys that serve to deliver and install moderators near the reactor core. The project is divided in three stages. In 2012 the first stage of development of complex of moderators will be finished. The water grooved moderator will be replaced with the new kombi-moderator for beams #7, 8, 10, 11. Main parameters of moderators for this direction will be studied then. The next stages will be done for beams #2-3 and for beams #1, 9, 4-6, consequently. Cold moderator chambers at the modernized IBR-2 reactor are filled with thousands of beads (~3.5 - 4 mm in diameter) of moderating material. The cold helium gas flow delivers beads from the charging device to the moderator during the fulfillment process and cools down them during the reactor cycle. The mixture of aromatic hydrocarbons (mesithylen and m-xylen) has been chosen as moderating material. The explanation of the choice of material for novel cold neutron moderators, configuration of moderator complex for the modernized IBR-2 reactor and the main results of optimization of moderator complex for the third stage of moderator development are discussed in the article.

1. Cold moderators at the modernized IBR-2 reactor

New complex of moderators of the modernized IBR-2 reactor will consist of three parts, or kombi-moderators. Each of them is going to provide neutrons for spectrometers situated onto the adjacent neutron beams (Fig.1.). The composition includes water pre-moderator and additional post moderators that are at ambient temperature, cold moderator with mixture of aromatic hydrocarbons with moderating media at about 20 K. Each of three cold moderators is going to be with the same design.

In the first stage of construction of moderator complex the part of the complex for beam 7, 8,10,11 will be installed [1].

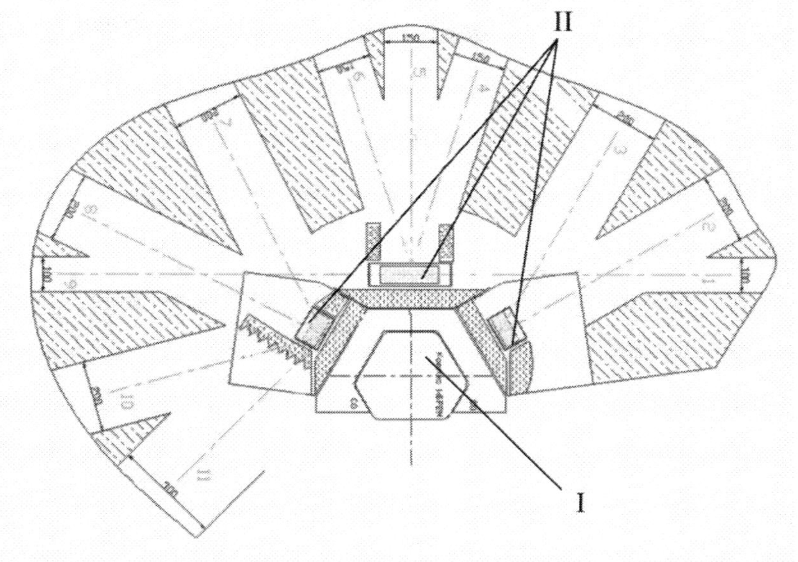

Fig 1. The schematic view of the modernized IBR-2 reactor with three parts of moderators surrounded it and 11 beam ports for neutron spectrometers (I-reactor core and stationary reflector, II - moderators)

The mixture of mesitylene and m-xylene has been chosen for some reasons:

- It has good moderating property [2-4];
- It suffers not dramatically of radiation [5];
- It is not explosive material;
- It might operate at wide range of low temperatures, etc.

It is well known, that hydrogenous materials are the most proper for solid cold neutron moderators [6-8]. The common way of cold moderator construction is that hydrogenous material is frozen to the very low temperature from a liquid phase and forms one bulk peace. The main problem with such a moderator is that molecules of moderating material are decomposed under the fast neutron radiation and hydrogen accumulates inside the bulk. Then, it builds up the pressure to the matrix of material in the process of heating up a moderator. The material expands and presses walls of a vessel that can be destroyed with such a pressure [9]. Another technical challenge for engineers is how to take off the heat induced by radiation from the bulk material with low thermal conductivity.

The main difference of the advanced cold moderator of the modernized IBR-2 reactor from existing solid cryogenic neutron moderators is in a fact that the cold moderator chamber is filled with thousands of separate beads (~3.5 - 4 mm in diameter) of moderating material occupying about 0.6-0.7 of volume of the chamber. The cold helium gas flow delivers beads from the charging device to the moderator before operation and then cools down them in the reactor cycle. It is possible to take off high radiation heat from beads due to their large surface area and small diameter. Free space between beads allows to expand material under the radiation without damaging of moderator vessel. With such a construction of a moderator and due to negligible radiation effects in aromatic hydrocarbons, thickness of pre-moderator can be decreased, and the cold moderator can be installed at the maximum of thermal neutron flux, opposite to the former solid methane cold moderators at the IBR-2 reactor. Broadly speaking that solid methane is the best moderating material for cold moderators [2-4], it is not absolutely true in case of the IBR-2 reactor. Cold moderator as a bulky solid methane shows even lower cold neutron production in comparison with a moderator of beads of aromatic hydrocarbons due to higher temperature of stable operation and thicker pre-moderator.

2. Optimization of moderator configuration for beams #1, 9, 4-6

The optimization of configuration of moderator complex for beam #1, 9, 4-6 (Fig.1.) has been done by Monte Carlo simulations. Some approaches have been applied to increase neutron flux such as utilization of additional water grooved reflector, surrounding of cold moderator by a water post moderator, a cold grooved beryllium filter, etc. The thickness of water pre-moderator was 5 cm.

Results of optimization show that the best thickness of cold moderator is 4 cm (Fig.2 (a,b). The flux distribution along the surface of such moderator is shown in the Fig 3 (a,b).

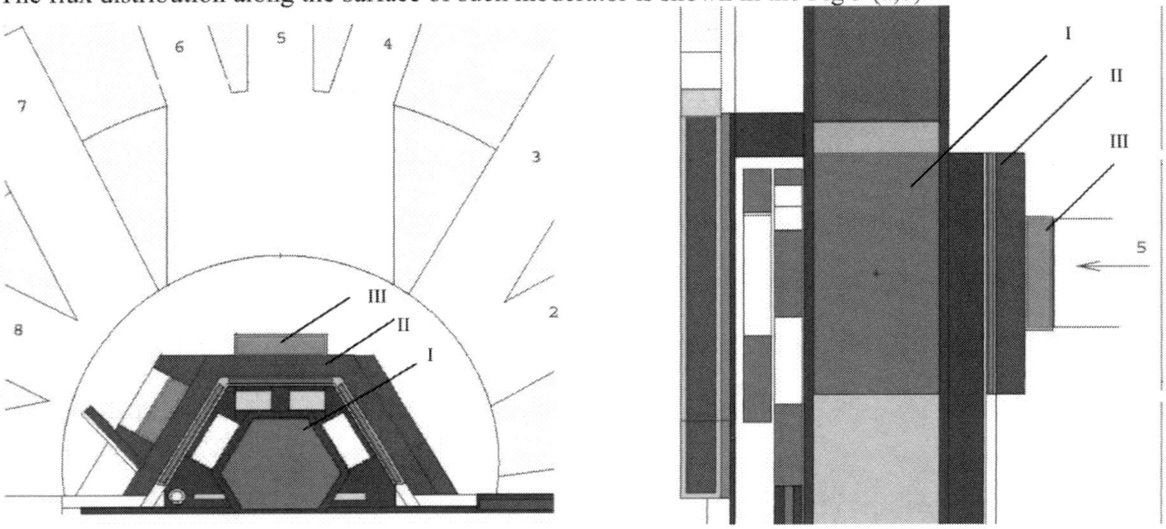

a) b)

Fig. 2. Horizontal (a) and vertical (b) cut-offs of the geometric model of modernized IBR-2 reactor used for neutron transport simulation (I - reactor core, II- water pre-moderator, III - cold moderator)

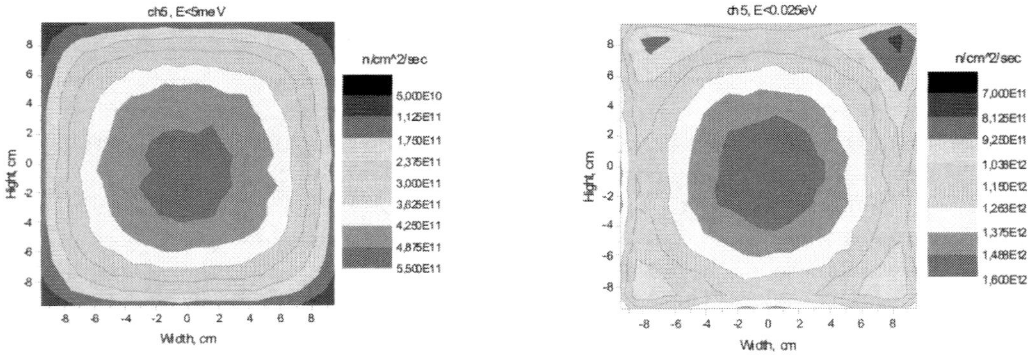

a) b)

Fig .3. Results of Monte-Carlo calculation of cold (a) and thermal (b) neutron fluxes distribution along the surface of cold moderator at the reactor power of 2 MW with combination of water pre-moderator (5 cm) + cold moderator 4 cm.

Distribution of thermal neutrons along the surface of moderator is not uniform (Fig. 3b). The additional water post-moderators above and below the cold moderator (Fig. 4.) give more bright pattern of thermal neutrons (Fig. 5a) and increase the area of high intensity of cold neutrons (Fig. 5b). Unfortunately, it is not possible to install water post-moderators at the left and right sides as spectrometers at the beams # 1, 9 view these sides of the cold moderator (Fig.1.). This result shows that a post-moderator additionally thermalizes and reflects neutrons back to a cold moderator and increases cold neutron flux. Otherwise, these neutrons would leak out of the moderating system and would be lost for spectrometers.

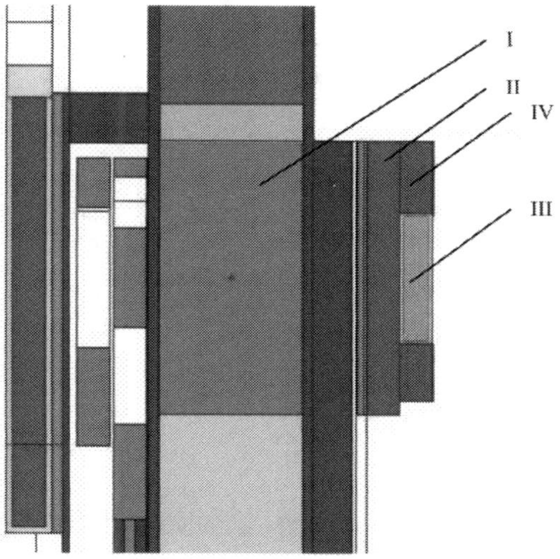

Fig. 4. Vertical cut-off of the model of modernized IBR-2 reactor used for neutron transport simulation (I - reactor core, II- water pre-moderator, III - cold moderator, IV-water post moderator)

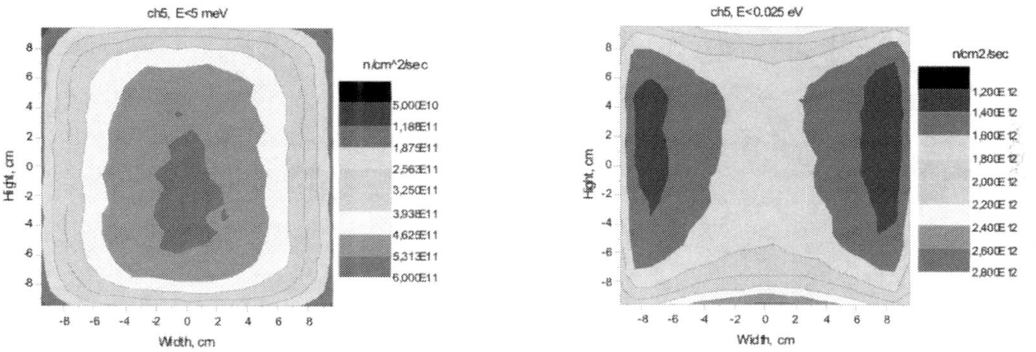

a) b)

Fig .5. Results of Monte-Carlo calculation of cold (a) and thermal (b) neutron fluxes distribution along the surface of the cold moderator at the reactor power of 2 MW with combination of water pre-moderator + cold moderator + water post-moderator (4 cm thick) above and below the cold moderator

This prospective result allows us to go further and increase (optimize) the thickness of post-moderator and add post-moderators at the left and right side of the cold moderator out of view of beams #1 and #9 (Fig. 6) then.

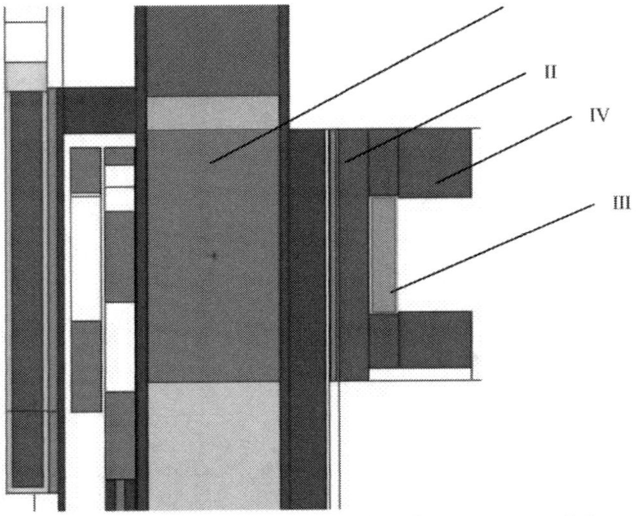

Fig. 6. Vertical cut-off of the model of the modernized IBR-2 reactor used for neutron transport simulation (I - reactor core, II- water pre-moderator, III - cold moderator, IV-water post moderator)

Optimized thickness of the post-moderator was calculated to be as high as 14 cm. It gives significant increase of the area with highest neutron fluxes both of cold and thermal neutrons. (Fig. 7 a,b)

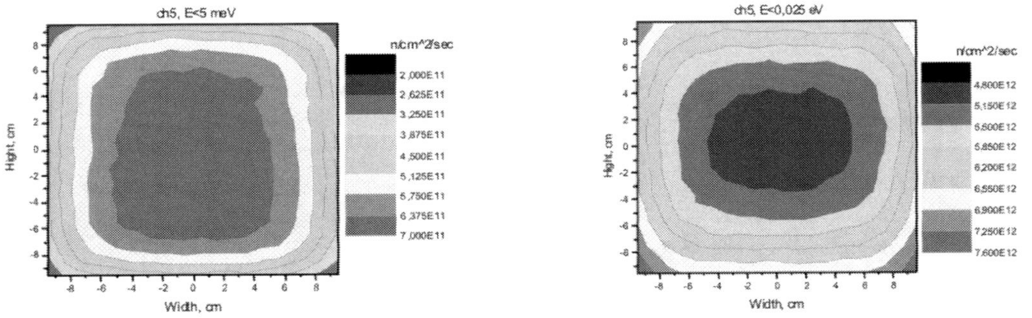

a) b)

Fig .7. Results of Monte-Carlo calculation of cold (a) and thermal (b) neutron fluxes distribution along the surface of the cold moderator at the reactor power of 2 MW with combination of water pre-moderator + cold moderator + 14 cm water post-moderator above and below the cold moderator

Another attempt to increase neutron fluxes has been done by utilization of grooved 10 cm thick Be reflector (T=77K) in front of the cold moderator. Gain factor of cold neutrons intensity appeared to be as high as 2.3 to a bare cold moderator and 1.6 to the best composition with water post-moderator. Nevertheless, this approach is not producible in this project due to effect of Be threshold and necessity of additional cooling to take off the heat from Be. Beryllium reflector might be optional for future improvements of moderators at the modernized IBR-2 reactor.

Generally speaking, neutron calculations for beams #1, 9, 4-6 have shown that utilization of optimized geometry will significantly increase cold neutron flux. The configuration has been chosen in such a way that the thermal neutron flux for beams #4-6 will not be dramatically decreased (Fig. 8).

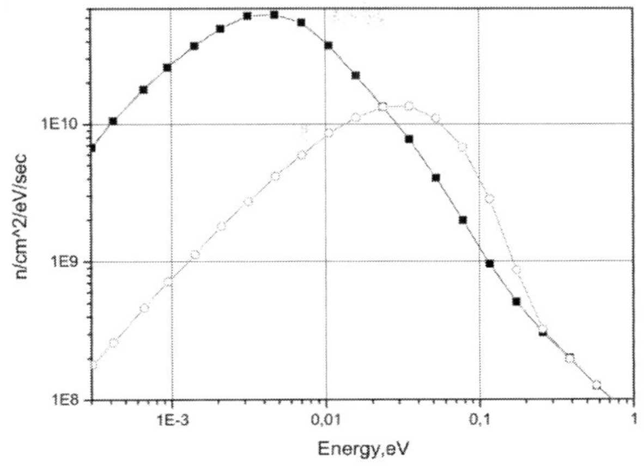

Fig. 8. The result of calculation of differential neutron spectra by virtual point detector installed at the distance of 4.5 m from the moderator surface in the direction of beam #5. The reactor power is 2 MW. (Black squares - the spectrum from cold moderator of the best composition with water post-moderator, open circles - neutron spectrum from the water flat moderator at ambient temperature.)

3. Conclusions

The modernized IBR-2M reactor will start its operation with water grooved moderators in 2011. In 2012 the first stage of development of complex of moderators will be finished. New kombi-moderator will replace a water ambient temperature grooved moderator for beams #7,8,10,11. Afterwards, the main parameters of the cold moderator and the complex of moderators for this direction will be studied. The next stage will be done for beams #2-3, and at the final stage - for beams #1, 9, 4-6.

The first results of computer optimization of configuration of moderators for beams #1, 9, 4-6 are optimistic. It appeared possible to increase cold neutron flux, not depressing significantly thermal neutron flux. As construction of moderators for this direction is planned to be done at the final stage of the project, it gives us a time to change slightly the geometry if experience with operation of the first two kombi-moderators shows it is necessary.

Acknowledges

The work has been supported by contracts of CRP of IAEA and Russian Ministry of Education and Science.

References

[1] Kulikov S., Shabalin E. 2009 Romanian Journal of Physics, New Complex of Moderators for Condensed Matter Research at the IBR-2M Reactor, 54, 3-4, Publishing House of the Romanian Academy, ISSN:1221-146X, 361-367

[2] Utsuro M., Sugimoto M, and Fujita Y. 1975 Report 8 Experimental study on a cold neutron source of solid methylbenzene. Ann. Rep. Res. Reactor Inst. Kyoto Univ, .pp. 17-25.

[3] Nunighoff K.N., Pohl Ch., Bollini V., at al. , 2004, European Physical Journal A - Hadrons and Nuclei, 22, Ice moderator experiments at very low temperatures, Springer-Verlag, ISSN:1434-6001 , eISSN:1434-601X , 519-528

[4] Natkaniec I., Shabalin E., Kulikov S., Holderna-Natkaniec K., June 2006, In Proc. of 17th Meeting of the International Collaboration on Advanced Neutron Sources, ICANS-XVII, Santa Fe, New Mexico, Comparison of Neutron Scattering and Radiation Properties of Methane and Water Ices with Methyl Derivatives of Benzene at Low Temperatures., Vol. II,

LA-UR-06-3904, p.519-529,

[5] Kulagin E., Kulikov S., Melikhov V., Shabalin E. 2004 Nuclear Instr. and Methods in Physics Research, B, Radiation Effects In Cold Moderator Materials: Experimental Study of Accumulation and Release of Chemical Energy, 215, 1-2, 181-186

[6] Carpenter J., Cold moderator for pulsed neutron sources, Proc.of Intern.Workshop on Cold Neutron Sources, LANSCE, Los-Alamos, USA, 1990, P. 131-153.

[7] Belyakov A.A., Solid methane cold moderator at the IBR-2 reactor // Journal of Neutron Research, 1996, N3, P. 209-221.

[8] Beliakov A.A., Tretiakov I.T., Shabalin E.P., Golikov V.V., and Luschikov V.I., 1998, Solid Methane Cold Moderator for the IBR-2 Reactor, Proc.of the International Workshop on Cold Moderators for Pulsed Neutron Sources, ANL, Argonne, Illinois, P. 73-78.

[9] Shabalin E., Fedorov A., Kulagin E., Kulikov S., Melikhov V., Shabalin D., 2008, Nuclear Instruments and Methods in Physics Research Section B: Beam Interactions with Materials and Atoms, Experimental study of swelling of irradiated solid methane during annealing, 266, Elsevier Science B.V., ISSN:0168-583X , eISSN:1872-9584, 5126-5131

Silver behenate and silver stearate powders for calibration of SAS instruments

M Nyam-Osor[1,2], D V Soloviov[1,3], Yu S Kovalev[1], A Zhigunov[4], A V Rogachev[1,5], O I Ivankov[1,3], R V Erhan[1,6], A I Kuklin[1,7]

[1] Joint Institute for Nuclear Research, Dubna, Russia
[2]National University of Mongolia, Ulaanbaatar, Mongolia
[3]Taras Shevchenko National University of Kyiv, Kyiv, Ukraine
[4]Institute of Macromolecular Chemistry of the Academy of Sciences of Czech Republic, Prague, Czech Republic
[5]Skobeltsyn Institute of Nuclear Physics of Moscow State University, Moscow, Russia
[6]Horia Hulubei National Institute for R&D in Physics and Nuclear Engineering, Bucharest-Magurele, Romania
[7]Bionanocenter of Moscow Institute for Physics and Technology, Dolgoprudny, Russia

Email: kuklin@nf.jinr.ru

Abstract. The possibilities of calibrations by silver behenate [$CH_3(CH_2)_{20}COOAg$] and silver stearate [CH_3 $(CH_2)_{16}CO_2Ag$] for the small angle scattering (SAS) technique are considered. It was shown that the long-period spacing of the silver compounds allow us to check the parameters of time-of-flight (TOF) method. Results of the data obtained from small angle neutron scattering (SANS) measurements performed with the silver behenate and the silver stearate powders are presented and compared with small angle X-ray scattering (SAXS) measurements.

1. Introduction

Small angle scattering of X-rays (SAXS) and neutrons (SANS) are two techniques that provide essential information about the structural properties of materials. The nature of the scattering in case of SAXS and SANS techniques are different (electron cloud interaction in case of SAXS and nuclei interaction in case of SANS), but nevertheless, the small angle scattering data treatment is similar. The module of q-vector is determined by the scattering angle and the wavelengths.

The traditional way of determining the wavelength and the resolution of the primary neutron beam produced by the monochromator of a SANS instrument is the time-of-flight (TOF) method, which is based on the principle to calculate the wavelength using the neutron speed given by the flight time of the neutrons. It is known that the TOF technique has high requirements for checking all of the parameters, such as the delay time (i.e. the time interval between the electronic start of the reactor and the real neutron pulse).

Therefore, we need a method for precise determination (checking) of the delay time (i.e. calibrations). One of the ways is the usage of known materials with ordered structure and long-period of the repetition distance. The methods for calibration of the SAXS and SANS instruments by using these materials, which can be so-called calibration standards, have been successfully tested [1-4]. In particular, silver behenate [$CH_3(CH_2)_{20}COOAg$] and silver stearate [CH3 (CH2)$_{16}CO_2Ag$] were used for the calibrations of the instruments, because these silver compounds are the most suitable for observing the Bragg peaks in the small q range.

In this paper, we present the results of the comparison of the data obtained by the SAXS and the SANS technique with using the silver behenate and silver stearate, as well as proposing a measurement time gaining method by using a mixed silver behenate and silver stearate sample.

2. Materials and methods

The silver behenate and silver stearate are known silver salts of the long-chain fatty acids. The long-period spacing of d_{001}=58.378 (8) Å for silver behenate and d_{001}=48.68 Å for silver stearate are known according to previous studies [1]. Silver behenate and silver stearate powders used in our study were purchased from Alfa Aesar, a Johnson Matthey Company [5].

SAXS experiments were performed using a pinhole camera Molecular Metrology SAXS System attached to a microfocused X-ray beam generator (Osmic MicroMax 002) that operates at 45 kV and 0.66 mA (30 W). The camera was equipped with a multiwire, gas-filled area detector with an active area diameter of 20 cm (Gabriel design). Two experimental setups were used to cover the q range of 0.0045 - 1.1 $Å^{-1}$ ($q=(4\pi/\lambda)\sin\Theta$, where λ is the wavelength and 2Θ is the scattering angle).

The neutron experiments were performed at YuMO spectrometer by two-detector system mode with ring wire detectors [6,7]. The beam was collimated to a diameter of 14 mm on the sample. The data treatment was performed by SAS program [8] with smoothing mode [9]. The control of devices for the YuMO spectrometer was performed using the SONIX Software Package [10]. For this measurements a standard thermobox and sample environment were used [11].

3. SAXS measurements

The first experiment was performed by SAXS in order to demonstrate that in our experiments no sample degradation was observed during the neutron scattering experiments even with the high intensity gamma flux [12].

Fig. 1 shows a scattering pattern of silver behenate. Background was fitted by nonlinear curve and was subtracted from the scattering pattern; we had obtained a picture for the Bragg peaks of silver behenate, as can be seen in Fig. 2.

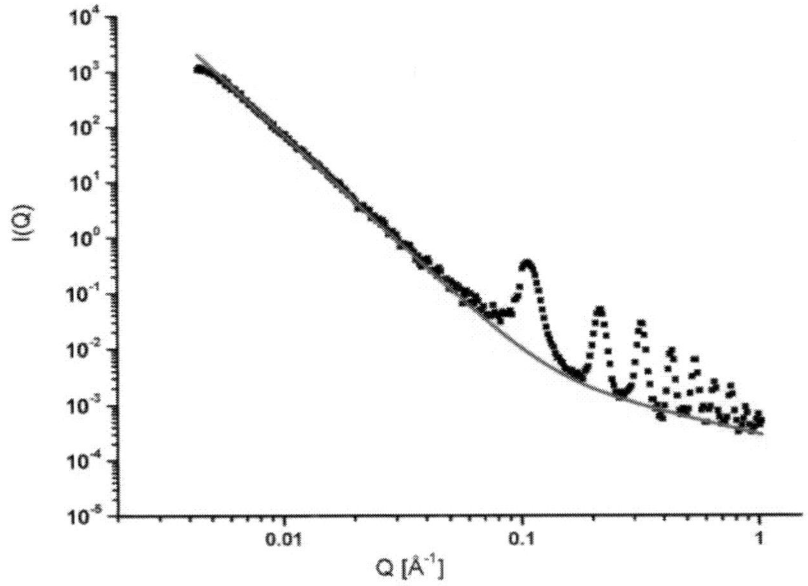

Figure 1. Scattering pattern from silver behenate obtained by SAXS.

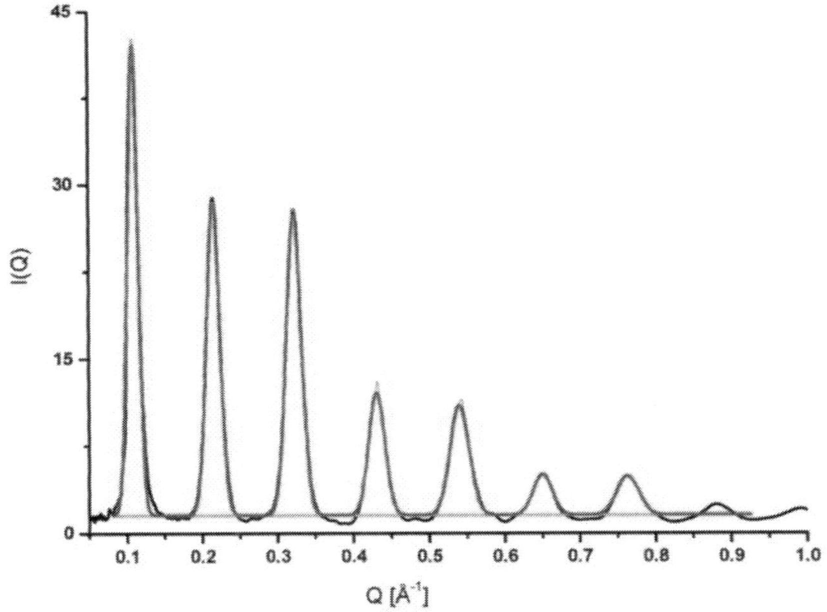

Figure 2. Bragg peaks from silver behenate obtained by SAXS.

These peaks were fitted by Gaussian peak functions. The fitting procedures and calculations were peformed for the first five peaks corresponding to 1-5th order of long-period spacing. An overview of the results of Bragg peaks of silver behenate powder is shown in Table 1.

Table 1. Bragg peaks of silver behenate measured by SAXS

Number of Bragg peaks	Q_{peak} [Å^{-1}]	d_{hkl} [Å]	Δd/d [%]	Δd [Å]	FWHM
1st	0.1086	57.9	0.9	0.5	0.016
2nd	0.2151	58.4	0.8	0.5	0.021
3rd	0.3222	58.5	0.2	0.1	0.023
4th	0.4307	58.4	0.4	0.2	0.023
5th	0.5399	58.2	0.3	0.2	0.027

Fig. 3 shows also a scattering pattern of silver stearate as another candidate sample for calibration standard. Typical picture of diffraction peaks of silver stearate is presented in Fig. 4. We should note that we have used same reduction procedure for silver stearate as before for silver behenate. Table 2 shows an overview of the results of Bragg peaks from the silver stearate powder.

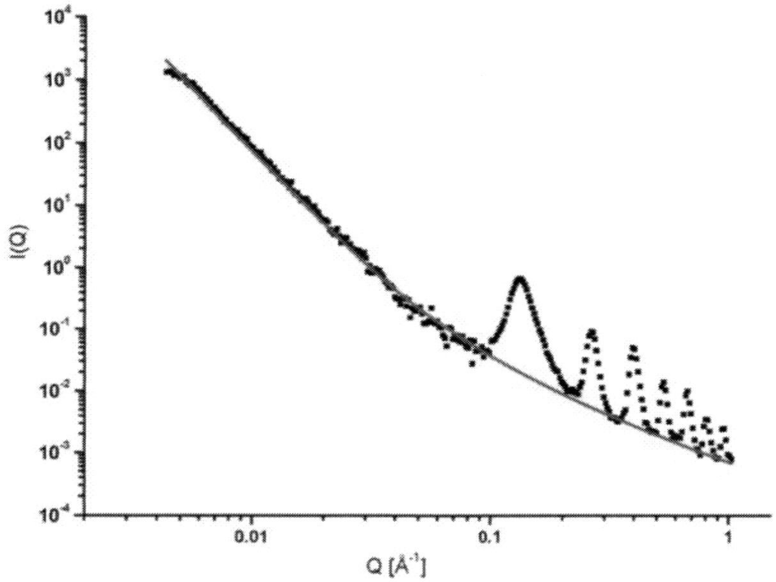

Figure 3. Scattering pattern from silver stearate obtained by SAXS.

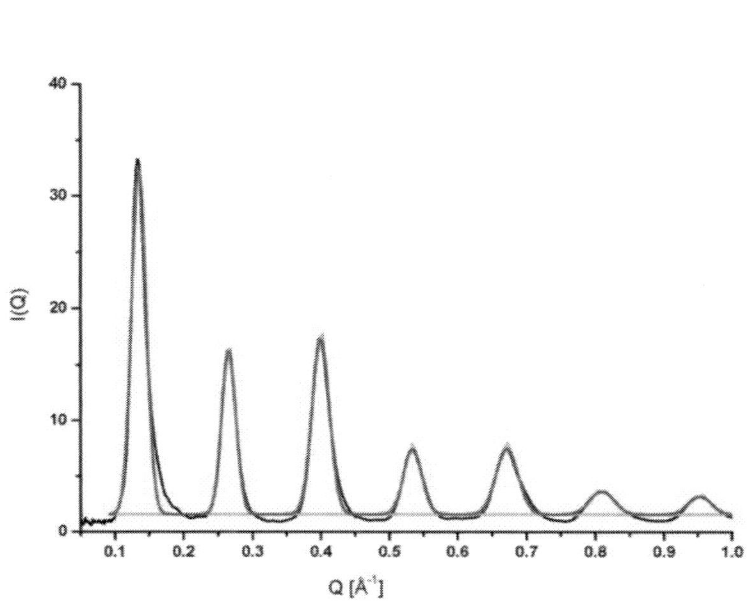

Figure 4. Bragg peaks of the silver stearate obtained from SAXS measurement.

Table 2. Bragg peaks of silver stearate measured by SAXS

Number of Bragg peaks	Q_{peak} [Å^{-1}]	d_{hkl} [Å]	Δd/d [%]	Δd [Å]	FWHM
1st	0.1351	46.5	4.5	2.1	0.025
2nd	0.2656	47.3	2.8	1.3	0.023
3d	0.3998	47.1	3.2	1.5	0.028
4th	0.5341	47.1	3.3	1.6	0.028
5th	0.6717	46.8	3.9	1.8	0.034

4. SANS measurements

SANS patterns of silver behenate and silver stearate obtained from the SANS YuMO spectrometer are shown respectively in Fig. 5 and Fig. 6. To provide the Bragg peak analysis on the SANS patterns, we have used Gaussian functions. Table 3 shows the results of the first and second Bragg peaks for silver behenate. For silver stearate, it is presented in Table 4.

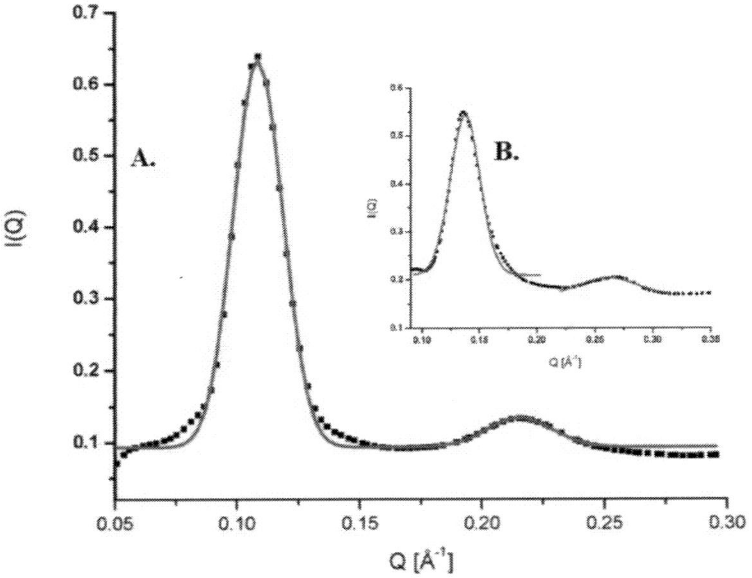

Figure 5. Neutron diffraction pattern from: A. silver behenate sample; B. silver stearate sample.

From the typical SANS diffraction pattern of silver behenate (see Fig. 5. A.) we can see that the first Bragg peak has sufficient intensity and second peak has lower intensity. Third, fourth and fifth peaks disappear in the incoherent background. However, to determine the wavelength from the Bragg reflections in the neutron diffraction pattern just one Bragg peak is enough [2, 3]. This statement is also valid for the silver stearate (Fig. 5.B).

Table 3. Bragg peaks of silver behenate measured by SANS

Number of Bragg peaks	Q_{peak} [Å^{-1}]	d_{hkl} [Å]	Δd/d [%]	Δd [Å]	FWHM
1st	0.1086	57.85	0.92	0.53	0.023
2nd	0.2154	58.34	0.09	0.05	0.031

Table 4. Bragg peaks from the silver stearate measurements by SANS.

Number of Bragg peaks	Q_{peak} [Å^{-1}]	d_{hkl} [Å]	Δd/d [%]	Δd [Å]	FWHM
1st	0.1332	47.17	3.1	1.46	0.032
2nd	0.2614	48.07	1.3	0.62	0.057

We have also measured using the YuMO spectrometer a mixed sample of two silver powders: silver behenate and silver stearate 50% + 50% wt%. The peak analyses of the mixture of silver behenate and silver stearate were carried out from the results of this measurement. Neutron diffraction patterns for the mixture of silver behenate and silver stearate powders are presented in Fig. 7 and Fig. 8. The difference between the two figures is that in the case of Fig. 7 the resulting curve was obtained by the measurement of the mixture of the compounds, while for the Fig. 8 we have combined the diffraction curves from pure silver behenate and silver stearate. In the Fig. 7 from the lower-to-higher q-range region in the neutron diffraction pattern, we can see the first peak Bragg peak from silver behenate, then first Bragg peak from silver stearate, then second Bragg peak from silver behenate and the last one is the second Bragg peak from silver stearate in the mixture of silver compounds. One can see that there is an overlapping between the silver behenate and silver stearate peaks due to the fact that the two silver powders have long period spacing in nearby q-range. Another reason why the overlapping occurs is because the SANS spectrometer has low resolution and it produces wider diffraction peaks than the peaks produced by SAXS instruments. An overview of results from peak analysis for the SANS measurement for the sample mixture are presented in Table 5.

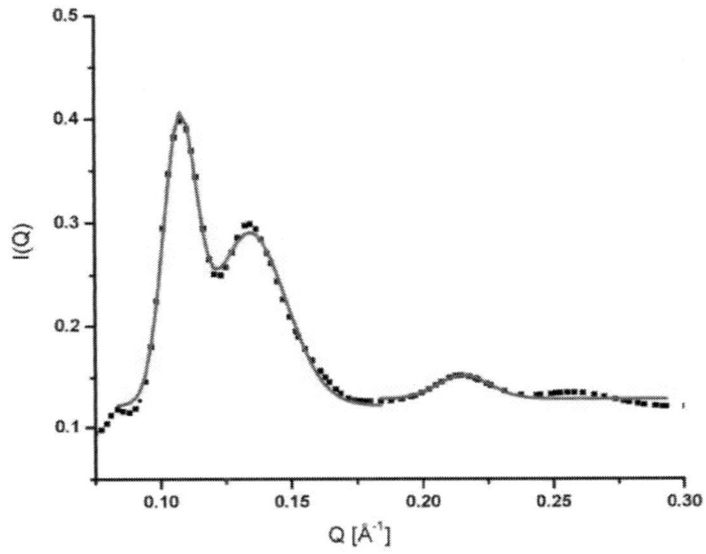

Figure 7. Neutron diffraction pattern for mixture from the silver behenate and silver stearate.

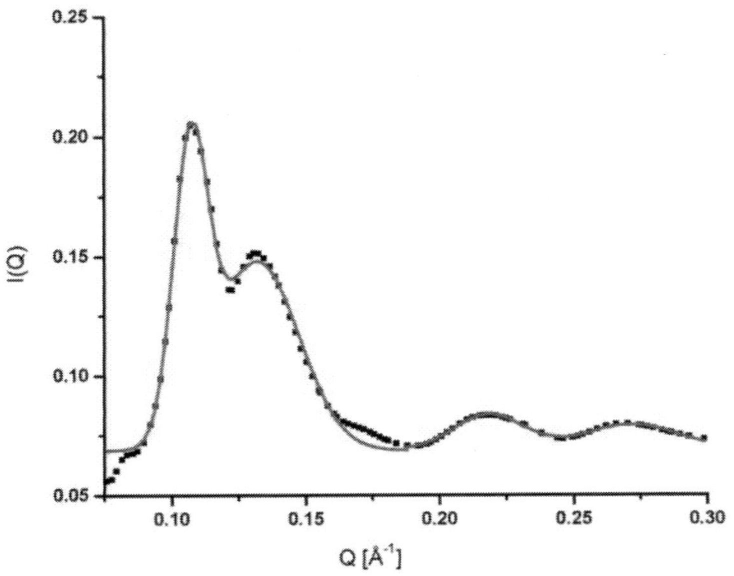

Figure 8. Neutron diffraction pattern for mixture from the silver behenate and silver stearate (combination of raw files).

Table 5. Results from the sample mixture measurement of silver stearate and silver behenate measured by SANS.

Powder	Q_{peak} [Å^{-1}]	d_{hkl} [Å]	Δd/d [%]	Δd [Å]	FWHM
Silver behenate	0.1075	58.45	0.09[a]	0.05[a]	0.015
Silver stearate	0.1343	46.82	3.9	1.85	0.033

[a] Uncertainty of delay time, only due to the delay time in accordance with lattice periodicity

From comparisons between full widths at half maximum (FWHM) results of the SAXS and the SANS measurements, we may conclude that the SANS YuMO spectrometer creates larger diffraction peak broadening than the SAXS instrument used in this study.

5. Conclusions

We considered the usage of the silver behenate and silver stearate sample measurements for the scattering vector or the wavelength calibration by the SAS technique. We have shown that the long-period spacing of the silver compounds allow us to check TOF method parameters. From comparisons between FWHM results of SAXS and SANS measurements, YuMO spectrometer has lower resolution than the pinhole camera Molecular Metrology SAXS System. For increasing the resolution of YuMO spectrometer we are planning to implement the new type of position sensitive detector with central hole in standard mode [13,14]. We have obtained some of the parameters of the YuMO spectrometer (namely: time of delay, correct wavelength etc.) using the measurements of silver behenate and silver stearate as calibrants. We suggest using a mixture of silver behenate and silver stearate for calibration of detectors on the YuMO spectrometer, because more peaks are visible in this case and in a single experiment we can obtain and check the calculation of the chosen parameters. The diffraction pattern of the mixture gives the same information with same accuracy as two separate measurements of silver compounds. SAXS and SANS measurement results of the silver behenate and the silver stearate powders are presented and compared.

Journal of Physics: Conference Series **351** (2012) 012024 doi:10.1088/1742-6596/351/1/012024

References

[1] Blanton T N, Huang T C, Toraya H, Hubbard C R, Robie S B, Louer D, Gobel H E, Will G, Gilles R and Raftery T 1995. *Powder Diffr.* **10** (2) 91-95

[2] Gilles R, Keiderling U and Wiedenmann A 1998 *J. Appl. Cryst.* **31** 957-959

[3] Keiderling U, Gilles R and Wiedenmann A 1999 *J. Appl. Cryst.* **32** 456-463

[4] Huang C, Toraya H, Blanton T N and Wu Y 1993 *J. Appl. Cryst.* **26** 180-184

[5] http://www.alfa.com/ru/gh100w.pgm

[6] Kuklin A I, Islamov A Kh and Gordely V I 2005 *Neutron News,* **16** (3) 16-18

[7] Kuklin A I, Islamov A Kh, Kovalev Y S, Utrobin P and Gordely V I 2006 *J. Surf. Investig-X-Ra* **6** 74-83

[8] Soloviov A G, Sioloviova T M, Stadnik A V, Islamov A K and Kuklin A I 2003, The Package for Small-Angle Neutron Scattering Data Treatment. *JINR Communication* P10-2003-86

[9] Soloviev A G, Litvinenko E I, Ososkov G A, Islamov A Kh and Kuklin A I 2003 *Nucl. Instrum. Methods Phys. Res. A* **502** (2-3) 498-500

[10] Kirilov A S, Litvinenko E I, Astakhova N V, Murashkevich S M, Petukhova T B, Yudin V E, Gordelii V I, Islamov A Kh and Kuklin A I 2004 *Instrum. Exp. Tech* **47** (6) 334-336

[11] Kuklin A I, Soloviov D V, Rogachev A V, Utrobin P K, Kovalev Yu S, Balasoiu M, Ivankov O I, Sirotin A P, Murugova T N, Petukhova T B, Gorshkova Yu E, Erhan R V, Kutuzov S.A, Soloviev A G and Gordeliy V I 2011 *JPCS* **291** (1) 012013

[12] Kuklin A I, Rogov A D, Gorshkova Yu E, Utrobin P K, Kovalev Yu S, Rogachev A V, Ivankov O I, Kutuzov S A, Soloviov D V and Gordeliy V I 2011 *Phys. Part. Nucl. Lett.* **8** (2) 119-128

[13] Kuklin AI, Kutuzov S A, Gabriel A, Eckold G, Utrobin P K, Smirnov A A, Ivankov A, Islamov A Kh, Kovalev Yu S, Feoktistov A, Rogachev A V, Kirilov A S, Gordeliy V I 2007 *Annual FLNP report 129-132*,

[14] Kuklin AI, Kutuzov S A, Gabriel A, Eckold G, Utrobin P K, Smirnov A A, Ivankov A, Islamov A Kh, Kovalev Yu S, Feoktistov A, Rogachev A V, Kirilov A S, Gordeliy V I 2003-2004 *LLB Scientific Report* 165

Practice of 2D data treatment in SAS

G Pépy

Research Institute for Solid State Physics and Optics, 1525 Budapest, Pf. 49. Hungary

E-mail: gpepy@laposte.net

Abstract. Many Small Angle Scattering experiments are one dimensional. Meanwhile the data of some experiments are 2 dimensional and the experimentalist may find advantages to make a full 2 dimensional data treatment or cannot avoid it. 2 dimensional data treatments present specific difficulties. The first one is an appropriate display of the whole data and of the most important part of them, in a way significant for the scientist. The second difficulty comes from the bad statistics of data which cannot be grouped together. In the following paper we describe various solutions to these problems, developed through a very long practice in LLB (Saclay, France) and SzFKI (Budapest, Hungary).

1. Introduction

The Small Angle Scattering (SAS) experiment data are usually obtained from a 2 dimensional Position Sensitive Detector (PSD); most of the time they depend of the scattering angle only; therefore the experimentalist is happy to gather these data versus the scattering angle (or scattering vector) in a 1 dimensional file and make the data treatment with one of the many appropriate softwares. However many experiments exhibit anisotropy or features not centered on the incoming beam: they cannot be reduced to 1 dimensional data because the model cannot be split in 1 D functions or it may be advantageous to make a full 2D treatment, for instance to have a single prefactor for the x dependent and the y dependent parts of the model.

Many fields provide anisotropic data: a non limitative list would include soft matter submitted to an external field, either shear, or stress, or magnetic field, of course magnetic materials ranging from nanolayers to nuclear industry alloys, and also stressed metallic alloys. A specific domain includes experiments on single crystals, which display non centered feature rather than centered anisotropy.

The domain of 2D data treatment has already been reviewed comparing the then existing softwares [1]. On the contrary this paper will be devoted to the main problems of 2D data treatment, and illustrated by examples treated with the PXY software developed by the author during a very long time [2].

In the next section we shall review how to display and select 2D data in order them to make sense.

In the following section we shall remind what means "fitting" a model and how to avoid some traps.

The problem of bad statistics (low counting rates) will be addressed next.

Finally we shall review a full data treatment case.

2. Display of 2D data

The first idea to display 2D data is to draw them in a 3D frame and to rotate the picture in order to appreciate its shape. This is certainly very attractive to the observer. Meanwhile it is a dead end once

Published under licence by IOP Publishing Ltd

one considers the next steps: how to select the most significant pixels and how to compare a model to the experimental data? Therefore we found that a better way to display 2D data is to make a flat map with various colors according to the intensity. For complex pictures a rainbow palette or a single color palette with shades are not appropriate; palettes with colors changing fast versus the intensity will much better enhance details.

At the beginning of a data treatment one will take care of a small part of the data, because it is easier to appreciate the intensity shape and also because the fit calculation will be faster. Indeed the author never found a good way to directly compare a model to 2D data. As a human being the experimentalist is so much used to a 1D representation to compare data points to a model curve that it seems impossible to escape it. For this purpose the flat color map allows easily to select the most relevant pixels in filters. The pixel intensity is then projected on one axis. The result is finally displayed on a specific drawing. Then a 2D model function can be calculated for each of the selected pixel and the projection process applied in the same way. In this manner it is easy to compare the model to the data. One should note that the projection process alters the function shape. For instance if the filter is a narrow rectangle it will be easy to recognize a Lorentzian or a Gaussian function; the shape will more and more differ from the classic shape as the rectangle is widened. The shape becomes not familiar at all if the rectangle includes the beam catcher hole!

Many 2D data exhibit different properties versus x and y axis. Therefore the most common representation uses rectangular filters (figure 1). They may be horizontal and/or vertical, possibly symmetric versus one axis; there may be several of them.

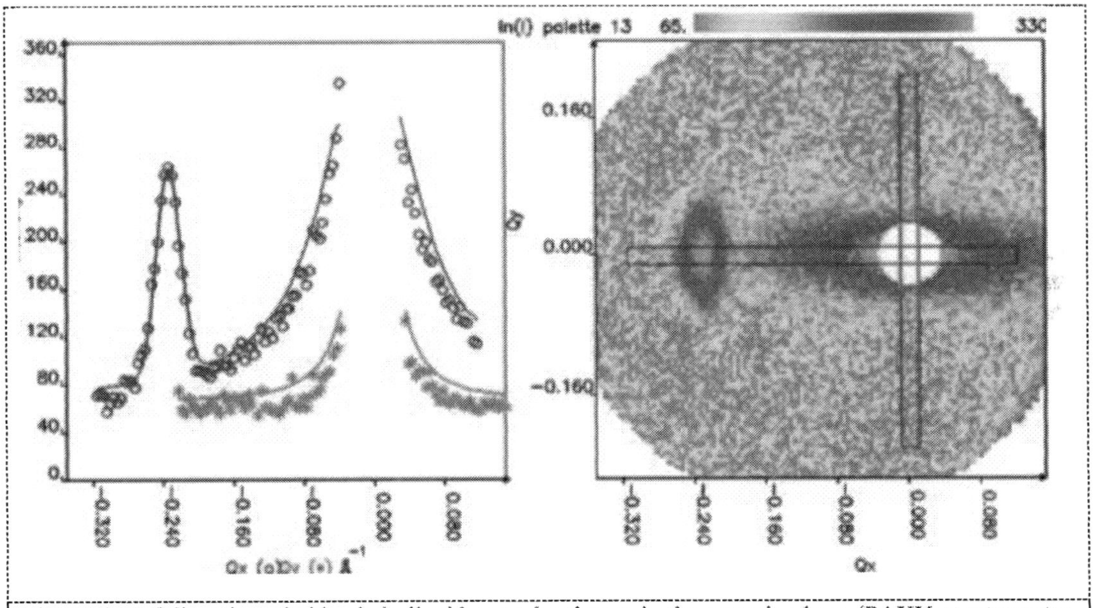

Figure 1. A fully oriented side chain liquid crystal polymer in the smectic phase (PAXY spectrometer, LLB, Saclay) [3]. The director is along the horizontal axis. On the right is the flat map of all the data. One observes an anisotropic central scattering due to the contrast between protonated and deuterated main chains and a Bragg peak due to the smectic structure. The intensity of the pixels inside two rectangular filters is projected on the relevant axis and displayed on the left. The continous line is the fitted model; a flat background + a double centered Lorentzian + a Gaussian*Lorentzian for the Bragg peak.

In many other cases a display with sector shape filters (figure 2) is appropriate, notably to exhibit anisotropy.

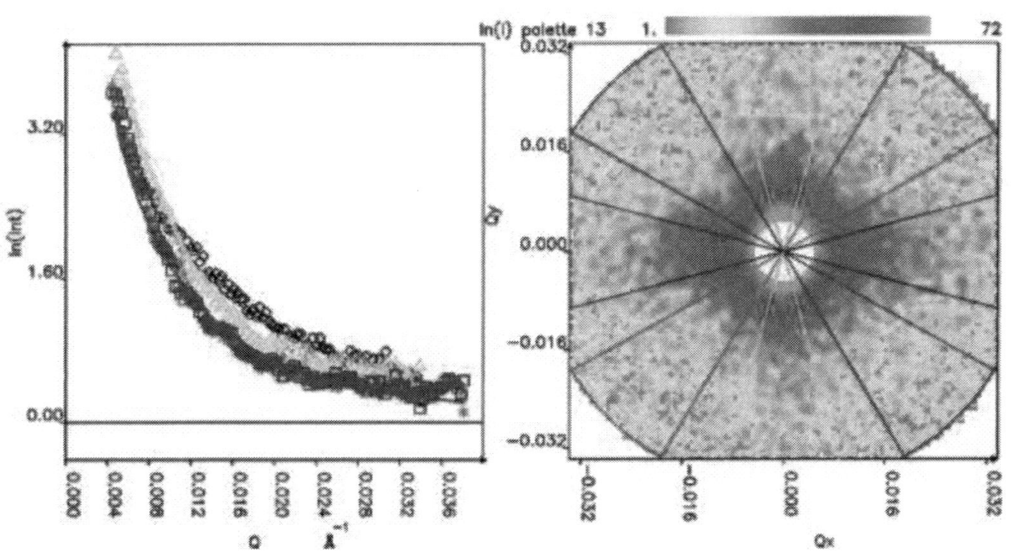

Figure 2. A smectic liquid crystal polymer (PAXY spectrometer, LLB, Saclay) [4]. 4 pairs of centro-symmetric sector filters are shown on the right. In this case the anisotropic Lorentzian characteristic of the main chain shape is polluted by a power law scattering due to oriented anisotropic catalyst fragments.

However it may happen that the classic display of the projected intensity is not enough to appreciate the quality of the fit. In figure 3, while the model is apparently good, the fit did not converge correctly. A representation of the data – calculated intensity (weighted by the uncertainty) exhibits a misset of the centre position. Using a free centred double Lorentzian (figure 4) allowed a better fit convergence.

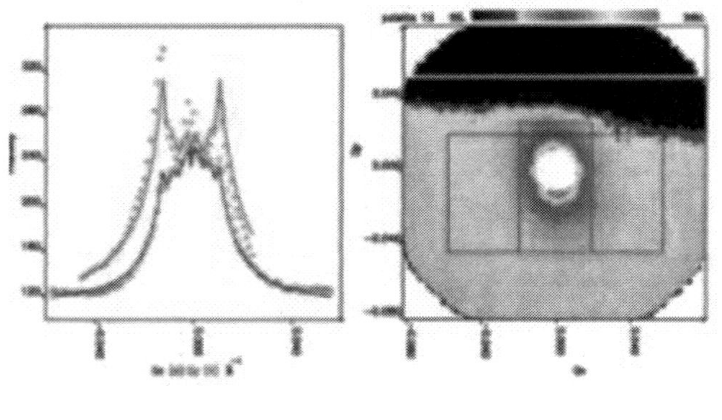

Figure 3. A nematic liquid crystal sheared in a cone-plane apparatus (which produces the shadow on top, PAXY spectrometer, LLB, Saclay) [5]. The model is a tilted (because of shear) double Lorentzian + flat background.

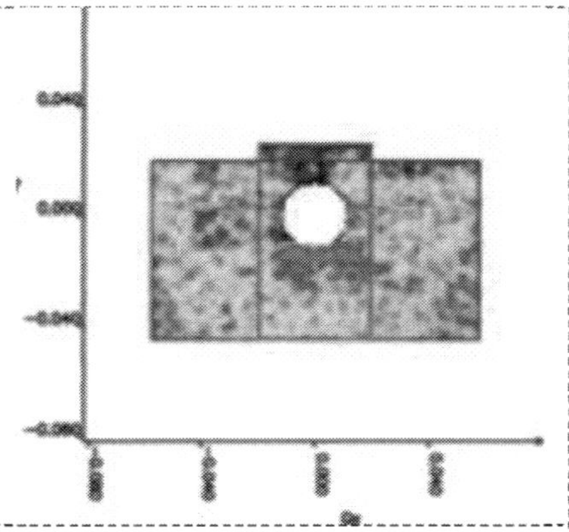

Figure 4. A display of the data – calculated intensity shows a vertical misset of the centre position.

In some cases it may be interesting to fit several files together, notably if only a few parameters differ. In figure 5 only the anisotropy changes versus temperature. One should note that a single prefactor is fitted for both files while it would be impossible to get a single value if 4 distinct

1D files were fitted. This feature of the 2D fitting software was developed to make more reliable the observation of the variation of a few parameters in a group of files.

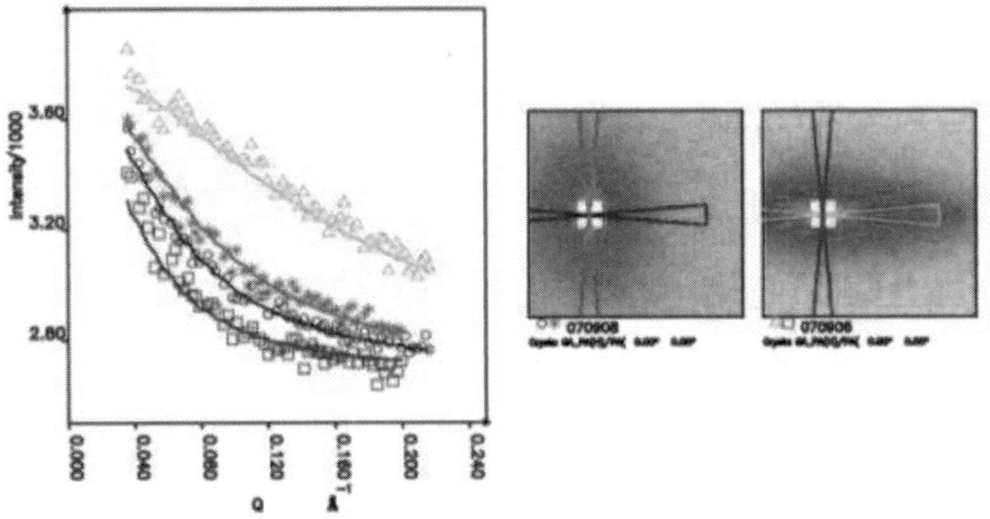

Figure 5. A nematic liquid crystal polymer (D11 spectrometer, ILL, Grenoble) at 2 different temperatures [6]. For each file 2 pairs of centro-symmetric sector filters are shown on the right. On the left the black and red data (1st file) exhibit small anisotropy, while the 2nd file, the lowest blue and highest green data, show a large anisotropy. The model is common to both files. Only the widths are independent. The prefactor is the same.

The polar geometry is valuable in very rare cases as shown on figure 6. While apparently attractive the polar geometry is difficult to use in practice. Displays versus various coordinates may be helpful.

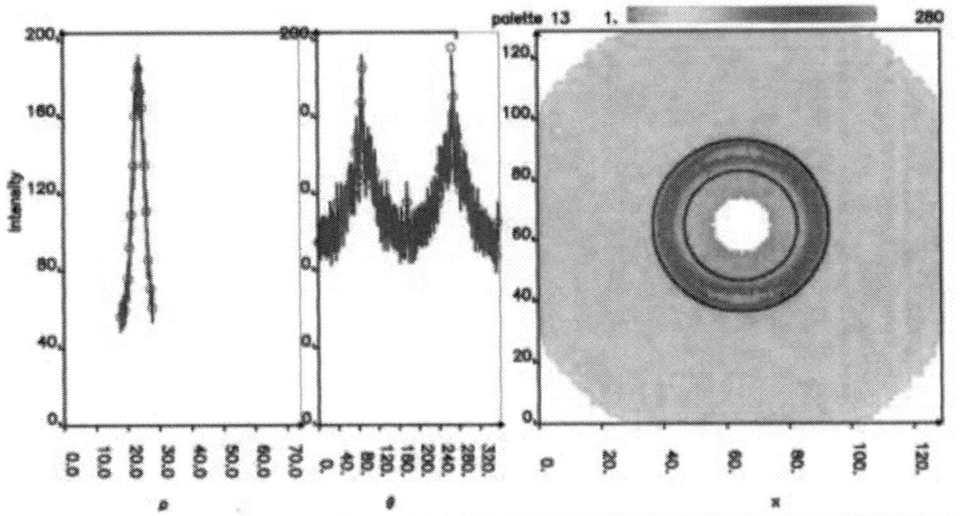

Figure 6. Fully oriented biaxial micels (PAXY spectrometer, LLB, Saclay) [7]. The model is the product of a Gaussian versus the radius and 2 symmetric Lorentzian versus the angle. This is a rare case when a polar representation is best. Pixels in the filter are shown from different points of view: radius and angle.

Sometimes the full Q range is obtained measuring at several distances or wave-lengths. For a better understanding the files should be embedded as on figure 7.

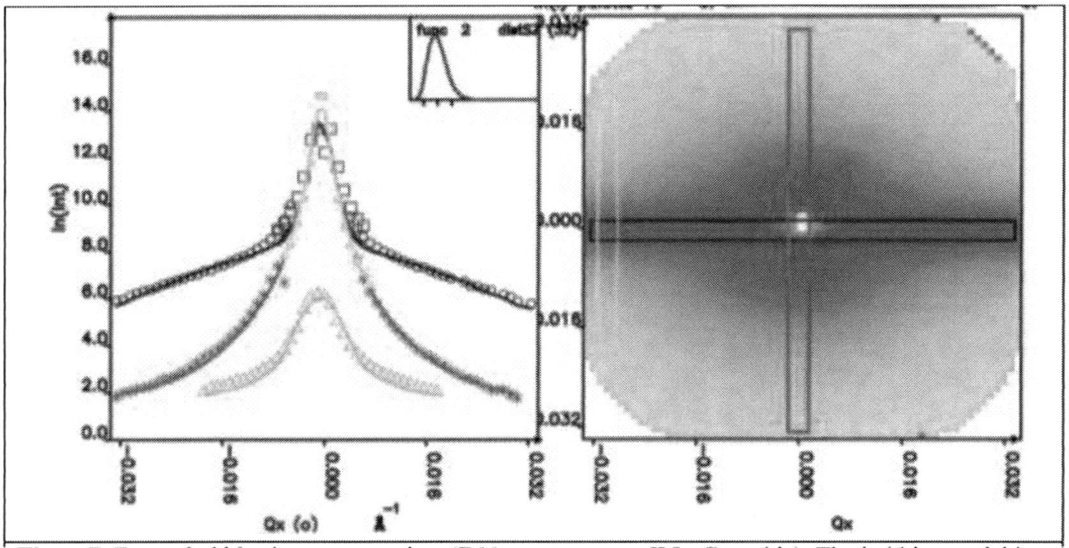

Figure7. Empty bubbles in tungsten wires (D11 spectrometer, ILL, Grenoble). The bubbles model is a set of ellipsoids with a Schultz-Zimm size distribution; it was necessary to add a centered power law [8]. The measurement took place at 2 different distances, 6 and 28m in order to increase the Q range. On the right, the map was built with 2 "embedded" files. To characterize the scattering 3 rectangular filters were necessary; the grouped pixels are displayed on the left.

In the case of X-ray pictures obtained with CCD cameras the pixel number is overwhelming. It is even more important to select the relevant parts of the picture than for pictures from gaz detectors (figure 8).

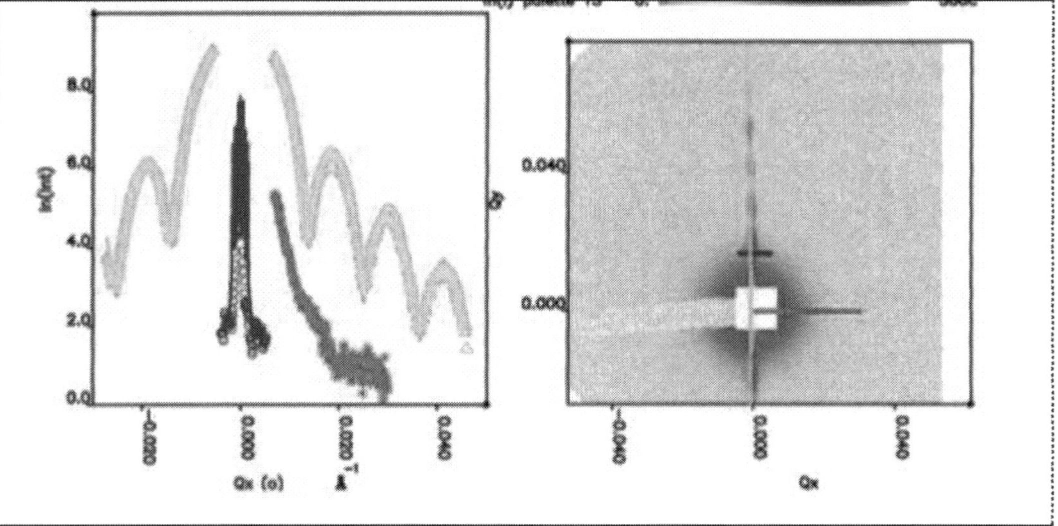

Figure 8. Nanochannels in polycarbonate foils (ID01 beam line, ESRF, Grenoble) [9]. A rough sample surface produces Porod scattering singled out in the red filter. The nanochannel model is a cylinder form factor, product of a Bessel function (green oscillating radial part) and a fast decreasing sine (blue longitudinal part).

3. Fitting process

There are several ways to fit a model to data. A very good introduction to data reduction and error analysis may be found in Bevington [10]. In the PXY software we use the mean-square minimization with steepest descent, which is indeed a very common process. For the clarity of this paper we consider useful to remind the main principles.

The distance between the data and the model is evaluated thanks to a "distance" function. Among several properties such a function should be a positive scalar function. Very often the distance function is a χ^2:

$$\chi^2 = \frac{1}{N-p}\sum_{i=1}^{i=N}\left(\frac{I_i - Y_i\{P\}}{\Delta I_i}\right)^2 \qquad (1)$$

where

Ii	intensity in pixel i
N	number of data points
p	number of free parameters
{P}	set of parameters
Yi	calculated intensity in pixel i
ΔIi	uncertainty of Ii

If the data probability law is a normal law χ^2 decreases down to 1 as the fit converges. This property is very useful to appreciate the quality of a fit. In particular it helps to detect systematic errors due to bad correction of the raw data.

The minimization process runs in cycles. Giving increments to the parameters during one cycle one calculates the χ^2 1st and 2nd derivatives versus the parameters. This allows to build an approximate set of linear equations. Solving this set of equations provides parameter increments. Assuming that the current position in the parameters space is close to the solution means that the χ^2 is close to the bottom and that the obtained increments will drive to it. Hopefully the χ^2 calculated for the new parameter set is smaller than the previous one. One has merely to cycle the process until the χ^2 does not decrease significantly.

Unfortunately the model functions of interest make the minimization process highly non linear. It is therefore important to start from a not too bad set of parameters. For this purpose it is helpful to select narrow data filters which will be broadened for the final fit. Also it is advisable to let free only a part of the parameters while the most obvious (background, centre position) are kept fixed. Indeed the more free parameters the highest the risk the minimization process falls in a "false" minimum from which it is impossible to escape. When this happens (a hint is a χ^2 far from 1), the only solution is to try better with other sets of start parameters...

Other minimization methods are available. One of the most interesting is the MINUIT package [11] developed at the CERN. It was recently included as a variant in the PXY software. The fitting results are the same as with our previous method. Sometimes MINUIT helps reaching a solution, sometimes our previous method is better (probably thanks to the χ^2 2nd derivatives calculation).

4. The problem of bad statistics

When dealing with 2D data, the problem of bad statistics occurs very often. Bevington [10] proposes two ways to improve the statistics: repeat the experiment increasing the measurement time or smooth out the data. We shall consider the smoothing in a subsequent section: before applying Bevington's advices one should consider the SAS raw data correction.

4.1. SAS raw data correction

Classically SAS raw data correction includes substracting a reference data file from the sample file and calibration by a specific incoherent scatterer (water, Plexiglas, vanadium) from which the background has been removed.

For each pixel the corrected scattered intensity is:

$$\left(\frac{d\Sigma}{d\Omega}\right) = F_w \frac{I_s/C_s - I_r/C_r}{I_w/C_w - I_c/C_c} \qquad (2)$$

$$C_s = \frac{(L_{SDs})^2}{A_s d_s M_s T_s} \qquad (3)$$

Where I is the intensity and C a coefficient.
The index mean

- s sample
- r reference
- w calibration (water…)
- c background (container…)

and the factors are

- A beam area
- d sample thickness
- M monitor
- T transmission
- L_{SD} sample to detector distance
- F_w normalization prefactor

Is, Ir, Iw, Ic and Cs, Cr, Cw, Cc are the pixel intensities and coefficients for sample, reference, calibration and background, respectively.

Of course bad statistics make the fit difficult because of the large associated uncertainties (error bars). It is trivial to observe that the uncertainty over the corrected file is the sum of the uncertainties of all the elements of equation (2). We shall assume that the coefficients uncertainty is negligible, including the transmissions. We shall therefore examine the usefulness of the correction files which are few and therefore much less time consuming than the data files acquisition. At this point we should remember Bacon's remark in is famous book about diffraction [12]: "the uncertainty of the correction files should not be worse than that of the measured files". He was hinting mostly to the problem of background measurement: this should be measured during a much longer time than the data in order to get the same level of uncertainty. In a modern SANS spectrometer at the end of a neutron guide, the background is very low indeed and experimentalists do not want to loose very much of their precious experimental time just to obtain good statistics… Meanwhile the calibration file usually exhibits small error bars; therefore it makes sense to forget the background correction just to avoid increasing a lot the final uncertainty!

The calibration file itself is extremely important. For gaz detectors the efficiency of each cell exhibits a slow drift versus time and it depends mostly of the preamplifier gain for the anode and cathode which define the cell. This last remark leads to a very useful trick. If one assume that the efficiency of a given cell is merely the product of the gain of the anode preamplifier and of the gain of the cathode preamplifier one can build a calibration matrix in the following way: calculate the average intensity of the calibration file, reject all pixels with intensity too far from the average, calculate again the average intensity for the remaining pixels, then calculate the average intensity of each line A_l and each column A_c and normalize them by the overall average intensity, finally build a new calibration matrix whose element A_{lc} will be the product of A_l and A_c. For a square matrix with n^2 elements the uncertainty of element A_{lc} will be smaller than that of the initial matrix by a factor 2/n. Of course this is not fully accurate but the fit will be more precise! In most cases this is a very efficient way to reduce the uncertainty.

Neither the reference file should be discarded at once as it sometimes contains useful information. For instance for protonated materials it allows to determine accurately the background. However for the data treatment one has to question whether it is better to prefer a single fitted background parameter. Assuming the error bars of the reference file are of the same order of magnitude as that of the sample files, replacing it by a single background parameter would divide the uncertainty by 2. Most often this is a worthwile transaction.

4.2. Smoothing

The true probability law of data obtained with gaz detectors is a Poisson law, which differs quite a lot from a normal law at the smallest counting rates. This takes place at the many pixels close to the background. And it has a strong influence on the fitting process: in equation (1) the data uncertainty ΔIi is usually approximated by √Ii, which is correct for a normal law but introduces a strong disturbance in case of a Poisson law. For instance the fit may converge towards an acceptable shape of the model but a much too high background parameter. To remedy this feature Bevington [10] advises to smooth the concerned data (it is enough to smooth only the smaller counting pixels).

The least disturbing method is the so-called "parabolic" smoothing where a pixel intensity is replaced by a weighted average over the neighbouring pixels. The weight of the relevant pixel is 1/4, while the weight for the 4 next pixels on the neighbouring lines and columns is 1/8 and the weight for the other 4 pixels on the diagonal is 1/16. This is very efficient to improve the fitting process. A single smoothing pass is enough.

Table 1

1/16	1/8	1/16
1/8	1/4	1/8
1/16	1/8	1/16

Meanwhile this manipulation has a drawback: it subtly changes the statistics and now the χ^2 decreases to a value lower than 1, which invalidates the convergence criterium and complicates the parameter error bar calculation. Papoular [13] pointed to a solution to this problem, namely replacing not only the intensity but also the associated uncertainty. Let Δ_j be the mean square deviation (uncertainty) of pixel j and w_j the relevant weight, after smoothing the uncertainty of pixel i should be:

$$\Delta \mathrm{Ii} = \sqrt{\sum_j \left(w_j . \Delta_j\right)^2} \qquad (4)$$

where the sum runs over the pixels as shown on table 1. This restores the behaviour of the χ^2.

4.3. Wild points

Wild points are pixels for which the intensity lies far away from the expected value. They are suspected of being linked to systematic errors. Authors of the FITTER 1 dimensional data treatment software [14] included a specific procedure. Several such procedures were developed in the literature under the name "robust fitting". Two of them (Huber, Tukey bisquares [15]) are now also included in the PXY software. In equation 1 is modified mutiplying the estimator e_i (6) in order to decrease the influence of the wild points in the χ^2. Equation 1 now becomes:

$$\chi^2 = \frac{1}{N-p}\sum_{i=1}^{i=N}(w_i e_i)^2 \qquad (5)$$

$$e_i = \frac{I_i - Y_i\{P\}}{\Delta I_i} \qquad (6)$$

The weights w_i are defined as:

Table 2

method	tuning constant	weight function
Huber	k = 1.345	$w = \begin{cases} 1 & for \quad \lvert e\rvert \le k \\ k/\lvert e\rvert & for \quad \lvert e\rvert > k \end{cases}$
bisquare	k = 4.685	$w = \begin{cases} \left[1-\left(\frac{e}{k}\right)^2\right]^2 & for \quad \lvert e\rvert \le k \\ 0 \quad for & \lvert e\rvert > k \end{cases}$

Meanwhile it seems that if systematic errors are carefully avoided these procedures are of limited interest.

5. Running a 2D data treatment

The experimentalist must be aware that no method will ever improve a bad experiment! The first task will be to read the data and use a representation best suited to the experiment. Do not forget to include a smoothing of small intensity pixels. Then a simple filter or multiple filters will be chosen, narrow enough to select the most significant pixels (in order to further limit the fitting time while getting a first approximation of the parameters). The intensity of the pixels inside the filter is displayed on a I = f(v) diagram, where v is one of the variable x, y, ρ, θ. The model function includes at least a background parameter and one or several functions. It is highly recommended to have a first look at the model shape produced by the input parameters; it should not be too far from the data. Once it is satisfactory one runs the fitting program. Hopefully it will converge. A χ^2 far from 1 is an unmistakable signature of a false minimum, one should start again the fit with another set of parameters. The calculation of the parameter uncertainty depends of the χ^2 being close to 1. Most often the results are good and it is not necessary to repeat the fit with a large filter only to somewhat improve the parameter error bars.

6. Conclusion

We have reviewed many practical aspects for a successful data treatment: the choice of the representation of the data and the filters, the criteria to include (or not) some corrections and smoothing of the raw data, not forgetting a trick to improve the calibration statistics, how to take care of wild points and, last but not least, how to identify a false minimum though the χ^2 observation.

7. Acknowledgements

Special thanks go to R Kahn (Laboratoire Léon Brillouin, Saclay) who provided the author with the first fitting kernel. The author is deeply indebted to two main users of the PXY software: L Noirez (Laboratoire Léon Brillouin, Saclay) and A Len (SzFKI). The PXY software was developed thanks to their many and always interesting requests. In the meantime they showed an extraordinary patience for the lengthy developments and an even more extraordinary patience for the many bugs which they contributed to identify and fix. It is also important to note the thorough discussions in Dubna with A Kuklin, A Islamov and A Soloviev about the MINUIT package and the wild points treatment. The play goes on…

References

[1] Pépy G, ECNS99 proceedings, Data imaging introductory course, 1999, KFKI-1999-04/E report Pépy G http://www-llb.cea.fr/cours/pepy/dl.htm
[2] Pépy G, *J. Appl. Cryst.* (2007). **40**, s433–s438
[3] Keller P, Carvalho B, Cotton J P, Lambert M, Moussa M and Pépy G 1985 *J. Physique Lett.* **46** L1065-L1071
[4] Vicentini F, Noirez L, Pépy G and Mauzac M 1995 *Europhys. Lett.* **32** (8) pp657-662
[5] Noirez L and Lapp A 1997 *Phys. Rev. Lett.* **78** 1 pp70-73
[6] Noirez L H. Mendil, L. Noirez,* and P. Baroni,I. Grillo PRL 96, 077801 (2006)
[7] Kiselev M A, Janich M, Lesieur P, Hoell A, Oberdisse J, Pépy G, Kisselev A M, Gapienko I V, Gutberlet T, Aksenov V I 2002 *J. Appl. Phys.* A **74** S1239-S1241
[8] Len A *submitted to Neutron News*
[9] Mendil H, Noirez N, Baroni P and Grillo I 2006 *PRL* 96, 077801
[10] Bevington P R 1969 *Data Reduction and Error Analysis for the Physical Sciences* (New Y ork: McGraw-Hill)
[11] James F 1994 Geneva http://wwwasdoc.web.cern.ch/wwwasdoc/minuit/minmain.html
[12] Bacon G 1955 *Neutron diffraction* (Oxford: Clarendon Press)
[13] Papoular R, LLB (Saclay), private communication
[14] Soloviev A, Stadnik A, Islamov A and Kuklin A 2007 Dubna http://wwwinfo.jinr.ru/programs/jinrlib/fitter/docs/html/index.html
[15] Huber P 1981 *Robust statistics* (New York: Wiley)
Fox J 2002 *An R and S-PLUS Companion to Applied Regression* (Thousand Oaks: Sage publications)

Journal of Physics: Conference Series **351** (2012) 012026 doi:10.1088/1742-6596/351/1/012026

KWS-2, the high intensity / wide *Q*-range small-angle neutron diffractometer for soft-matter and biology at FRM II

A Radulescu[1], V Pipich, H Frielinghaus and M-S Appavou

Forschungszentrum Jülich GmbH, Jülich Centre for Neutron Science (JCNS), Outstation at FRM II, 85747 Garching, Germany

E-mail: a.radulescu@fz-juelich.de

Abstract. The KWS-2 small-angle neutron diffractometer operated by JCNS at FRM II is upgraded and optimized towards high intensity and wide *Q*-range studies of mesoscopic structures and structural changes due to rapid kinetics and becomes a dedicated SANS instrument to soft-matter and biology. The high intensity permits fast measurement of small or weak scattering samples and time resolved structural studies with a time resolution up to 100ms. The possibility to cover up to four decades in *Q* will soon enable structural investigation over a wide length scale, between several Å and 1μm. The characteristics and performance of the instrument in the conventional pinhole mode is detailed presented and the new upgrades currently in progress and aiming for boosting the instrument performance towards higher intensity and wider *Q*-range are reported.

1. Introduction

After the shut-down of the FRJ-2 "DIDO" research reactor in Jülich, the small-angle neutron diffractometers KWS-1 and KWS-2 have been rebuilt by the Jülich Centre for Neutron Science (JCNS) at the neutron source Heinz Maier-Leibnitz (FRM II reactor) in Garching [1]. Both instruments are 42m long and based on the same concept: by the pinhole geometry the exploration of a wide momentum transfers *Q*, between 7×10^{-4} and $0.5Å^{-1}$, is possible by the variation of the sample-to-detector distance between 1m and 20m and of the wavelength between 4.5Å and 20Å. At FRM II the combination of the high flux supplied by the cold neutron source (CNS) and the newly designed neutron guide system [2] supported the major upgrade of KWS-2 towards particular scientific topics: with a new "fast" detection electronics accepting high count rates up to 0.6 MHz, a new collimation system allowing for larger experimental flexibility and the routine use of large beam-sizes, KWS-2 became a dedicated SANS diffractometer to high-intensity / wide *Q*-range investigation of mesoscopic structures and structural changed due to rapid kinetics in soft-matter, chemistry and biology.

In the middle of 2010 the instrument entered a second phase of major upgrade aiming for boosting the instrument performance towards higher intensities and wider *Q*-range by means of focusing elements – the MgF_2 lenses [3-5]. The neutron lenses are planned for the high intensity mode (up to 15 times gain in intensity compared to the conventional pinhole mode by increasing the sample area while holding the resolution) and the high resolution mode (exploration of low *Q* range up to 1×10^{-4}

[1] To whom any correspondence should be addressed.

$Å^{-1}$) when they will be used in the combination with a double-disk chopper with variable opening and a high resolution position-sensitive detector.

In this paper we will present the design and characteristics of the KWS-2 SANS diffractometer in the standard pinhole mode including some data for demonstrating its performance.

2. Instrument components and characteristics

The 42m long instrument KWS-2 occupies an end-of-guide beam position (50x50mm^2 cross section) on the newly built NL3-ao vertically "S-shaped" neutron guide [2,6] in the cold-neutron-guides West-Hall of FRM II. A conceptual design of the KWS-2 instrument is shown in figure 1 while the key components and working procedure for the pinhole standard mode are discussed in turn below. A detailed description and characterization of the new components that are currently in construction (the double-disc chopper with variable opening, the neutron lenses and the high-resolution detector) makes the subject of a forthcoming publication.

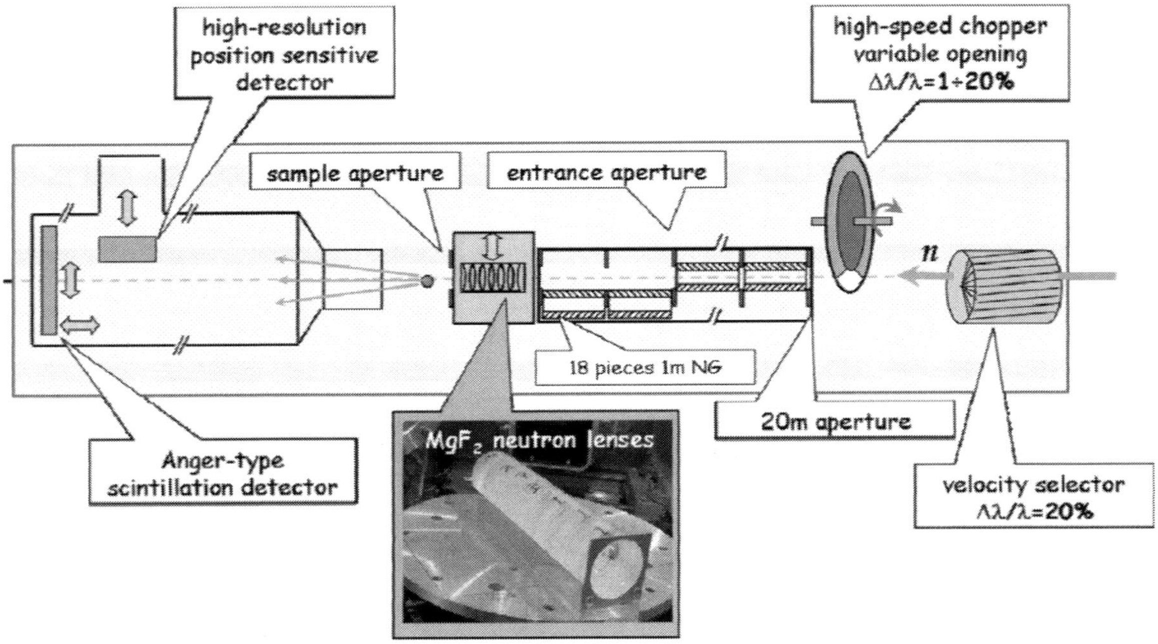

Figure 1. Schematic diagram of the KWS-2 SANS instrument.

2.1. Wavelength selection and collimation system

The incident beam from the neutron guide is monochromatized by a mechanical slit selector (ASTRIUM GmbH) with a wavelength resolution $\Delta\lambda/\lambda$=0.20 in a wavelength range of $\lambda = 4.5 \div 20$Å. The monochromated beam is collimated by a system of moveable guides and fixed position apertures. The 18 neutron guide elements, each 1m long with 50x50mm^2 cross-section and m=1.4 non-magnetic coating, allow for the variation of the effective source-to-sample distance from 2m to 20m in 1m increments to vary the beam divergence and flux on the sample. Each guide is followed by a squared source-defining aperture (entrance aperture). The apertures placed at 2, 4, 8, 14 and 20m in front of the sample and the sample aperture have variable opening, between 1x1mm^2 and 50x50mm^2, allowing for either symmetric or slit-like beam geometry; the other apertures have a 50x50mm^2 fixed opening. All apertures are equipped with ^{10}B enriched ceramic blades with sharp edges and large BC_4 masks. The variable opening apertures are equipped with additional elements that increase their performance (transducers, special driving systems, special positioning and alignment systems). Thus, collimation of the incident beam is defined by the entrance aperture (typically 50x50mm^2 for standard pinhole mode) and the sample aperture (typically 10x10mm^2); this is the pinhole standard working mode.

2.2. The sample area

The KWS-2 SANS instrument has a sample area that allows for the accommodation of large sample environments such us cryostats, furnaces, rheometers, shear cells, stretching devices, pressure cells, humidity cells, etc. A sample stage that allows for the xyzθ positioning of sample holders and can support heavy sample environments is typically used. In this case, an evacuated tube with variable length, the so-called "collimation nose", enables the positioning of the sample aperture very close to the sample or the entrance window of the sample environment, diminishing the air flight path of neutrons. For measurements at room temperature or within a typical temperature range for soft-matter and biology topics (-20°C÷150°C) a multi-position / multi-level thermostated sample holder with dry-air flow is routinely used. For special cases, an evacuated sample chamber allowing for the use of a multi-position thermostated sample holder with sealed sample cells can be used. The sample area is massively shielded by 1.5cm thick inner layer of borated plastic and 5cm thick outer layer of lead. The access of users at the sample position is allowed through a motorized door commanded by the instrument controlling computer and coordinated with the status of the instrument shutters.

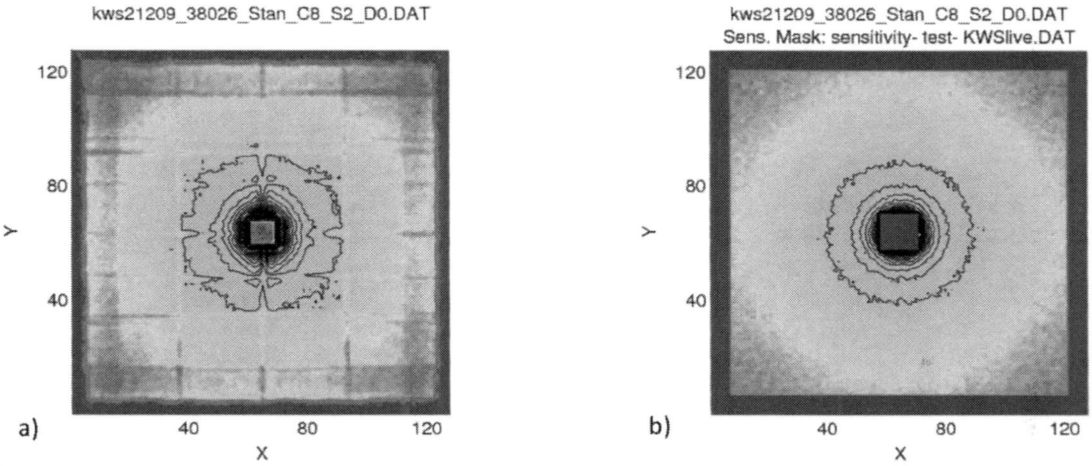

Figure 2. Typical scattering pattern collected with the scintillation detector of KWS-2 as raw data (a) and corrected data for detector sensitivity (b).

2.3. Main detector

The KWS-2 main detector is based on the Anger principle [7] with an array of 8x8 photomultipliers, a 60x60cm^2 ^{6}Li glass scintillator with an efficiency >95% for λ>4.5Å neutrons [8] and a light disperser glass, which spread the light produced by a neutron event to an array of 3x3 photomultipliers [9]. The readout electronics enables fast signal and data processing (about 0.6μs statistical dead-time) and allows for the acquisition of stable and reliable count rates up to 0.6MHz. Because of the photomultipliers array and the scintillator active area made of 4x4 plates glued together, a grid-like structure corresponding to detector regions with lower sensitivity characterizes the raw scattering patterns (figure 2a). Therefore, corrections that take into account the detector sensitivity measured with an incoherent standard sample (Plexiglas) must be always applied to the raw data (figure 2b).

The determination of the system dead-time was carried out by comparing the detector count rates under uniform neutron illumination to that of a fission chamber. Because the count rate of the fission chamber is linearly related to the true count rate n of the scattered neutrons, the dead time of the detector may be derived from the deviations of the measured count rates n_m from the ideal behavior (figure 3). The recorded data can be fully corrected for detector dead-time losses for count rates up to 150kHz, when a 10% dead-time effect occurs in the measured rates n_m compared to the true rates n (figure 4). The ^{6}Li scintillator is sensitive for γ-radiation originating directly from the reactor or from activated materials in the surroundings. In order to suppress the γ-background discriminators and filtering algorithms have been implemented that reduce the γ-to-neutron ratio to 10^{-5}. The spatial resolution of the detector determined using an absorbing boron coated diaphragm with holes placed in front of the scintillator is 5.25mm.

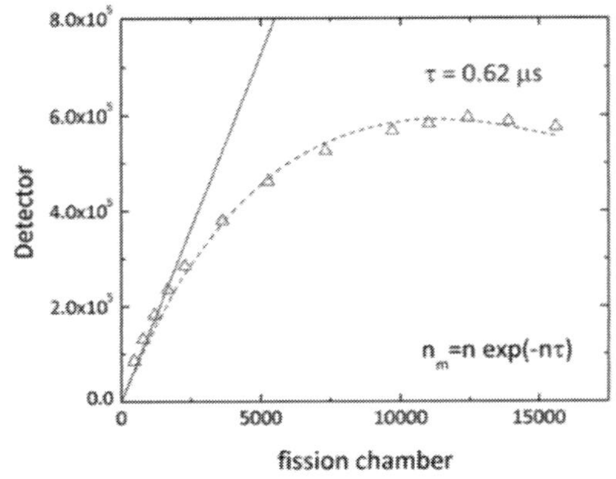

Figure 3. Measurement of the detection electronics dead-time using a "zero dead-time" fission chamber; the fit of the data with the paralysable model (curve) delivered the statistical dead-time; the ideal behavior of the detection electronics is indicated by the straight line.

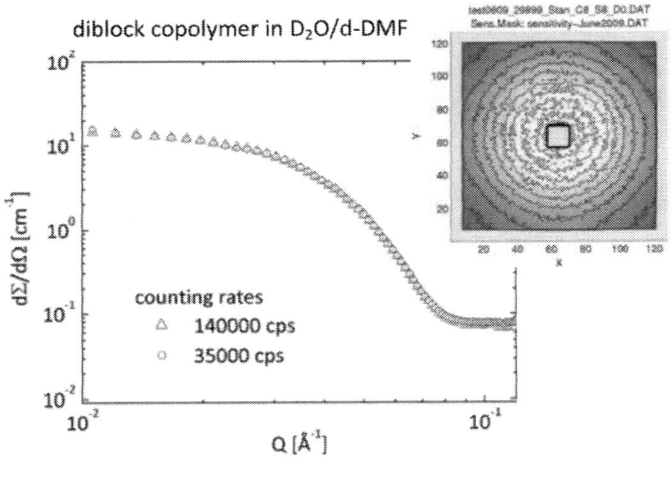

Figure 4. Radially-averaged scattering patterns from a diblock copolymer in solution, measured for different count rates; the data were corrected for the detector sensitivity, scattering from empty cell and instrument background and calibrated on the absolute scale using a Plexiglas secondary standard; the inset show the two-dimensional scattering pattern recorded with a count rate of 140 kHz; the quality of data in both presentations proves the reliability of the dead-time and detector sensitivity corrections applicable up to high count rates.

The main detector is equipped with a system of three mobile rectangular beam-stop masks with sizes of 5x5, 7x7 and 9x9 cm^2, which can be separately brought in beam in order to mask the direct beam and to tune the Q_{min} resolution by matching different beam size and divergence conditions. The detector can be moved between 1m and 20m after the sample and positioned vertically and laterally in beam, in order to catch on the beam-stop the direct beam for different wavelengths.

3. Instrument performance

3.1. Flux at the sample position

The flux at the sample position as a function of the incident wavelength and collimation configuration has been measured by standard gold foil activation method [10]. The results are shown in figure 5.

The magnitude of the neutron flux strongly depends on the amount of liquid deuterium in the cold neutron source (CNS) of FRM II: the full filling of the CNS requires about 15l of it. Detailed flux measurements have been carried out for the partial CNS filling of 10.6l. The highest flux for the shortest collimation length L_c=2m and λ=4.5Å is about $2x10^8$ n/s cm^2. Later measurements using the Plexiglas secondary standard performed for a better CNS filling (13l) have shown a considerable increase in the neutron flux at the sample position by a factor of 1.7. Under the optimal conditions of CNS the neutron flux is comparable to that of the SANS instruments at the high-flux reactor of the Institute Laue-Langevin in Grenoble.

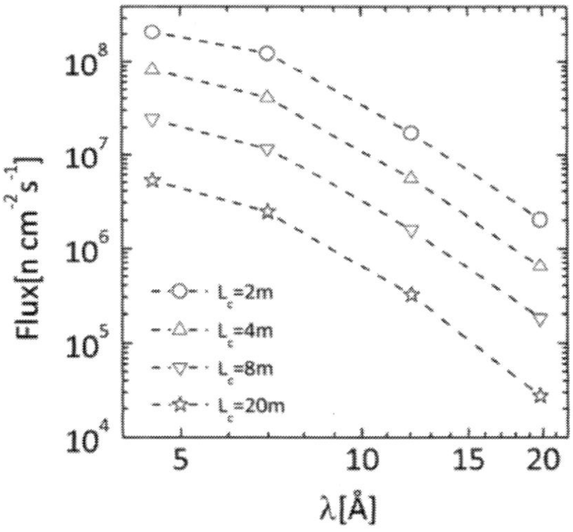

Figure 5. Absolute neutron flux at the sample position as a function of wavelength λ for different collimation lengths L_c, typical opening of the entrance and sample apertures and 10.6l CNS filling.

3.2. Available Q-range

The momentum transfer Q at a given point on the detector is given by

$$Q(r)=4\pi(\sin\theta/2)/\lambda=4\pi\{\sin[(\tan^{-1}(r/L_D))/2]\}/\lambda \qquad (1)$$

where r is the radial distance of a point on the detector from the beam axis, L_D the sample-to-detector distance, θ the scattering angle and λ the neutron wavelength. If we considered the closest distance at which reliable data may be recorded given by the smallest beam-stop and the largest distance given by the detector corners, the Q-range available for a selection of instrument configuration is shown in figure 6.

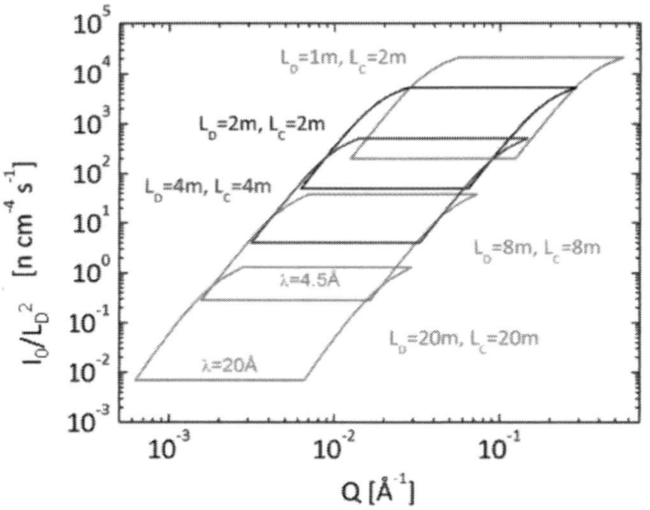

Figure 6. Available range of momentum transfer Q for a selection of instrument configurations at KWS-2: the neutron flux at the sample position divided by the square of the detector distance is plotted against Q for different combinations of collimation length L_C and detection length L_D and for neutron wavelengths in the range 4.5Å ≤ λ ≤ 20Å.

3.3 Beam Profile

A very important quantity in the characterization of the instrument is the contrast ration at the detector. This is defined by the peak intensity in the center of the beam stop versus the flux at twice the half-widths of the primary beam. As can be seen from the intensity distribution in figure 7, the contrast ratio is of the order of 10^5, which allows measurements of very small scattering cross-sections.

3.4. Instrument resolution

One of the elements enabling for the achievement of a high neutron flux on the sample is the large wavelength spread used, $\Delta\lambda/\lambda$=0.20. The large $\Delta\lambda/\lambda$ and the limited geometric resolution of a SANS instrument (finite collimation, detector resolution, gravity) lead to smearing that can cause to a certain extent loss of the structural information contained in the SANS curve. In order to recover at least part of this information lost as a result of the limited resolution, the actual instrumental resolution parameters have to be known a priori or determined independently. The structural information is then retrieved usually by implementing the resolution function in the fitting procedure [11] or, less used, by desmearing the measured data [12]. The resolution of pinhole SANS instruments with area detectors is well understood and can be modeled according to [11, 13, 14]. At KWS-2 the wavelength distribution measured by time-of-flight [6] agrees well with that resulting from the setting of the mechanical velocity selector, a Gaussian-like distribution with $\Delta\lambda/\lambda$=0.20. For long wavelengths (λ>10Å), the wavelength distribution deviates slightly from the triangular form and becomes gradually asymmetric as a consequence of combined effects arising from the Maxwell speed distribution of the incoming "white" neutron beam and the deterioration of the transmission and resolution of the velocity selector for large beam divergence characteristic for long neutron wavelengths [15].

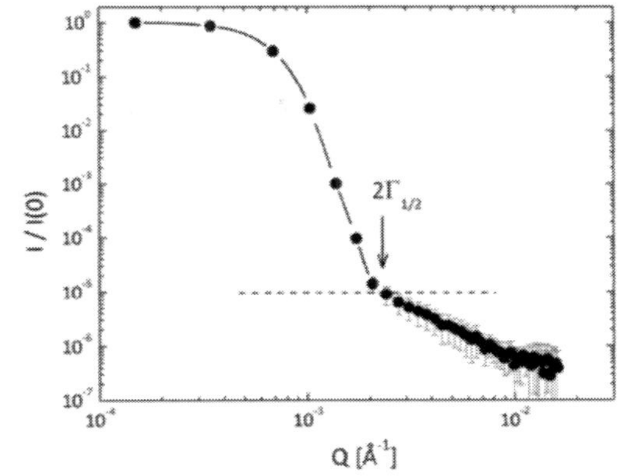

Figure 7. Intensity profile measured at KWS-2 for the collimation length L_C=8m, detection distance L_D=8m and λ=12Å.

A dilute dispersion of charged-stabilized, monodisperse silica nanoparticles has been used to verify the instrumental resolution function based on the wavelength spread provided by the velocity selector, the beam divergence determined by the collimation system and the cell size of the detector. The data measured with λ=4.5Å, corrected for detector sensitivity, instrument background, scattering from empty cell and solvent and calibrated on the absolute scale are displayed in figure 8 over the typical *Q*-range explored at KWS-2 by using different L_C-L_D combinations.

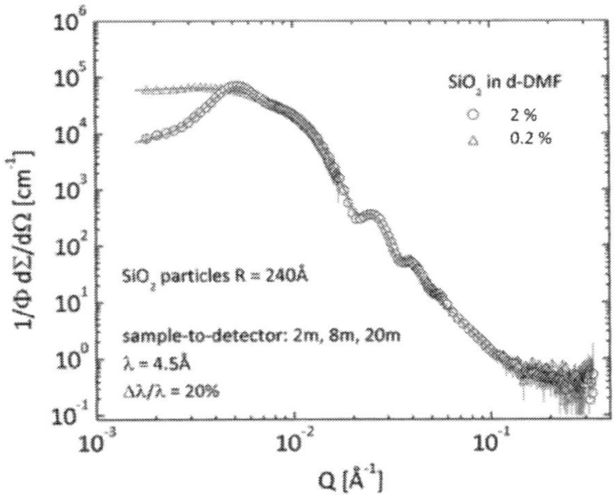

Figure 8. SANS patterns from silica particles in deuterated DMF for different particle volume fractions described by model function convoluted with the instrument resolution.

The experimental data were fitted by a model function characteristic for a system of hard spheres [16] convoluted with the resolution function of the instrument. The fit of the data was successful and the structure parameters of the reference samples determined from a SAXS experiment [17] were recovered. One should remark that, even with a significant higher wavelength spread such as

$\Delta\lambda/\lambda$=0.20, the structural features of the form factor can be observed up to the third oscillation, which enables a quite reliable interpretation of the measured data in terms of structural models.

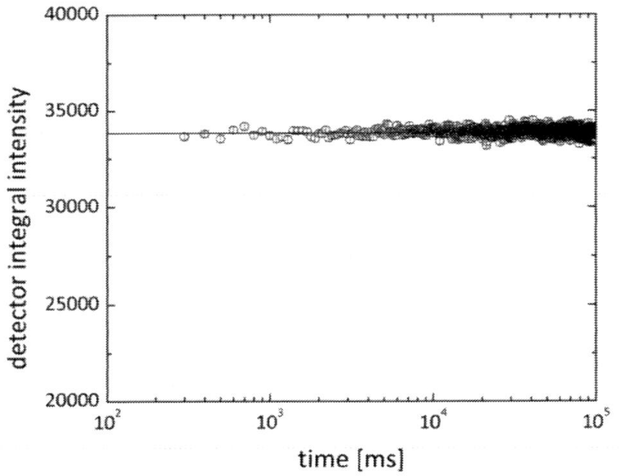

Figure 9. Fast SANS measurements of Plexiglas standard sample (isotropic constant pattern) performed with a time-resolution of 100ms as a test of the real-time data acquisition mode; final upgrade of the detection electronics allows for the fast start of data acquisition within 100ms after the opening of the instrument shutter.

3.5. Time-resolved SANS with 100ms time-resolution

KWS-2 was optimized for real time measurements of fast structural changes due to rapid kinetics. Data acquisition on sub-second time scale allows for the exploration of the early stages of formation of structures. Valuable new information are obtained in various soft matter topics like, for example, micellization behavior of block copolymers [18], fast crystallization of polymeric systems [19], deformation of elastomers [20] or application of magnetic fields [21]. At KWS-2, the detector electronics correlated with the experimental conditions (neutron wavelength and flight-path) allows for time-resolutions of the order of 100ms (figure 9). An example of time-resolved SANS study of the chain exchange kinetics in block copolymer micelles of spherical and cylindrical geometry is presented in [22].

4. Calibration and measurement methods

4.1. Wavelength calibration

The wavelength calibration is carried out by using Bragg reflection from silver behenate (AgBE powder supplied by Chemos GmbH, Germany), which is one of the very few materials that feature cold neutron Bragg reflections in the SANS angular range. These reflections (figure 10) can be analysed with Bragg's law to determine the wavelength of the primary beam delivered by the velocity selector [23,24], knowing the long-period spacing d=58.380Å of AgBE. The large particle size of the AgBE powder yields the scattering observed at low Q. The three Bragg peaks at high Q observed using different L_D's and various velocity selector speeds were analysed and the absolute wavelengths were determined.

4.2. Absolute intensity calibration

One of the advantages of the SANS technique is that the scattered intensity can be obtained on the absolute scale [25] and thus quantitative information such as molecular weight of polymers or aggregation number of micelles can be directly extracted from the "forward scattering" I($Q \rightarrow 0$). The absolute calibration of the scattering from a sample of interest is performed by scaling the recorded intensity with that from a sample (standard) with well known scattering properties. One of the typical standard samples used in neutron scattering is vanadium. Because of the weak scattering of vanadium, vanadium-calibrated materials are typically used as secondary standards for SANS. The secondary standard sample used at KWS-2 is a Plexiglas slab with a thickness of 1.5mm.

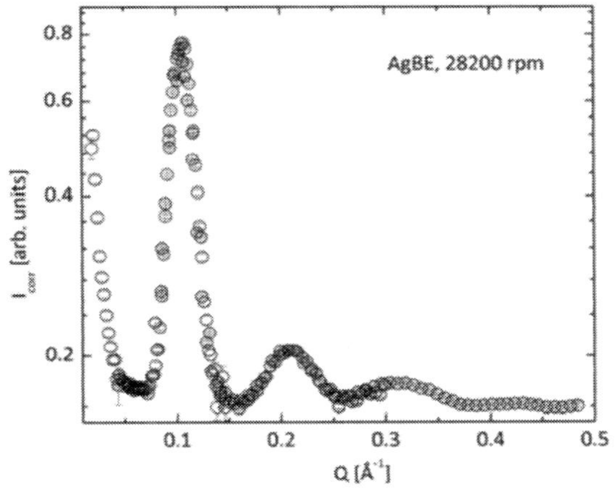

Figure 10. Scattering pattern from AgBE measured for the selector speed of 28200 rpm at three detection distances and corrected for detector sensitivity and background (instrument and cell).

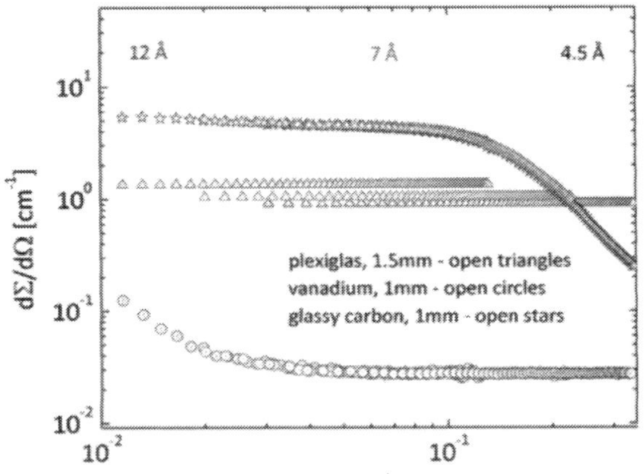

Figure 11. Absolute scattering cross-section from KWS-2 Plexiglas secondary standard for three different wavelengths: the data were calibrated with vanadium and cross-checked with glassy carbon.

The Plexiglas standard is calibrated with vanadium and, in order to account for multiple scattering effects, a cross-check with a glassy carbon standard sample [26] was done. This complex calibration procedure applied to the Plexiglas secondary standard allowed for a precise determination of the wavelength dependent calibration factor [24], which is further used for the absolute calibration of the scattering from the studied samples.

At KWS-2 the Plexiglas secondary standard is used also to normalize the individual detection cells of the main detector for local variations in efficiency (the detector sensitivity measurement).

4.3. Data reduction

The scattering intensities recorded at KWS-2 from studied samples are corrected, calibrated and, in the case of isotropic scattering, radial averaged using standard procedure [25] by means of the QtiKWS data analysis software [27].

5. Future upgrades

The boosting of the instrument performance towards higher intensities and wider Q-range by means of focusing elements – the MgF_2 aspherical lenses [3-5] and tilted velocity selector is aimed in the near future. For this purpose the neutron lenses are currently installed inside the last collimation segment towards the sample position, between 2m and 1m in front of the sample. The lenses have been tested and the experimental arrangements for the high resolution and high intensity modes were optimized [5]. A set of 26 lenses grouped in three packages that can be automatically brought in beam will be used either separately or in a combined way, to match the focusing conditions for different neutron wavelengths and collimation-detection conditions. In order to suppress the scattering on phonons and to increase the transmission by a factor of 2 the lenses will be cooled at 70K. For the high resolution mode the lenses will be used in combination with a high resolution position-sensitive scintillation detector and a double-disc chopper. The high resolution detector with space resolution of 0.5mm is installed in a fixed position at 17m after the sample. It is designed to be parked in a housing in top of the main detector evacuated tube (see figure 1) and to be brought automatically in beam when needed and the main detector is parked in the end position at 20m after the sample. The housing and positioning system of the high resolution detector are currently in construction.

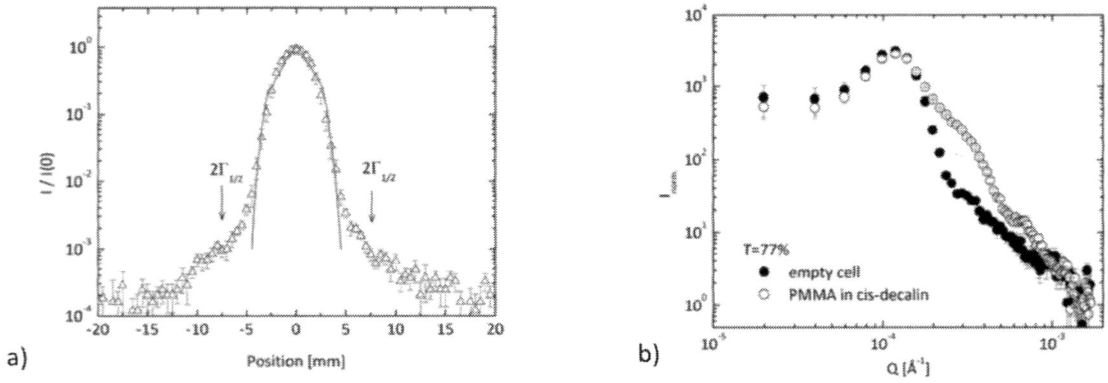

Figure 12. (a) – Focused beam profile obtained at KWS-2 using seven aspherical lenses as measured (symbols) and simulated (line) results; (b) – the scattered intensity from 0.25% PMMA in cis-decalin solution obtained by focusing-lenses at KWS-2; a beam-stop attached on the high resolution detector masks to a large extent the transmitted beam while the scattering from the sample is clearly observed above the background (empty cell) at Q>2x10^{-4}Å^{-1}.

Journal of Physics: Conference Series **351** (2012) 012026 doi:10.1088/1742-6596/351/1/012026

The high resolution mode combining the high resolution detector with aspherical MgF_2 lenses was tested at KWS-2. Figure 12a shows the beam profile measured while using seven lenses at room temperature and a neutron wavelength λ=18Å. The high resolution detector was placed at 8.9m distance from the lenses (7.1m after the sample) and the collimation length used was 20m (18.2m in front of the lenses) with an entrance aperture of 8x8 mm^2 and a sample aperture of 10x20 mm^2. The measured profile agrees very well in the central part with the results of simulations carried out using the McStas package with focusing lens compound [28] included. The contrast ratio obtained in these conditions is about 10^{-3} while the significant background observed towards the edge of the detector may arise from chromatic aberration (due to the large wavelength spread used) and scattering from the instrument windows. Definitely, these imperfections remain to be further improved.

Figure 12b shows in the radial averaged presentation the scattering from a sample of PMMA colloids in cis-decalin solution (1mm thickness) compared with the scattering from the empty sample cell. The signal from the sample is clearly visible above the background at $Q>2\text{x}10^{-4}$Å^{-1}. Corrected data are shown in figure 13 in parallel with the results obtained at KWS-2 with λ=7Å using the pinhole mode and at KWS-3 mirror-focusing instrument [29] with λ=12.7Å. The minimum Q value reached at KWS-2 with neutron lenses goes down to 10^{-4}Å^{-1}, which shows that in the near future more than three decades in Q range, from 0.5Å^{-1} up to 10^{-4}Å^{-1}, can be covered at a single instrument. A fit with the solid sphere form factor including the polydispersity σ(R)/R of the particle radii reported by static light scattering (σ_0=0.06 of the log-normal size distribution), which is depicted by the solid curve in figure 13, delivered a colloidal radius of about 8100Å,, a value similar to that of 7900Å obtained by static light scattering. Because of finite resolution, the minima of the form factor are smeared out in the experimental data measured using both focusing methods.

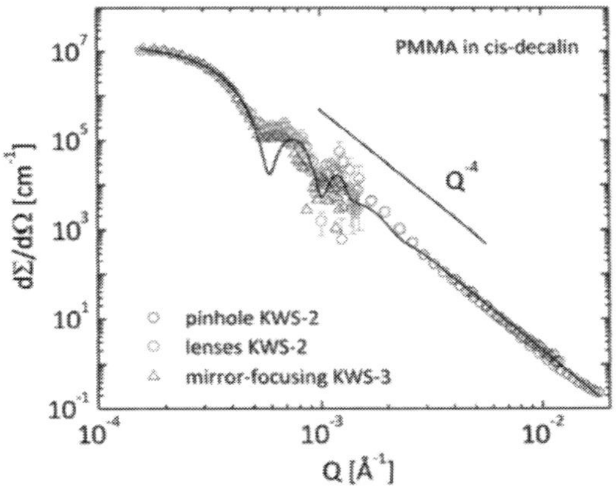

Figure 13. Intensity versus scattering vector of the 0.25% PMMA in cis-decalin solution obtained at KWS-2 by combining the classical pinhole and lenses-focusing modes. The data measured on the same sample at KWS-3 mirror-focusing instrument are shown in parallel. The solid curve represents the fit of the experimental data with the form factor of spheres (see text).

In order to have thus the possibility to tune the wavelength resolution for the conventional pinhole mode and to avoid the chromatic aberration effects while using the focusing lenses the analysis of the wavelength distribution by time-of-flight using a double-disc chopper with variable opening is planned to be used. The installation of the chopper in front of the collimation is currently in progress. The two discs with two diametrically opposed windows (90° sectors) are placed very close to each other (5cm). Thus, a variable opening of the chopper can be achieved by changing the phase between the discs which will allow for a high flexibility with respect to defining the time resolution and intensity on the sample.

Finally, the possibility to explore larger Q domains, up to 1Å^{-1}, is targeted in close connection to the small correlation lengths typically exhibited by biological structures.

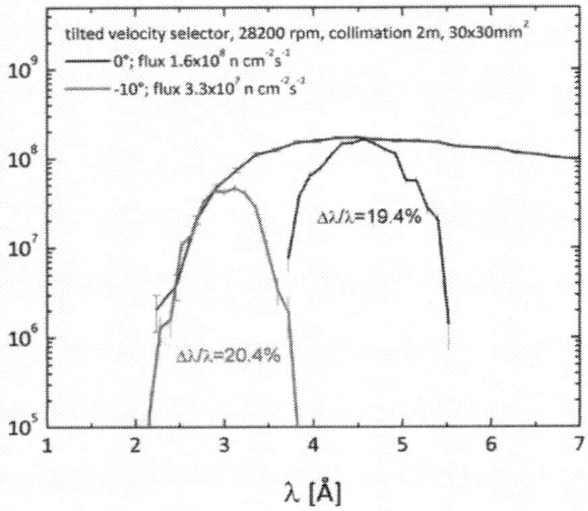

Figure 14. Simulated results at the end of the vertical "S-shaped" NL3a-o guide obtained for the white flux (blue line) and the monochromatic flux (black and red lines) using the velocity selector at a speed of 28200 rpm in two tilting situations (0° and -10° with respect to the beam axis).

For this purpose a vertical tilting of the velocity selector up to -10° with respect to the beam axis is intended. One should mention that the velocity selector is placed in the middle of the vertical "S-shaped" guide with the entrance window in a "*three o'clock*" position. Simulations carried out with McStas package [30] have shown that the neutron wavelength of λ=3Å ($Q_{max}\rightarrow$1Å^{-1}) can be selected with certain advantages prevailing over the disadvantages: due to the cut-off of the "S-shaped" neutron guide the wavelength spread (20%) is held while the price in intensity is acceptable for the low resolution–high intensity (shortest sample-to-detector distance) configuration aimed in this situation. The mechanism for the controlled vertical tilting of the selector is in construction phase and will be installed in the summer of 2012.

6. Conclusions

KWS-2 instrument operated by the JCNS at FRM II is a high intensity / wide Q-range small-angle neutron diffractometer optimized for studies of soft-matter and biological systems. The high intensity permits on one hand rapid measurement of small or weak-scattering samples and, on the other hand, time-resolved structural investigations with a time resolution up to 100ms. The possibility to cover up to four decades in Q, between 1x10^{-4}Å^{-1} and 1Å^{-1}, will enable structural studies over a wide length scale spanning between several Å and 1μm.

Acknowledgements

The constant help for operating, optimizing and upgrading the KWS-2 instrument from M.Heiderich, R.Hanslik, S.Staringer, H.Schneider, T.Kohnke, U.Bünten, R.Engels, G.Kemmerling, M.Wagener, M.Drochner, L.Fuss-Fleischhauer, R.Möller, F.Suxdorf and H.Kleines (all Forschungszentrum Jülich

GmbH) is gratefully acknowledged. We thank J.Stellbrink (Forschungszentrum Jülich GmbH) for the kindness to provide us the KWS-3 and pinhole KWS-2 SANS data on PMMA. Kind supply of the glassy carbon standard sample characterized with SAXS by J.Ilavsky (Argonne National Laboratory) and of SiO_2 size-standard sample by W.Sager (Forschungszentrum Jülich GmbH) is acknowledged. We thank D.Schwahn (Forschungszentrum Jülich GmbH) for useful discussions and his valuable advices with respect to the operation and development of KWS-2.

References

[1] Gläser W and Petry W 2000 *Physica B* **276-278** 30
[2] Radulescu A and Ioffe A 2008 *Nucl. Instr. and Meth. A* **586** 55
[3] Choi S M, Barker J G, Glinka C J, Cheng Y T and Gammel P L 2000 *J. Appl. Cryst.* **33** 793
[4] Koizumi S, Iwase H, Suzuki J, Oku T, Motokawa R, Sasao H, Tanaka H, Yamaguchi D, Shimizu H M and Hashimoto T 2007 *J. Appl. Cryst.* **40** s474
[5] Frielinghaus H, Pipich V, Radulescu A, Heiderich M, Hanslik R, Dahlhoff K, Iwase H, Koizumi S and Schwahn D 2009 *J. Appl. Cryst.* **42** 681
[6] Radulescu A, Pipich V and Ioffe A. *Nucl.Instr. and Meth. A* submitted
[7] Heiderich M, Reinartz R, Kurz R and Schelten J 1991 *Nucl. Instr. And Meth. A* **305** 423
[8] Engels R, Clemens U, Kemmerling G, Nöldgen H and Schelten S 2009 *Nucl. Instr. And Meth. A* **604** 147
[9] Kemmerling G, Bünten U, Clemens U, Engels R, Heiderich M, Pyckhout-Hintzen W, Rongen H, Schelten J, Schwahn D and Zwoll K 2004 *IEE Trans. Nucl. Sci.* **51** 1098
[10] Zeitelhack K et al. 2006 *Nucl. Instr. and Meth. A* **560** 444
[11] Pedersen J S, Posselt D and Mortensen K 1990 *J.Appl.Cryst.* **23** 321
[12] Kline S R 2006 *J.Appl.Cryst.* **39** 895
[13] Pedersen J S 1997 *Adv. Colloid Interface Sci.* **70** 171
[14] Barker J G and Pedersen J S 1995 *J. Appl. Cryst.* **28** 105
[15] Friedrich H, Wagner V and Wille P 1989 *Physica B* **156-157** 547
[16] Mortensen K 1993 *Prog. Colloid Pol. Sci.* **93** 72
[17] Vad T, Sager W F C, Zhang J, Buitenhuis J and Radulescu A 2010 *J. Appl. Cryst.* **43** 686
[18] Willner L, Poppe A, Allgaier J, Monkenbusch M and Richter D 2001 *Europhys. Lett.* **55** 667
[19] Wang H 2006 *Polymer* **47** 4897
[20] Burford R P, Markotsis M G and Knott R B 2006 *Physica B* **385-386** 766
[21] Wiedenmann A, Keiderling U, Meissner M, Wallacher D, Gähler R, May R P, Prevost S, Klokkenburg M, Erne B H and Kolbrecher J 2008 *Phys. Rev. B* **77** 184417
[22] Lund R, Willner L, Pipich V, Grillo I, Lindner P, Colmenero J and Richter D 2011 *Macromolecules* **44** 6145
[23] Keiderling U, Gilles R and Wiedenmann A 1999 *J. Appl. Cryst.* **32** 456
[24] Okabe S, Karino T, Nagao M, Watanabe S and Shibayama M 2007 *Nucl. Instr. and Meth.* **572** 853
[25] Lindner P 2002 Neutrons, X-Rays and Light: Scattering Methods Applied to Soft-Condensed Matter, ed. Lindner P and Zemb Th (Elsevier – North Holland Delta Series, Amsterdam)
[26] Zhang F, Ilavsky J, Long G, Quintana J, Allen A and Jemian P 2010 *Metall. Mater. Trans.* **41** 1151
[27] http://iffwww.iff.kfa-juelich.de/~pipich/dokuwiki/doku.php/qtikws
[28] http://www.mcstas.org/download/components/contrib/Lens.html
[29] Kentzinger E, Dohmen L, Alefeld B, Rücker U, Stellbrink J, Ioffe A, Richter D and Brückel T 2004 *Physica B* **350** e779
[30] Willendrup P, Farhi E and Lefmann K 2004 *Physica B* **350** 735

Erratum

Figure 5 in the original paper incorrectly showed the absolute neutron flux for a CNS filling of 13l, which results in an increased flux at the sample position with a factor of up to 1.7. The correct figure and caption are shown below.

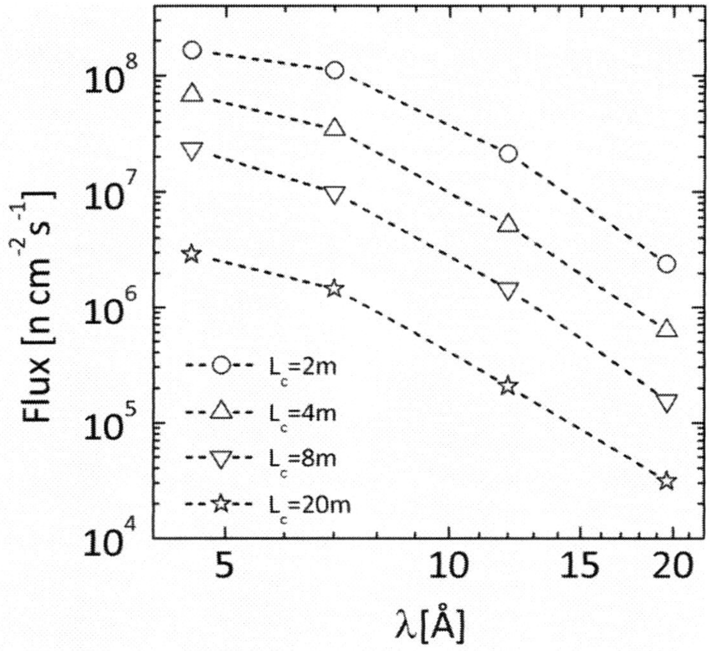

Figure 5. Absolute neutron flux at the sample position as a function of wavelength λ for different collimation lengths L_C, typical opening of the entrance and sample apertures and 10.6l CNS filling.

FITTER. The package for fitting a chosen theoretical multi-parameter function through a set of data points. Application to experimental data of the YuMO spectrometer

A G Soloviev[1], T N Murugova[2,3], A H Islamov[3] and A I Kuklin[2,3]

[1] Laboratory of Information Technologies, JINR, Joliot-Curie-6, Dubna 141980 Russia

[2] Frank Laboratory of Neutron Physics, JINR, Joliot-Curie-6, Dubna 141980 Russia

[3] Bionanocenter of Moscow Institute for Physics and Technology, Dolgoprudny, Russia

E-mail: solovjev@jinr.ru

Abstract. Fitter is a C++ program aimed to fit a chosen theoretical multi-parameter function through a set of data points. The method of fitting is chi-square minimization. Moreover, the robust fitting method can be applied in Fitter. Fitter was disigned to be used for a small-angle neutron scattering data analisis. Respective theoretical models are implemented in it. Some common used models (Gaussian and polynomials) are also implemented for wider aplicability. The programm include the option the resolution of spectrometer. The investigation has been performed at the Laboratory of Information Technologies and at the Frank Laboratory of Neutron Physics, JINR. Fitter can be downloaded from [1].

1. Introduction

Fitter, the package for fitting a chosen theoretical multi-parameter function through a set of data points, has been successfully used in small-angle neutron scattering data analysis since 2003 [2; 3]. The several examples of researches using Fitter are given below. This handbook gives a long write-up description of the new version of Fitter. The method of small-angle neutron scattering (SANS) is frequetly used for research of fundamental and applied problems in condensed matter physics, biology and chemistry.This method is often an only way to obtain direct structural information oo systems with disordered and partly ordered arrangement of the nuclear density with charachteristic sizes of about 20-1000 angstroms.

The small-angle neutron scattering (SANS) method has long history. So, the earliest references to SANS can be traced back to 1948 by D.J.Huges et al. [4]. Starting with classical work of A.Guinier, who suggested the so called ``Guinier approximation" to powerfull package of D.Svergun team [5-11], who developed and brogth SANS method up to present moment. At the same time, in spite of the wide SANS method application and already more than 50-years history of its development, the problem of interpreting the data obtained on spectrometers, still remain open. It is related, first of all, to the ill-posedness of the solved problem and, on the other hand, to specifics of existing SANS spectrometers. Practically each spectrometer is provided by not only software on initial processing experimental data, but also by a complex of programs for fitting the curves measured in coordinates scattering intensity/q, where $q = (4\pi / \lambda)\sin(\theta / 2)$.

A modernization of the time-of-flight SANS YuMO spectrometer at the high flux pulse IBR-2 reactor (JINR, Dubna) [12] was launched in 2000. One of the goals of the improvement was to increase the dynamic Q-range of the instrument. It was realized by using a two-detector system with central holes at the YuMO instrument. For this purpose "old" and "new" detectors were used. The main view of the setup of the modernized YuMO instrument as well as an examples of a SANS experiments are presented in [13]. The use of the two detectors YuMO set up for the users have clearly demonstrated its high efficiency [14-30]. That are only several examples. In 2006 the position sensitive detector started at the YuMO spectrometer [31]. For mono-disperse, homogeneous, globular particles, general intensity of scattering can be expressed by a number of particles N, structural factor S(q) and form-factor F(q). However, for many practically important applications, influence of the structural factor on experimental results is insignificant. It is, for example, for strongly diluted not charged samples or even for the charged samples but with good charge shield [32]. For more complicated model it is necessary the new programm for segment [33] or a long helical model was created [34]. Neveretheless, easy development and quickly programm necessary to have during modernization of spectrometer.

The new Fitter version is better suit requirements of new detector setup. The main changes are:

- fitting in a given range;
- improving the robust approach;
- choosing the MINUIT strategy;
- function in STD models;
- new visualization and GUI;

This paper is a long write-up description of the current version of program for fitting a chosen theoretical multi-parameter function through a set of data points. Examples of application to experiments on the small-angle spectrometer YuMO are considered.

2. Long Write-Up

The Fitter is designed to be used for SANS data processing first of all. Thus, SANS theoretical models are implemented in it. Moreover, some standard mathematical models are added for wider applicability. Besides the implemented theoretical models, Fitter has a minimization module. It provides a safe call of MINUIT [35] procedures in the current version. The important feature of Fitter's design is its expandability: both new models and new minimizing algorithms can easily be added to the existing ones.

2.1. Theoretical Models

Fitter's model module is designed as follows (figure 1). Abstarct base class Model provides a common interface, used by a minimization module. Thus, any theoretical model class inherits from it. Model classes currently implemented in the Fitter are: STDModel, SANSModel, YuMOModel. Each concrete class provides several theoretical functions. All of them are described below.

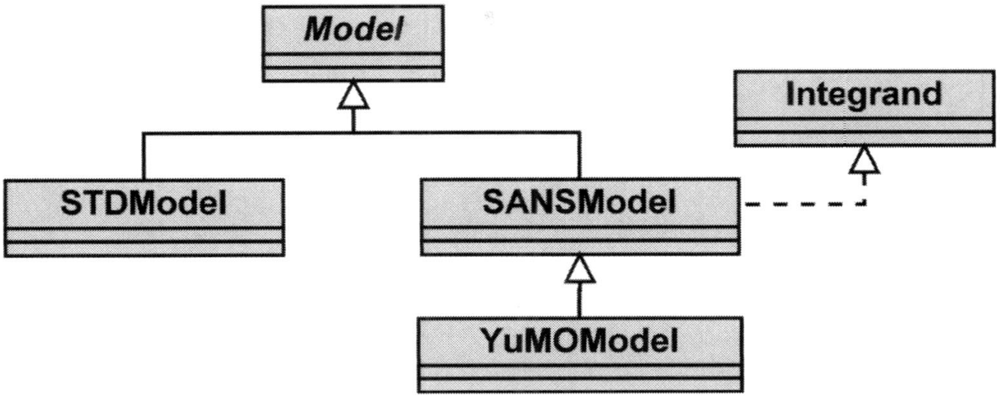

Figure 1. Fitter's model module class hierarchy.

2.1.1. Standard Models. *Standard mathematical models implemented in the Fitter are:*

- **Gaussian**

$$G(x) = A\exp\left\{-\frac{1}{2}\left(\frac{x-x_0}{\sigma}\right)^2\right\} + B$$

- **Exponential**

$$E(x) = A\exp\{mx+b\} + B$$

- **Polynomials** up to the eighth order

$$P_n(x) = \sum_{k=0}^{n} p_k x^k$$

2.1.2. SANS (Small-Angle Neutron Scattering) Models. Determination of invariants for small-angle scattering curves allows one to analyze the structure of a particle under study. Upon the first step of this analysis the particle form is approximated by simple geometrical bodies - ellipsoids, cylinders, prisms.

Thus, SANS models implemented in the Fitter are:

- **Ball** of the radius R

$$I(Q) = A\Phi^2(QR) + B$$

$$\Phi(t) = 3\frac{\sin t - t\cos t}{t^3}$$

- **Ellipsoid of revolution** with half-axises a and av

$$I(Q) = A\int_0^1 \Phi^2\left[Qa\sqrt{1 + x^2(v^2 - 1)}\right]dx + B$$

$$\Phi(t) = 3\frac{\sin t - t\cos t}{t^3}$$

- **Cylinder** of the radius R and the length H

$$I(Q) = A\int_0^1 \Lambda_1^2\left(QR\sqrt{1 - x^2}\right)S^2(QHx/2)dx + B$$

$$\Lambda_1(t) = 2J_1(t)/t$$

$$S(t) = \sin t/t$$

- **Elliptical cylinder** with half-axises a, av and the length H

$$I(Q) = A\int_0^1 \Psi_{EC}(Q, a\sqrt{1 - x^2})S^2(QHx/2)dx + B$$

$$\Psi_{EC}(Q,a) = \frac{1}{\pi}\int_0^\pi \Lambda_1^2\left(Qa\sqrt{\frac{1 + v^2}{2} + \frac{1 - v^2}{2}\cos y}\right)dy$$

$$\Lambda_1(t) = 2J_1(t)/t$$

$$S(t) = \sin t/t$$

- **Parallelepiped** with the ribs a, b, c

$$I(Q) = A\int_0^1 \Psi_p\left(Q, b\sqrt{1 - x^2}, a\right)S^2(Qbcx/2)dx + B$$

$$\Psi_p(Q,b,a)=\frac{2}{\pi}\int_0^{\pi/2} S^2(Qa\sin y/2)S^2(Qb\cos y/2)dy$$
$$S(t)=\sin t/t$$

- **Spherical shell** of the outer radius R_1 and the inner radius R_2

$$I(Q)=A\left[\Phi(QR_1)-\left(\frac{R_2}{R_1}\right)^3\Phi(QR_2)\right]^2+B$$
$$\Phi(t)=3\frac{\sin t-t\cos t}{t^3}$$

- **Triaxial ellipsoid** with half-axises a, b, c:

$$I(Q)=A\int_0^1\int_0^1\left[\Phi[Q\sqrt{(a^2\cos^2\frac{\pi}{2}x+b^2\sin^2\frac{\pi}{2}x)(1-y^2)+c^2y^2}\,]\right]^2 dxdy+B$$
$$\Phi(t)=3\frac{\sin t-t\cos t}{t^3}$$

- **Two axis ellipsiodal shell** with half-axises a, a, va and thickness t:

$$I(Q)=A\int_0^1\left[V_t\Phi\left[Q\sqrt{(a+t)^2(1-x^2)+x^2(av+t)^2}\right]-V_c\Phi\left[Qa\sqrt{1+x^2(v^2-1)}\right]\right]^2 dx+B$$
$$V_c=\frac{4}{3}\pi abc \qquad \text{(Volume of core)}$$
$$V_t=\frac{4}{3}\pi(a+t)^2(va+t) \quad \text{(Volume of core with shell)}$$

- **Triaxial ellipsiodal shell** model

$$I(Q)=A\int_0^1\int_0^1[V_t\Phi(Q\sqrt{((a+t)^2\cos^2\frac{\pi}{2}x+(b+t)^2\sin^2\frac{\pi}{2}x)(1-y^2)+(c+t)^2y^2)}$$
$$-V_c\Phi(Q\sqrt{(a^2\cos^2\frac{\pi}{2}x+b^2\sin^2\frac{\pi}{2}x)(1-y^2)+c^2y^2})]^2dxdy+B$$
$$\Phi(t)=3\frac{\sin t-t\cos t}{t^3}$$
$$V_c=\frac{4}{3}\pi abc \text{ (Volume of core)}$$

$$V_t = \frac{4}{3}\pi(a+t)(b+t)(c+t) \text{ (Volume of core with shell)}$$

More detailed information about SANS models is available in [36].

2.1.3. SANS Models with YuMO spectrometer resolution. These models are implemented to fit data measured on the YuMO spectrometer [37] operated on the 4-th channel of the fast pulsed reactor IBR-2 [12]. YuMO models are the same as SANS, but they take into account the spectrometer resolution.

- **SANS Model with resolution**

$$I(Q) = \frac{1}{\sqrt{2\pi}\sigma}\int_{Q-3\sigma}^{Q+3\sigma} I_{SANS}(q)\exp\left\{-\frac{1}{2}\left(\frac{q-Q}{\sigma}\right)^2\right\}dq$$

$$\sigma = \sqrt{2\Delta Q}$$

2.2. Minimization

To find theoretical model parameters, one should minimize a functional, which is a measure of deviation between a theoretical curve and experimental data. In a common case of a least-squares fit, the functional under minimization is a chi-square:

$$\chi^2 = \frac{1}{N-N_{parms}}\sum_{i=1}^{N}\left(\frac{f(x_i)-y_i}{\Delta y_i}\right)^2$$

2.2.1. Minuit. We are using ROOT::TMinuit class to perform a minimization. This package was originally written in Fortran by Fred James and part of PACKLIB (patch D506) and has been converted to a C++ class by R.Brun. The current implementation in C++ is a straightforward conversion of the original Fortran version. The main changes are:

- The variables in the various Minuit labeled common blocks have been changed to the TMinuit class data members.
- The internal arrays with a maximum dimension depending on the maximum number of parameters are now data members arrays with a dynamic dimension such that one can fit very large problems by simply initializing the TMinuit constructor with the maximum number of parameters.
- The include file Minuit.h has been commented as much as possible using existing comments in the code or the printed documentation
- The original Minuit subroutines are now member functions.
- Constructors and destructor have been added.
- Instead of passing the FCN function in the argument list, the addresses of this function is stored as pointer in the data members of the class. This is by far more elegant and flexible in an interactive environment. The member function SetFCN can be used to define this pointer.
- The derived class TMinuitOld contains obsolete routines from the Fortran based version.

Additional modifications were made to separate TMinuit class from the ROOT package (Fitter is ROOT independent indeed).

MINUIT offers the user a choice of several minimization algorithms. The MIGRAD algorithm is, in general, the best minimizer for nearly all functions. It is a variable-metric method with inexact line search, a stable metric updating scheme, and checks for positive-definiteness. Its main weakness is that it depends heavily on knowledge of the first derivatives, and fails miserably if they are very inaccurate.

For further details see MINUIT documentation [35].

2.2.2. Robust Fitting. The least-squares fitting involves the minimization of the sum of the squared residuals. There are two instances where this minimization produces less than satisfactory fit:

1. The significant outliers are present in data. In this case, the square of the residuals of these out-lier points may, within a given region, significantly shift the fitted curve away from the bulk of the data.

2. The the Y-data span more than several orders of magnitude. The squared residuals of the largest valued Y-points can overwhelm the influence of the squared residuals of the smallest Y-valued points, causing the smallest Y-value points to either be poorly fitted or not fitted at all. Data that require a logarithmic Y-scale to see all of the points may be a good candidate for robust fitting, especially if four or more major log divisions take place.

Robust estimates designed to be successful in such cases. The essence of robust fitting is to use a minimization that is less influenced by outliers and the dynamic range of the Y-variable. Each data measured point took into account with it own weight, which indicate influence of given point.

It is based on so-called M-estimates which follow from maximum like-hood approach, M-estimates are usually the most relevant for model fitting. Robust approach uses gross-error model (Hubert). Probability distribution function of measurement errors d_i suggested as a superposition of two distributions: basic $g(d_i)$ and distribution of big errors $h(d_i)$

$$D(d_i) = (1-\varepsilon)g(d_i) + \varepsilon h(d_i), \quad \varepsilon \in [0,1]$$

$$g(d_i) = \frac{1}{\sigma\sqrt{2\pi}}\exp\left\{\frac{-d_i^2}{2\sigma^2}\right\}$$

$$h(d_i) = \alpha\exp\{-\beta\,|\,d_i\,|\}$$

In practice, robustness works also in cases of other distributions $h(d_i)$, for example, normal one like $g(d_i)$ is, but with bigger RMS.

Application of maximum like-hood approach and few simplifications and the fact that RMS of the distribution $g(d_i)$ can be approximated by chi-square of data points, give us well known Tukey bi-square weights, which is suitable in most cases.

$$w_i = \begin{cases} \left(1 - \dfrac{|\,f(x_i) - y_i\,|^2}{c_T\chi^2}\right)^2, & |\,f(x_i) - y_i\,| \le c_T\chi^2 \\ 0, & |\,f(x_i) - y_i\,| > c_T\chi^2 \end{cases} \tag{1}$$

In chi-square function data values are multiplied by their weights:

$$\chi^2 = \frac{1}{\sum_{i=1}^{N} w_i - N_{parms}} \sum_{i=1}^{N} w_i \left(\frac{f(x_i) - y_i}{\Delta y_i} \right)^2 \tag{2}$$

The initial values of the weights are equal to 1. In the following iterations, weights are recalculated after each procedure of minimization and calculation of the new value of chi-square. Iterations are repeated until convergence (until chi-square value is stabilized within a predefined accuracy).

In general, the robust approach looks like:

- setting $w_i = 1$, calculating χ^2 by (2), setting $c_T = 5$;
- while $c_T > 1$ doing the loop:

  - calculating w_i by (1) using current values of χ and c_T;
  - minimizing χ^2 given by (2);
  - reducing c_T by $c_T / = 2$.

  For detailed explanation see [38-40].

3. Installation

3.1. Installation on UNIX systems

In order to compile and install Fitter on your system, type the following in the base directory of the Fitter distribution:

```
% ./configure
% make
% make install
```

To build Fitter with GUI you should install FLTK [41] first. Second, you should install Gluplot ploting package [42], since it is used for visualization. Finally, enable using FLTK libraries while Fitter configuring:

```
% ./configure –enable-fltk
```

Since Fitter uses autoconf you should have not trouble compiling it. Should you run into problems please report them to the the authors Alexei Soloviev and Alexei Stadnik.

3.2. Installation on Windows

Fitter provides an user-friendly setup procedure for Windows users. Just run setup and follow the instructions.

4. Usage

4.1. General Usage

Fitter is a C++ program aimed to fit a chosen theoretical multi-parameter function through a set of data points. The method of fitting is chi-square minimization. Moreover, the robust fitting method can be applied in Fitter. Fitter was designed to be used for a small-angle neutron scattering data analysis. Respective theoretical models are implemented in it. Some commonly used models (Gaussian and polynomials) are also implemented for wider applicability.

4.1.1. Synopsis.

fitter **--** { **std** | **sans** | **yumo** } *model* [**-s** *name*] [**-r**] [**-R** *x1:x2*] [**-v** [*file*]] [**-o** *file*] [**-H** *file*] [**--rc-read**] [**--rc-write**] *file*
fitter [**-h**] [**-V**]

4.1.2. Input. Input data file (ASCII) for fitter should have at least three columns: the first one would be interpreted by fitter as "X"-samples, second one - as "Y"-samples and the third one - as "Y"-errors. For YuMO models (see below) should be one more column, which is interpreted as "X"-errors.

4.1.3. Options.

MODEL OPTIONS

–std *model* Standard models. One of the following (see '–std help' for list):
- *g* Gaussian
- *e* Exponential
- *p0* Polynomial of the order 0
- *p1* Polynomial of the order 1
- *p2* Polynomial of the order 2
- *p3* Polynomial of the order 3
- *p4* Polynomial of the order 4
- *p5* Polynomial of the order 5
- *p6* Polynomial of the order 6
- *p7* Polynomial of the order 7
- *p8* Polynomial of the order 8

–sans *model* SANS models. One of the following (see '–sans help' for list):
- *b* Ball
- *c* Cylinder
- *c2* Elliptical cylinder
- *e2* Two axis ellipsoid
- e2s Two axis ellipsiodal shell
- e3 Three axial ellipsoid
- e3s Triaxial ellipsiodal shell
- *p* Parallelepiped
- *ss* Spherical shell

–yumo *model* SANS models for YuMO spectrometer. One of the following (see '–yumo help' for list):
- *b* Ball with resolution
- *c* Cylinder with resolution

c2 Elliptical cylinder with resolution
e2 Two axis ellipsoid with resolution
e2s Two axis ellipsiodal shell with resolution
e3 Three axial ellipsoid with resolution
e3s Triaxial ellipsiodal shell with resolution
p Parallelepiped with resolution
ss Spherical shell with resolution

PROCESSING OPTIONS

-s, –strategy *name* Change MINUIT strategy. MINIMIZE is a default one.
-r, –robust Robust fitting. Useful for highly noised data.
-R, -range *x1:x2* Fitting in range from x1 to x2 . Note that no fit is applied if a range appearers to be empty.
-v, –verbose [*file*] Output MINUIT processing information. No output is produced by default. Repeating this option determines how much output will be produced:
(a) minimum MINUIT output
(b) normal MINUIT output
(c) additional output giving intermediate results
(d) maximum output, showing progress of minimizations

OUTPUT OPTIONS

-o, –output *file* Output data file with the resulting theoretical curve. No file is created by default.
-H, –header *file* Output resulting parameters and chi-square value to file (standard output by default). File may be the same as for theoretical curve, parameters and chi-square will be put to its beginning in this case.
–rc-read Read rc-file. If specified, this option cause reading parameters staring values from rc-file. Some times this can highly speed-up minimization. Rc-file .fitter should exist in $HOME (in C: on Windows).
–rc-write Write rc-file. If specified, this option cause writing resulting parameters to rc-file. Useful for creating/repairing rc-file. Rc-file .fitter is placed in $HOME (in C: on Windows).

HELP OPTIONS

-h, –help Display short help message and exit.
-V, –version Display version information and exit.

4.2. GUI

Besides the command-line interface the Fitter offers an user friendly GUI (graphical user interface). Fitter's main window, shown in figure 2, is implemented in the standard manner with menu-bar, tool-bar and status-bar. All menu items exactly correspond to the command-line options described above. The command-line to be executed is just formed by choosing corresponding menu items and is displayed on the status-bar. Some of the most frequently used menu items are duplicated as buttons or combo-boxes on the tool-bar. A window with fitting results arises after fit is complete.

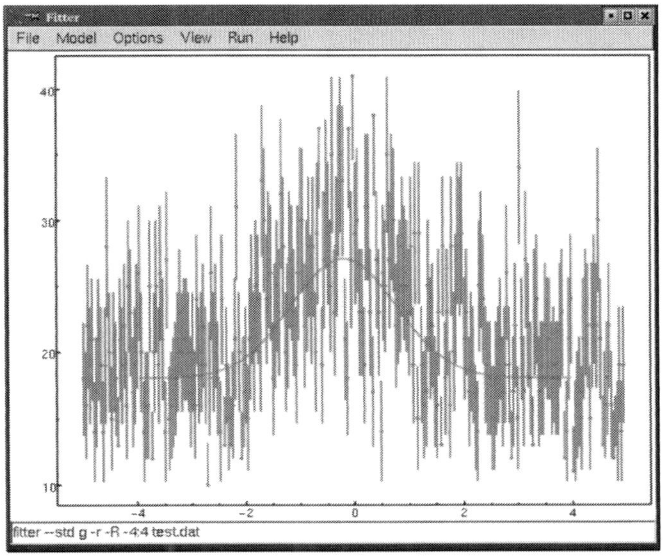

Figure 2. Fitter's main window.

5. Examples

5.1. Robust feature example

Here is a simple example of the Robust feature usage. File test.dat is an example of input data for fitter. It is just a generated gaussian distribution, contaminated by uniformly distributed random noise.

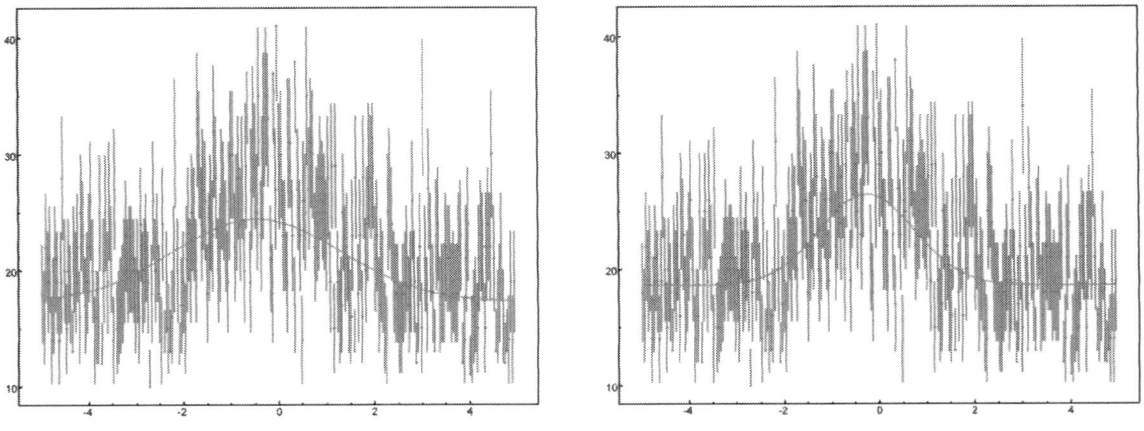

Figure 3. Highly noisted data fitting example.

Typing in command line " % fitter test.dat –o fit1.dat –std g"
we get the following result:

Model..Gaussian

Mean..-0.4576 +- 0.1719
RMS..1.7571 +- 0.3680
Amplitude..7.1333 +- 0.5719
Background..17.3032 +- 0.6111

Chi square..1.1068

---

and the fitting curve shown in figure 3 (left).

This fit is not good enough, thus, we try "% fitter test.dat –o fit2.dat –std g -r" invoking robust fitting. This gets a better result:

Model..Gaussian

Mean..-0.2564 +- 0.1595
RMS..0.9985 +- 0.1638
Amplitude..7.8680 +- 1.2633
Background..18.6319 +- 0.5641

Chi square..1.0708

The fitting curve (figure 3, right) is much better this time.

5.2. SANS/YuMO example

To demonstrate the fitter usage, we have used the results of neutron measurements, obtained on the apoferritin protein sample obtained from Aldrich corporation. The file apdn.dat containing a SANS spectra is used as fitter input. A spherical shell model, which is one of the most adequate for apoferritin, is chosen. The corresponding command line is the following: "% fitter apdn.dat –o apdn.fit –yumo ss".

This gives the following result:

Model..Spherical shell with resolution

Outer radius..59.6547 +- 0.0576
Inner radius..40.1314 +- 0.0700
Amplitude..28.3414 +- 0.1721
Background..0.0191 +- 0.0003

Chi square31.2243

---

and the fitting curve shown in figure 4.

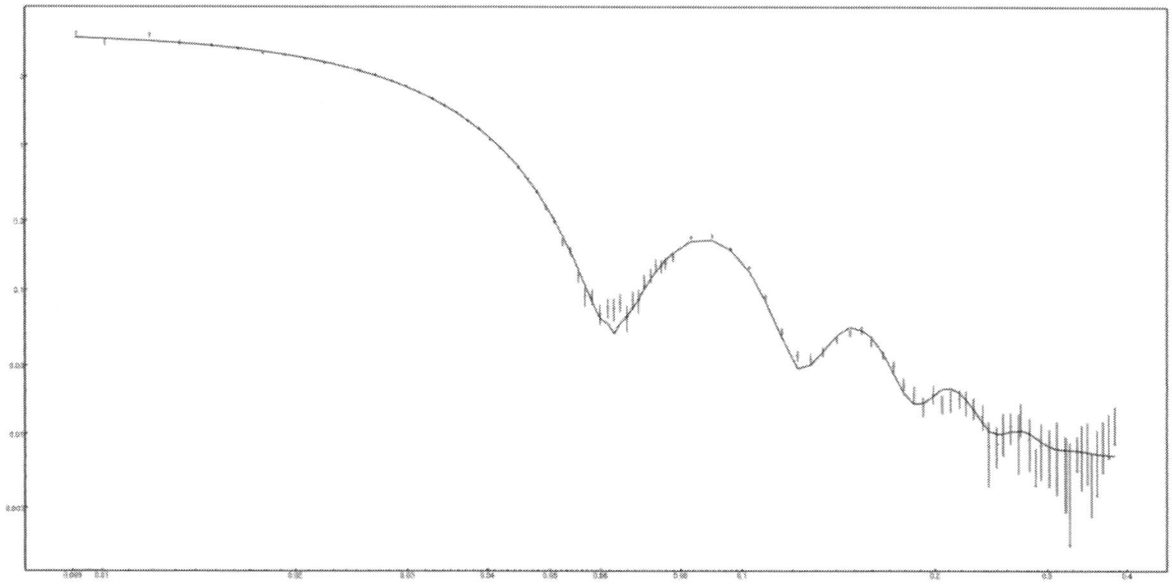

Figure 4. Apoferritin data fitting example.

6. Conclusion

Fitter is used as a part of a program complex on spectrometer YuMO. It is designed to be easily is understood by the experimenters. Both command line interface and GUI are provided. Despite of simplicity of offered models wide usage of fitter for many major appendices is supposed. The further development of the program is connected to expansion of offered models and introduction in the program of processing of the account of the structural factor. Fitter's design allows the simple adding of new models. The programm is demanded by users due to using the resolution of YuMO spectrometer also.

References

[1] *Library "JINRLIB". FITTER.* http://wwwinfo.jinr.ru/programs/jinrlib/fitter/indexe.html

[2] Soloviev A G, Stadnik A V, Islamov A Hand Kuklin A I 2003 *FITTER: The package for fitting experimental data of the YuMO spectrometer by theoretical form-factors. Version 1.0. Long Write-Up and User's Guide.* (Dubna: JINR) p

[3] Soloviev A G, Stadnik A V, Islamov A Kand Kuklin A I 2008 *Fitter. The package for fitting a chosen theoretical multi-parameter function through a set of data points. Application to experimental data of the YuMO spectrometer. Version 2.1.0. Long Write-Up and User's Guide.* (Dubna: JINR) p

[4] Hughes D J, Wallace J Rand Holtzman R H 1948 *Physical Review* **73** 1277

[5] Svergun D 1993 *Journal of Applied Crystallography* **26** 258

[6] Svergun D 1994 *Acta Crystallographica Section A* **50** 391

[7] Svergun D 1992 *Journal of Applied Crystallography* **25** 495

[8] Svergun D, Barberato Cand Koch M H J 1995 *Journal of Applied Crystallography* **28** 768
[9] Svergun D I 1999 *Biophys J* **76** 2879
[10] Konarev P V, Petoukhov M Vand Svergun D I 2001 *Journal of Applied Crystallography* **34** 527
[11] Kozin M Band Svergun D I 2000 *Journal of Applied Crystallography* **33** 775
[12] *Frank laboratory of neutron physics*. http://flnp.jinr.ru/34/
[13] Kuklin A I, Islamov A Kand Gordeliy V I 2005 *Neutron News* **16** 16
[14] Kovalev Y S, Kuklin A I, Novikov A G, Savostin V V, Shimkevich A Land Jadrovsky E L 2007 *Journal of Non-Crystalline Solids* **353** 3532
[15] Murugova T N, Solodovnikova I M, Yurkov V I, Gordeliy V I, Kuklin A I, Ivankov O I, Kovalev Y S, Popov V I, Teplova V Vand Yaguzhinsky L S 2011 *Neutron News* **22** 11
[16] Molchanov V S, Philippova O E, Khokhlov A R, Kovalev Y Aand Kuklin A I 2006 *Langmuir* **23** 105
[17] Kosacheva E, Kudryavtsev D, Bakeeva R, Kuklin A, Islamov A, Kudryavtseva L, Sopin Vand Konovalov A 2006 *Colloid Journal* **68** 713
[18] Haramagatti C R, Islamov A, Gibhardt H, Gorski N, Kuklin Aand Eckold G 2006 *Physical Chemistry Chemical Physics* **8** 994
[19] Islamov A, Haramagatti C R, Gibhardt H, Kuklin Aand Eckold G 2006 *Physica B: Condensed Matter* **385-386, Part 1** 791
[20] Andreeva S A, Fomenkov I, Islamov A K, Kuklin A I, Filippova O Eand Khokhlov A R 2005 *Polymer Science A* **47** 194
[21] Ozerin A N, Svergun D I, Volkov V V, Kuklin A I, Gordelyi V I, Islamov A K, Ozerina L Aand Zavorotnyuk D S 2005 *Journal of Applied Crystallography* **38** 996
[22] Kříž J, Pleštil J, Pospíšil H, Kadlec P, Koňák Č, Almásy Land Kuklin A I 2004 *Langmuir* **20** 11255
[23] Lebedev D, Filatov M, Kuklin A, Islamov A, Stellbrink J, Pantina R, Denisov Y, Toperverg Band Isaev-Ivanov V 2008 *Crystallography Reports* **53** 110
[24] Kovalev Y, Levkovich N, Kuklin Aand Apel' P 2009 *Colloid Journal* **71** 634
[25] Bakeeva R, Kosacheva E, Razina I, Kudryavtsev D, Kovalev Y, Kuklin A, Kudryavtseva Land Sopin V 2008 *Kinetics and Catalysis* **49** 631
[26] Herec M, Islamov A, Kuklin A, Gagos Mand Gruszecki W I 2007 *Chem Phys Lipids* **147** 78
[27] Anitas E M, Balasoiu M, Bica I, Osipov V Aand Kuklin A I 2009 *Optoelectronics and Advanced Materials - Rapid Comunications* **3** 621
[28] Balasoiu M, Anitas E M, Bica I, Osipov V A, Orelovich O L, Savu D, Savu S, Erhan Rand Kuklin A I 2008 *Optoelectronics and Advanced Materials - Rapid Comunications* **2** 730
[29] Balasoiu M, Craus M, Anitas E, Bica I, Plestil Jand Kuklin A 2010 *Physics of the Solid State* **52** 917
[30] Balasoiu M, Craus M L, Kuklin A I, Plestil J, Haramus V, Islamov A H, Erhan R, Anitas E M, Lozovan M, Tripadus V, Petrescu C, Savu D, Savu Sand Bica I 2008 *Journal of Optoelectronics and Advanced Materials* **10** 2932

[31] Kuklin A I, Kutuzov S A, Gabriel A, Eckold G, Utrobin P K, Smirnov A A, Ivankov A, Islamov A K, Kovalev Y S, Feoktistov A, Rogachev A V, Kirilov A Sand Gordeliy V I. 2007. Measurements with New Type of Position Sensitive Detector in Backscattering and Small-angle Setup, Dubna
[32] Eckold Gand Gorski N 2001 *Colloids and Surfaces A: Physicochemical and Engineering Aspects* **183-185** 361
[33] Rogachev A V, Cherny A Y, Ozerin A N, Gordeliy V Iand Kuklin A I 2007 *Crystallography Reports* **52** 500
[34] Lebedev D V, Baitin D M, Islamov A K, Kuklin A I, Shalguev V K, Lazov V Aand Isaev-Ivanov V V 2003 *FEBS Letters* **537** 182
[35] *CERN Program Library Long Writeup D506 MINUIT* http://wwwasdoc.web.cern.ch/wwwasdoc/minuit/
[36] Feigin L Aand Svergun D I 1987 *Structure Analysis by Small-Angle X-Ray and Neutron Scattering* (New York: Plenum Press) p 335
[37] Ostanevich Y M 1988 *Makromolekulare Chemie. Macromolecular Symposia* **15** 91
[38] Huber P 1981 *Robust statistics* (New-York: John Wiley and Sons) p
[39] Ososkov G. Robust regression for the heavy contaminated sample. *Proc. 2nd International Tampere Conference in Statistics, Tampere, Finland, 1987*:
[40] Chernov N, Kolganova Eand Ososkov G 1999 *Nuclear Instruments and Methods in Physics Research Section A: Accelerators, Spectrometers, Detectors and Associated Equipment* **433** 274
[41] *Fast Light Toolkit*. http://www.fltk.org
[42] *The Gluplot Plotting Package*. http://wwwinfo.jinr.ru/programs/jinrlib/gluplot/indexe.html

AUTHOR INDEX